Hui-lian Xu, PhD
James F. Parr, PhD
Hiroshi Umemura, PhD
Editors

Nature Farming and Microbial Applications

Nature Farming and Microbial Applications has been co-published simultaneously as *Journal of Crop Production*, Volume 3, Number 1 (#5) 2000.

Pre-publication REVIEWS, COMMENTARIES, EVALUATIONS . . .

T0371978

"*Nature Farming and Microbial Applications* authored by Xu, Parr and Umemura is of great interest to agriculture specialists, plant physiologists, microbiologists and entomologists as well as soil scientists and environmentalists. It gathers very original and innovative data on organic farming obtained by . . . scientists all over the world. It constitutes an excellent reference to both graduate and undergraduate students. I sincerely congratulate all authors for this remarkable book."

Dr. André Gosselin
Professor
Department of Phytology
Center for Research in Horticulture
Université Laval, Quebec, Canada

Nature Farming and Microbial Applications

Nature Farming and Microbial Applications has been co-published simultaneously as *Journal of Crop Production*, Volume 3, Number 1 (#5) 2000.

The *Journal of Crop Production* Monographic "Separates"

Below is a list of "separates," which in serials librarianship means a special issue simultaneously published as a special journal issue or double-issue *and* as a "separate" hardbound monograph. (This is a format which we also call a "DocuSerial.")

"Separates" are published because specialized libraries or professionals may wish to purchase a specific thematic issue by itself in a format which can be separately cataloged and shelved, as opposed to purchasing the journal on an on-going basis. Faculty members may also more easily consider a "separate" for classroom adoption.

"Separates" are carefully classified separately with the major book jobbers so that the journal tie-in can be noted on new book order slips to avoid duplicate purchasing.

You may wish to visit Haworth's website at . . .

http://www.HaworthPress.com

. . . to search our online catalog for complete tables of contents of these separates and related publications.

You may also call 1-800-HAWORTH (outside US/Canada: 607-722-5857), or Fax 1-800-895-0582 (outside US/Canada: 607-771-0012), or e-mail at:

getinfo@haworthpressinc.com

Nature Farming and Microbial Applications, edited by Hui-lian Xu, PhD, James F. Parr, PhD, and Hiroshi Umemura, PhD (Vol. 3, No. 1 #5, 2000). *"Of great interest to agriculture specialists, plant physiologists, microbiologists, and entomologists as well as soil scientists and evnironmentalists. . . . very original and innovative data on organic farming." (Dr. André Gosselin, Professor, Department of Phytology, Center for Research in Horticulture, Université Laval, Quebec, Canada)*

Water Use in Crop Production, edited by M.B. Kirkham, BA, MS, PhD (Vol. 2, No. 2 #4, 1999). *Provides scientists and graduate students with an understanding of the advancements in the understanding of water use in crop production around the world. You will discover that by utilizing good management, such as avoiding excessive deep percolation or reducing runoff by increased infiltration, that even under dryland or irrigated conditions you can achieve improved use of water for greater crop production. Through this informative book, you will discover how to make the most efficient use of water for crops to help feed the earth's expanding population.*

Expanding the Context of Weed Management, edited by Douglas D. Buhler, PhD (Vol. 2, No. 1 #3, 1999). *Presents innovative approaches to weeds and weed management.*

Nutrient Use in Crop Production, edited by Zdenko Rengel, PhD (Vol. 1, No. 2 #2, 1998). *"Raises immensely important issues and makes sensible suggestions about where research and agricultural extension work needs to be focused." (Professor David Clarkson, Department of Agricultural Sciences, AFRC Institute Arable Crops Research, University of Bristol, United Kingdom)*

Crop Sciences: Recent Advances, Amarjit S. Basra, PhD (Vol. 1, No. 1 #1, 1997). *Presents relevant research findings and practical guidance to help improve crop yield and stability, product quality, and environmental sustainability.*

Nature Farming and Microbial Applications

Hui-lian Xu, PhD
James F. Parr, PhD
Hiroshi Umemura, PhD
Editors

Nature Farming and Microbial Applications has been co-published simultaneously as *Journal of Crop Production*, Volume 3, Number 1 (#5) 2000.

FPP

Food Products Press
An Imprint of
The Haworth Press, Inc.
New York • London • Oxford

Published by

Food Products Press®, 10 Alice Street, Binghamton, NY 13904-1580 USA

Food Products Press® is an imprint of The Haworth Press, Inc., 10 Alice Street, Binghamton, NY 13904-1580 USA.

Nature Farming and Microbial Applications has been co-published simultaneously as *Journal of Crop Production*, Volume 3, Number 1 (#5) 2000.

Cover design by Thomas J. Mayshock Jr.

Library of Congress Cataloging-in-Publication Data

Nature farming and microbial applications/Hui-lian Xu, James F. Parr, Hiroshi Umemura, editors.
p. cm.
"Nature farming and microbial applications has been co-published simultaneously as Journal of crop production, volume 3, number 1 (#5) 2000."
ISBN 1-56022-082-1 (hardcover: alk. paper)–ISBN 1-56022-083-X (pbk.: alk. paper)
1. Organic farming. 2. Organic fertilizers. 3. Microbial inoculants. I. Xu, Hui-lian. II. Parr, J. F. (James Floyd), 1929- III. Umemura, Hiroshi. IV. Journal of crop production.

S605.5 .N36 2000
631.5′84–dc21 00-056193

INDEXING & ABSTRACTING

Contributions to this publication are selectively indexed or abstracted in print, electronic, online, or CD-ROM version(s) of the reference tools and information services listed below. This list is current as of the copyright date of this publication. See the end of this section for additional notes.

- *AGRICOLA Database*
- *BIOBASE (Current Awareness in Biological Science) <URL: http://www.elsevier.nl>*
- *Chemical Abstracts*
- *CNPIEC Reference Guide: Chinese National Directory of Foreign Periodicals*
- *Crop Physiology Abstracts <www.cabi.org/>*
- *Derwent Crop Production File*
- *Environment Abstracts. Available in print–CD-ROM–on Magnetic Tape. For more information check: www.cispubs.com*
- *Field Crop Abstracts <www.cabi.org/>*
- *FINDEX <www.publist.com>*
- *Foods Adlibra*
- *Food Science and Technology Abstracts (FSTA)*
- *Grasslands & Forage Abstracts <www.cabi.org/>*
- *PASCAL, ℅ Institute de L'Information Scientifique et Technique http://www.inist.fr>*
- *Plant Breeding Abstracts <www.cabi.org/>*
- *Referativnyi Zhurnal (Abstracts Journal of the All-Russian Institute of Scientific and Technical Information in Russian)*
- *Seed Abstracts <www.cabi.org/>*
- *Soils & Fertilizers Abstracts <www.cabi.org/>*
- *South African Assn for Food Science & Technology (SAAFOST)*
- *Weed Abstracts <www.cabi.org/>*

(continued)

Special Bibliographic Notes related to special journal issues (separates) and indexing/abstracting:

- indexing/abstracting services in this list will also cover material in any "separate" that is co-published simultaneously with Haworth's special thematic journal issue or DocuSerial. Indexing/abstracting usually covers material at the article/chapter level.
- monographic co-editions are intended for either non-subscribers or libraries which intend to purchase a second copy for their circulating collections.
- monographic co-editions are reported to all jobbers/wholesalers/approval plans. The source journal is listed as the "series" to assist the prevention of duplicate purchasing in the same manner utilized for books-in-series.
- to facilitate user/access services all indexing/abstracting services are encouraged to utilize the co-indexing entry note indicated at the bottom of the first page of each article/chapter/contribution.
- this is intended to assist a library user of any reference tool (whether print, electronic, online, or CD-ROM) to locate the monographic version if the library has purchased this version but not a subscription to the source journal.
- individual articles/chapters in any Haworth publication are also available through the Haworth Document Delivery Service (HDDS).

Nature Farming and Microbial Applications

CONTENTS

ALL FOOD PRODUCTS PRESS BOOKS
AND JOURNALS ARE PRINTED
ON CERTIFIED ACID-FREE PAPER

ABOUT THE EDITORS

Hui-lian Xu, PhD, is Senior Crop Physiologist, the research station Deputy Director and a center board Director at the International Nature Farming Research Center, Japan. Dr. Xu is a productive young scientist with numerous research publications in several fields of crop science, horticulture, soil physics, soil microbiology, plant physiology and sustainable agriculture. He studied his university courses in China Laiyang Agricultural University, did his PhD from the University of Tokyo, and worked with Canada Laval University for five years. With generous international experiences, Dr. Xu is a member of several professional societies in the United States, Canada and Japan.

James F. Parr, PhD, received his doctorate degree in 1961 from Purdue University in soil microbiology and soil physics. Dr. Parr has more than 35 years of research experience with the U.S. Department of Agriculture, the Tennessee Valley Authority, and U.S. land-grant Universities in aspects of soil water and crop management systems. He has served as a technical consultant to FAO, UNDP and USAID on international development projects worldwide. He has published more than 200 papers on soil water/crop management systems, environmental quality, organic farming, rainfed farming and sustainable agriculture. Dr. Parr is a fellow of the American Society of Agronomy and the Soil Service Society of America. He retired from USDA in 1994 but continues to serve as a research consultant and technical editor.

Hiroshi Umemura, PhD, worked from 1953 to 1989 with Nagano Prefecture Agricultural Experiment Station on soil survey, soil improvement and pollution control. He received his doctorate degree from the University of Tokyo with his thesis on sustainability of the soil productivity of cool uplands. Since 1989, he has been Director of the International Nature Farming Research Center and has taken part in several international development projects.

Preface

Recent concerns over the environmental pollution and food quality degradation by excessive input of chemicals have prompted scientists and policymakers to reevaluate our modern agriculture and find alternatives to produce safe and nutritious food and protect our environment. In this regard, there is a growing interest in organic farming. Nature farming, a concept similar to organic farming, was proposed by Mokichi Okada, a Japanese naturalist and philosopher, more than 50 years ago when chemical inputs to agriculture were first started in Japan. Many farmers have practiced nature farming following Okada's concept. With support from Japanese authorities, farmers and some financial groups, the International Nature Farming Research Center was founded to foster the organic agriculture movement in Japan and worldwide. This special issue summarizes the research achievements by scientists at this center and collaborators from universities and research institutes in Japan and other countries.

The papers are grouped into the following sections: (1) Nature Farming, and (2) Microbial Applications. The first section covers the historical aspects of nature farming and some examples of classical farming practices. The second section reports on nature/organic farming and applications of organic fertilizers and microbial products including "effective microorganisms," mycorrhizae and rhizobia in crop production and environmental protection. Some papers present unique and original methodologies, including mathematical and physical modeling, to analyze photosynthesis, transpiration, plant-water relations and stress resistance of crops.

We have made considerable effort to improve the readability of papers by scientists who are not native speakers of English. Thus, we assume responsibility for any mistakes due to translation and editing. We would like to thank Setsunori Shinoda, Chairman of the Board, International Nature Farming Research Center, for his support of this editorial work. We would like to

[Haworth co-indexing entry note]: "Preface." Xu, Hui-lian, James F. Parr, and Hiroshi Umemura. Co-published simultaneously in *Journal of Crop Production* (Food Products Press, an imprint of The Haworth Press, Inc.) Vol. 3, No. 1 (#5), 2000, pp. xxv-xxvi; and: *Nature Farming and Microbial Applications* (ed: Hui-lian Xu, James F. Parr, and Hiroshi Umemura) Food Products Press, an imprint of The Haworth Press, Inc., 2000, pp. xv-xvi. Single or multiple copies of this article are available for a fee from The Haworth Document Delivery Service [1-800-342-9678, 9:00 a.m. - 5:00 p.m. (EST). E-mail address: getinfo@haworthpressinc.com].

express our thanks to the Editor, Dr. Amarjit S. Basra, for his encouragement and the staff of The Haworth Press Inc., for their careful work in producing this special issue.

Hui-lian Xu
James F. Parr
Hiroshi Umemura

PART I: NATURE FARMING

Nature Farming: History, Principles and Perspectives

Hui-lian Xu

SUMMARY. The history, principles and perspectives of nature farming, as advocated by Mokichi Okada, a Japanese philosopher in 1935, are described. According to Okada, the principles of nature farming must fulfill five requirements: (1) produce safe and nutritious food that ensures good health; (2) be economically and spiritually beneficial to both producers and consumers; (3) be sustainable and easily practiced; (4) conserve and protect the environment; and (5) produce sufficient food of high quality for an expanding world population. In practice, both synthetic chemicals and raw waste from animals without treatment are prohibited as fertilizers or soil amendments for crop production. Composts from plant materials are recommended. This is the main difference with the principles of organic farming, which allows the use of animal manure, untreated or composted. Although there are strong

Hui-lian Xu is Senior Crop Scientist, International Nature Farming Research Center, 5632 Hata, Nagano 390-1401, Japan (E-mail: huilian@janis.or.jp).

[Haworth co-indexing entry note]: "Nature Farming: History, Principles and Perspectives." Xu, Hui-lian. Co-published simultaneously in *Journal of Crop Production* (Food Products Press, an imprint of The Haworth Press, Inc.) Vol. 3, No. 1 (#5), 2000, pp. 1-10; and: *Nature Farming and Microbial Applications* (ed: Hui-lian Xu, James F. Parr, and Hiroshi Umemura) Food Products Press, an imprint of The Haworth Press, Inc., 2000, pp. 1-10. Single or multiple copies of this article are available for a fee from The Haworth Document Delivery Service [1-800-342-9678, 9:00 a.m. - 5:00 p.m. (EST). E-mail address: getinfo@haworthpressinc.com].

requirements for nature farming principles, it is one of the most idealistic and practicable farming methods to ensure human health and environmental protection. *[Article copies available for a fee from The Haworth Document Delivery Service: 1-800-342-9678. E-mail address: getinfo@haworthpressinc.com <Website: http://www.HaworthPress.com>]*

KEYWORDS. Environmental protection, food quality, nature farming, organic farming

INTRODUCTION

Today, modern industrialized conventional agriculture is heavily dependent on inputs of chemical fertilizers and pesticides. The excessive use of such synthetic agricultural chemicals has caused many problems including environmental pollution and degradation, impaired food safety and quality and adverse effects on human and animal health (Musa, 1976). Moreover, chemical farming or conventional agriculture often creates an unstable ecosystem in which the potential for maximum yield is inevitably associated with risks due to ecosystem instability (Vogtmann, 1984). These concerns prompted scientists to reevaluate our modern chemical farming systems and seek appropriate alternative practices to ensure a more sustainable agriculture. Organic farming is a method of crop production that has been utilized in the early history of agriculture. "Nature farming" may appear as a new terminology for the agriculturists, although its principles are similar to those of organic farming. It was proposed by Mokichi Okada, a Japanese philosopher, more than 60 years ago, when chemical agriculture began. Some consider that organic farming or nature farming are reversions to the old agriculture of the 1940s (Vogtmann, 1984) and nature farming may be a reversion to the agriculture earlier than the 1940s, according to the industrialization of Japan. The purpose of this paper is to review and discuss the history, principles and perspectives of organic farming and nature farming as well as their differences and to provide a better understanding of these two non-chemical farming methods.

CONCEPT AND PRINCIPLES

The main objective of nature farming is to develop and practice an ideal agriculture according to the principles of Mokichi Okada (1953), which advocate the following requirements:

1. produce safe and nutritious food that ensures good health;
2. be economically and spiritually beneficial to both producers and consumers;
3. be sustainable and easily practiced;
4. conserve and protect the environment; and
5. produce sufficient food of high quality for an expanding world population.

No synthetic chemical fertilizers and pesticides are used in nature farming systems. Human and animal manure, urban sewage and other untreated animal wastes are also prohibited from use as organic amendments in nature farming. Instead, farmers can use composted crop/plant residues and industrial processing wastes (rice husks and bran, oil mill sludge) as organic amendments to maintain soil organic matter and supply plant nutrients (Minamino, 1994). Recently, some nature farming practitioners have advocated the use of treated animal wastes and urban garbage (e.g., fermented through composting) as organic amendments and soil conditioners. However, the treated materials must not contain any dangerous pathogens, harmful chemical residues or toxic levels of heavy metals.

THE BACKGROUND

Mokichi Okada proposed the concept of nature farming in 1935 because of his concern for food safety and quality, protecting the environment and ensuring human health and nutrition. To fulfill his goals, he established an organization, which later developed into Sekai Kyuse Kyo (SKK). Mokichi Okada's philosophy is respectful of nature and nature's laws. This is reflected in one of his typical slogan, "Nature is God."

In 1941, Mokichi Okada did some experiments and concluded that the chemical farming method was neither compatible nor in accordance with nature. "*If we grow crops with true love and respect towards the natural power of the soil, the soil will function to an astonishing degree. All the difficult problems and troubles that harass both farmers and consumers today can be solved through this method of cultivation.*" He also wrote a poem criticizing chemical agriculture for environmental pollution and degradation:

What a foolish thing!
Today's man is polluting
The earth's precious soil
Which produces the treasure
Vitally important food.

Mokichi Okada came to understand the vicious circle that accompanied the chemical farming method, i.e., that the application of chemical fertilizers resulted in the generation of harmful and destructive insects which necessitate the application of toxic pesticides that would often cause insects to develop even stronger resistance. Then, in turn, the application of still more toxic chemicals would ultimately lead to further deterioration of the soil, the growth of still more pests, and the production of unsafe food that could endanger human health and nutrition. Consequently, a number of experimental farms to demonstrate the nature farming method were established throughout Japan by his followers after the Nature Farming Society was founded in 1953.

Some original essays written by Mokichi Okada in 1953 and translated into English in 1987 (Okada, 1987) are presented here, so that readers can fully comprehend his philosophy on nature farming.

The Adverse Effect of Fertilizers

Over ages, agriculture has steadily, though imperceptibly, fallen into evil ways. Unestimating the power of the soil, farmers have accepted the necessity of using manure or chemical fertilizers for the improvement of crops. As a result the soil has degenerated and altered and its generative power has weakened. Unaware of this and suffering under the delusion that insufficient use of fertilizers causes poor crops, farmers resort to still more fertilization, with the result that soil declines even further.

Fertilizers produce quite good results for a while but they gradually have adverse effects when used over long periods because, forced to rely on these artificial substances, crop plants change so that they become incapable of drawing substances from the soil. The situation is similar to that of drug addiction. The fertilizer addiction which is destroying soil throughout Japan is equally terrible. Farmers have blindly trusted in fertilizers for such a long time that they are unable to open their eyes to the truth.

One of the great agricultural worries today is insect pests. However, instead of finding out what causes them, farmers concentrate all their efforts on exterminating them. In other words, unable to find out the reasons why harmful insects occur, they take what seems to be the next-best alternative. In fact, fertilizers generate harmful insects, and the number of such pests has increased in recent years as the number of fertilizers used has increased. Furthermore, people are unaware that the insecticides used to exterminate pests penetrate the soil and cause it to degenerate, thus leading to the appearance of still more insects. The application of fertilizers gravely weakens crops. Heavily fertilized plants fall and break in wind and rain. They bear few fruit because their blossoms drop. In the case of rice, wheat, and beans, the plants grow too tall and their leaves are too big, so that crops are shaded, husks are thick, and seeds are small and sick. Vegetables and grains raised by nature farming are better tasting, larger, and faster growing, and crops are bigger than those produced with fertilizers.

Low Nutrient Availability at Beginning But Long Nutrient Sustainability of Organic Materials

When people change from the fertilizer method to nature farming, they generally experience the following series of developments. When the seedlings are first transplanted from beds to the paddy fields, for a while the color of their leaves will be poor and their stalks thin. Indeed, by comparison with seedlings in other fields, they will look so scrawny that neighboring farmers will smirk, and the farmers trying the natural method will become worried enough to return to God in prayer. After two or three months, plant conditions will improve, and, at blooming time, the farmers' brows will un-knitted a little. They will finally be put at ease when, just before harvest time, they see that their crops are as good as or better than ordinary. When harvest comes around, they will be astounded to see not only that their crops are big, but also that the grain is of high quality, lustrous, strongly adhesive, and delicious.

Organic Materials Are Decomposed to Humus and Remain in the Soil

It is important to use large quantity of natural compost, which I shall now discuss. In stimulating plant growth it is of the greatest importance to allow the root hairs at the ends of the large roots to develop. This means preventing the soil from compacting and hardening. If natural compost has decomposed too much, it tends to harden easily; about half decomposed is best. This kind of remaining organic matter increases the soil's power to hold water and nutrients. If the soil is loose enough to aerate the roots, the root hairs have plenty of room to grow. This roominess is the reason why plants grow well in such soil. To soften soil for dry field cultivation thus to permit roots to grow and spread, allow dried leaves and grasses to decompose until soft, then mix this compost with the soil. Recently, it has been said that ground should be soft in order that air may penetrate the roots. The composts are employed to produce three effects; to prevent the soil from hardening, to warm the soil, and to retain moisture around plant roots.

Chemical Fertilizations Increase Insects and Pesticides Increase Poison Resistance of Insects

Though insect pests may not totally vanish when the nature farming method is used, it does reduce them to a fraction of what they are when fertilizers are employed. Farmers often say that too much fertilizer increases the number of insects that plague the crops. I have heard from experts that Manila and Havana tobaccos, which are used in the finest cigars and are famous for their superb aroma, are never troubled with insect pests, because the plants are never fertilized with chemicals. In particular, there is the fact that insects do not damage weeds; and the special fragrance of the wild parsley and that edible chrysnthemum picked in the country in springtime were never damaged by insects is due to the fact they have had no fertilizer.

Soil Recovery from Degradation

For the first year or two after the shift is made from fertilized agriculture to nature farming, results will be poor. This phenomenon occurs because the soil has become addicted to fertilizers. . . . It is imperative to be patient and wait two or three years until the soil and seeds are freed of the fertilizer poison and land can manifest its great power.

DIFFERENCES BETWEEN NATURE FARMING AND ORGANIC FARMING OR OTHER FARMING SYSTEMS

There are several kinds of farming systems now practiced in the world. They are intensive farming or intensive agriculture, comprehensive agriculture, ecological agriculture, controlled agriculture, organic farming or organic agriculture, and nature farming. Often, their definitions overlapped with each other. They are briefly summarized as follows:

- *Intensive farming.* In economically developed regions, large quantities of chemical fertilizers, pesticides, and herbicides are used in both field and equipped facilities.
- *Comprehensive agriculture.* Chemicals are reduced to 70% of intensive agriculture. Legume crops and green manure crops are adopted.
- *Controlled agriculture.* Chemical fertilizers, pesticides and herbicides are not used on the crops for direct eating such as wheat for everyday bread and vegetables for everyday dining tables, but used for silage and industrial crops such as corn and sugarcane. Pesticides are prohibited for vegetable and fruit production although chemical fertilizers are used.
- *Organic farming.* Chemical fertilizers, pesticides and herbicides are not used but animal manure and urban sewage are allowed for use as fertilizers and soil conditioners.
- *Nature farming.* All synthesized chemicals, animal manures and urban sewage are prohibited. Composts fermented using plant organic materi-

als are used to increase soil fertility and improve the soil physical properties.

Many believe that nature farming is the most ideal and appropriate farming system to safeguard human health and to protect the environment. However, nature farming is difficult to practice because of certain restrictions. Organic farming allows the practitioners to use animal manure and raw animal products. However, principles of nature farming prohibit the use of not only chemical fertilizers and pesticides, but also untreated animal manures, urban sewage and animal products. Since the beginning, some people have not understood why animal manures and animal products are prohibited by nature farming. Now people do understand that without treatment many disease pathogens would persist in the animal manures, for example, the roundworm, which can infect humans. Without treatment, some kinds of hormones and antibiotics, used to hasten animal growth and prevent diseases, would remain in the manures and could be absorbed by crops especially vegetables. These hormones and antibiotics are potentially harmful to human health. Another problem is that of heavy metals in the animal manures.

CURRENT STATUS OF NATURE FARMING IN JAPAN

The current status of organic farming and nature farming have been reviewed in recent books written by Kumazawa (1996), Hasumi (1991), Nakasuji (1991), and Minamino (1994). The following are some key points discussed by these authors as well as some relevant issues that do not appear in these books.

Mokichi Okada's Followers

As mentioned earlier, there are many of Mokichi Okada's followers who are practicing nature farming successfully on a farm scale. Mokichi Okada's followers have also founded several organizations relevant to nature farming according to Mokichi Okada's philosophy. The International Nature Farming Research Center is also financially supported in part by Mokichi Okada's followers. Most of these organizations are supported one way or another by Sekai Kyusei Kyo (SKK), which was founded by Mokichi Okada and is also known as the Church of World Messianity. Before 1997, some organizations including SKK itself were separated into several groups, which advocated different ideological and organizational aspects. Consequently, prior to 1997, Okada's followers were not able to focus adequately on nature farming research and development needs. Now that the followers have re-united, many expectations should be realized in the future on much-needed research and development for nature farming methods.

The Technology of Effective Microorganisms or EM

Because some of Okada's followers are not scientists, some activities and propositions diverge from modern scientific principles. This makes it difficult for these people to collaborate with scientists in universities and national research institutes. This phenomenon is especially reflected in the research and applications of a technology called "Effective Microorganisms" or EM (Higa, 1994). The necessary fundamental research on this technology has been inadequate. Many benefits were claimed and conclusions reached about this technology without valid scientific support. This has resulted in aversion and criticism of the technology by scientists. Nevertheless, the microbes involved in this technology have demonstrated their effectiveness on ecosystem improvement of the soil microflora (Fujita, 1995), promotions of crop root development (Xu et al., 1998), and recovery of polluted environments (Li and Ni, 1996). Research is needed to elucidate the mechanisms or modes-of-action that elicit these beneficial effects.

Practical Scale of Nature Farming

Several nature farms are operated on relatively large scales by organizations associated with the International Nature Farming Research Center and SKK. On these farms, the principles of nature farming proposed by Okada are strictly followed. Cereals, vegetables and fruits are produced without the use of chemical fertilizers and pesticides, and even without animal manures and urban wastes. Although nature farming has not been adopted by farmers on production scale levels, most farmers are practicing nature farming on a small scale for their own consumption. Some farmers never eat cereals, vegetables and fruits that they produce with chemical fertilizers and pesticides. They know how poisonous these chemicals are to humans, animals and wildlife. These farmers are not lacking morality when they produce agricultural products with poisonous chemicals for sale to consumers. It is society and the consumers who push the farmers to do this. If the farmers do not use chemicals, they can not gain enough income to sustain themselves and their families. Society does not insure their normal life without appropriate income from the high yield of chemical agriculture. The consumers select agricultural products, especially those such as fruits and vegetables, by shape, size and color. One will be surprised to hear that the apples under 270 g are not marketable in Japan. Even a small scar caused by insects on the fruit surface renders them un-marketable. Without chemical fertilizers and pesticides, the farmers cannot produce larger attractive products that are demanded by consumers today. Thus, farmers are forced to depend on poisonous chemicals even though they are aware of their adverse effects to societies and the consumers. The society's policies and consumers' habits make it difficult for nature farming to be adopted on large scales.

The Future of Nature Farming

The future of nature farming or organic farming has been reviewed and discussed by many scientists (Lockeret et al., 1981; Harwood, 1984; Vogtmann, 1984; Hasumi,1991; Nakasuji, 1991; Minamino, 1994; Kumazawa, 1996). Detailed descriptions are omitted here. Nature farming and organic farming should have a bright future because people worldwide come to realize that they must respect nature and nature's laws in our modern society. If chemical agriculture is allowed to continue, the environment in which we live will be severely threatened. Therefore, it is concluded that nature farming or organic farming has a good future although there are many problems yet to be overcome by practitioners.

REFERENCES

Fujita, M. (1995). An approach of apple cultivation towards nature farming. The 4th Conf. Technol. of Effective Microorganisms, Saraburi, Thailand, Nov. 20, 1995.

Harwood, R.R. (1984) Organic farming research at the Rodale Research Center. In *Organic Farming: Current Technology and Its Role in a Sustainable Agriculture*, D.F. Bezdicek, J.F. Power, D.R. Keeney, and M.J. Wright (eds.). Madison: American Society of Agronomy, pp. 1-18.

Hasumi, T. (1991). *Devotions to Organic Agriculture*. Tokyo: Nippon Keizai Hyoronsha, 357 p.

Higa, T. (1994). *The Completest Data of EM Encyclopedia*. Tokyo (in Japanese): Sogo-Unicom, 385 p.

Kumazawa, K. (1996). *What Is the Environment Sound Agriculture*. Tokyo: Agricultural Statistics Association, 315 p.

Li, W.J. and Y.Z. Ni. (1996). *Research and Applications of EM Technology*. Beijing: China Press of Agricultural Science and Technology, pp. 42-102.

Lockeretz, W.G. Shearer and D.H. Kohl. (1981). Organic farming in the Corn Belt. *Science* 211: 540-547.

Minamino, Y. (1994). *Manifesto of Nature Farming. In the Front of Organic Agriculture*. Tokyo: Fumin-Kyokai, 195 p.

Musa, S. (1976). *Horrible Food Pollution*. Tokyo: Kodansha, 220 p.

Nakasuji, F. (1991). *Nature, Organic Agriculture and Pests*. Tokyo: Tojusha, 292 p.

Okada, M. (1987). *True Health*. Atami, Japan: Church of World Messianty, pp. 147-172.

Vogtmann H. (1984). Organic farming practices and research in Europe. In *Organic Farming: Current Technology and Its Role in a Sustainable Agriculture*, D.F. Bezdicek, J.F. Power, D.R. Keeney, and M.J. Wright (eds.). Madison: American Society of Agronomy, pp. 19-36.

Xu, H.L., N. Ajiki, X. Wang, C. Sakakibara and H. Umemura. (1998). Water retention in excised leaves of sweet corn grown under organic and chemical fertilizations with or without effective microbe applications. *Pedoshpere* 8: 1-8.

Classical Farming Systems of China

Zitong Gong
Putian Lin
Jie Chen
Xuefeng Hu

SUMMARY. During the long and dramatic history of agricultural civilization, a variety of technological methods have been developed and many valuable experiences accumulated on the farming practices. This review describes the progression of the agricultural development in ancient China. The practices and philosophies of classical farming in different historical periods are expounded and the characteristics of the classic farming practices in China are summarized. The paper also describes regeneration of the organic farming system, the development of the ecological and sustainable agricultural systems, the realistic significance of the knowledge of the classic farming methods such as application of organic manure and the systems of crop rotation and intercropping. *[Article copies available for a fee from The Haworth Document Delivery Service: 1-800-342-9678. E-mail address: getinfo@haworthpressinc.com <Website: http://www.HaworthPress.com>]*

KEYWORDS. Agricultural history, ancient China, classical farming systems, cropping systems, organic farming, organic manure, soil fertilization

---

Zitong Gong is Senior Soil Scientist, Jie Chen and Xuefeng Hu are Soil Scientists, Institute of Soil Science, Chinese Academy of Sciences, 71 East Beijing Road, P.O. Box 821, Nanjing, 210008 China. Putian Lin is Professor, Loudi Agricultural School, Loudi, Hunan, China.

Address correspondence to: Zitong Gong at the above address (E-mail: xzshi@ns.issas.ac.cn).

[Haworth co-indexing entry note]: "Classical Farming Systems of China." Gong, Zitong et al. Co-published simultaneously in *Journal of Crop Production* (Food Products Press, an imprint of The Haworth Press, Inc.) Vol. 3, No. 1 (#5), 2000, pp. 11-21; and: *Nature Farming and Microbial Applications* (ed: Hui-lian Xu, James F. Parr, and Hiroshi Umemura) Food Products Press, an imprint of The Haworth Press, Inc., 2000, pp. 11-21. Single or multiple copies of this article are available for a fee from The Haworth Document Delivery Service [1-800-342-9678, 9:00 a.m. - 5:00 p.m. (EST). E-mail address: getinfo@haworthpressinc.com].

INTRODUCTION

During the last half century, crop production has been intensified by abundant uses of chemical fertilizers and pesticides. Environmental pollution has greatly increased because of excessive use of chemical fertilizers, pesticides and petroleum energy. Soils are being increasingly degraded by unbalanced inorganic fertilization, monoculture croppping, and drought or flooding caused by a deteriorating environment. These problems not only make agriculture unsustainable but also jeopardize the existence of the earth on which humankind resides and depends. Therefore, concerns about environmental protection, agricultural sustainability, food quality and human health have prompted agricultural scientists and policymakers to reevaluate the modern chemical farming methods and to seek alternatives. Many alternatives have been proposed and practiced throughout the world, involving the substitution of biological alternatives for chemical inputs. These alternatives as a whole constitute "organic farming" although other similar terminology such as "nature farming," "sustainable agriculture," and "ecological agriculture," are used in different cases. Organic farming practices comprise a system of farming that harmonizes the relations between land use and land maintenance and guarantees stable harvests that ensure both product quantity and quality. Organic farming is not a new concept but an agricultural method that has been used throughout the history of mankind. Organic farming is actually the agricultural system that was utilized before the 1940s in developed industrial countries and before the 1960s in China. For the purpose of organic farming research, one needs to review the history of agriculture. China has a long history of civilization including agriculture and many experiences by farmers are recorded in documents. The treasure of classical farming knowledge accumulated by the Chinese in their history can no doubt be helpful to the further advancement of organic farming research and practices. In the present review, the practices and philosophies of classical farming systems in different historic periods of China are described. Information and ideas are extracted from the ancient agriculture and subsistence documents, which are expected to be useful to the regeneration of modern organic farming in China and in the world.

AGRICULTURAL DEVELOPMENT IN ANCIENT CHINA

The agricultural development of ancient China covers a very long history of thousands of years. From the "Fire Tillage" and the "Si Tillage" of the early farming practices to the establishment of crop rotation systems, and from the combination of land use and land maintenance to the maturity of the classical farming systems, the agricultural civilization of ancient China has witnessed several important stages of development.

The Periods of the Pre-Qin and Han Dynasty (200 B.C.)

The early farming practices are grouped into two stages: "Fire Tillage" and "Si Tillage" (Si is a spade-shaped farm tool used in ancient China). "Slash-and-burn" and "lying waste" are primitive farming methods, which are versions of "Fire Tillage." The main characteristic of fire tillage is that it is practiced without farming tools. The practices involve cutting down forests, burning them and sowing after rain. Because the plant ash after burning provides plant-nutrients, good harvests could be possible in the beginning. However, the harvests always declined after one or two years because of the lack of knowledge about fertilization, intertillage and irrigation. The only option was that ancient Chinese farmers had to abandon the used land after several years and find another block of wasteland for cultivation, which is referred to as lying waste. Thus, the farming practice described here is called Fire Tillage (Guo, 1981).

China entered the stage of Si Tillage as early as the Shen-Nong Period, about 8 or 9 thousand years ago. Si Tillage is a practice by which people loosen soils and sow seeds with primordial tools such as sharp wooden sticks, stones and bones after burning wasteland. Because the soil was loosened, the conditions of moisture, nutrition, aeration and temperature changed, and its fertility increased. The yields, however, were not high enough because farmers knew little about soil fertility and fertilization (Li, 1981; Tang, 1985).

In the New Stone Age, 4 to 6 thousand years ago, agriculture along the upper and middle reaches of the Yellow River developed gradually. During this time, the three slavery dynasties of Xia, Shang and Zhou were established successfully and bronze tools were invented. In the Shang Dynasty, the dominant farming practice was "Cooperative Tillage," namely, three slaves were organized into one team and worked in fields cooperatively. In the West Zhou Dynasty, three thousand years ago, farming tools were improved, and the farming practice was also changed from three-person to two-person teams, called "Mate Tillage." The ridged tillage could be used as an example. When tilling fields, two persons worked side by side, and one turned earth over to the left, the other to the right, resulting in the formation of furrows and ridges. With accumulated experience, the farming practices led to the development of the fallow system, which was more advanced than the lying waste system. One cycle of fallow system is a period of three years: one year for fallow and two years for cultivation (Research Department of Agricultural Heritage of China, 1984a, b).

In the Spring and Autumn Period (770-476 B.C.), China had already entered the feudal society, when widespread use of iron farming tools and animal traction, i.e., oxen tillage, gave special impetus to agricultural production. At that time, the fallow system had gradually developed into the crop succession system, which gave rise to crop rotations. A document referred to as *Mr. Shang's Works* indicates that Shang-Yang encouraged the reclamation of wasteland and utilization of fallow fields when he carried out the political reforms in the Qin Dynasty (Hu, 1959; Lin, 1996). The utilization of fallow fields gave an impetus to the popularization of the crop succession system. During the successive cultivation, the system of crop rotations occurred and developed as a result of the measures to adjust manpower and control plant diseases and insect pests. The multiple crop indexes were raised by crop rotations. The combination of land use and land maintenance received increasing attention and intensive farming systems began to develop.

In the period of the Qin and Han Dynasties (221 B.C.-220 A.D.), the system of crop succession already had a fixed model, the crop rotations had been recorded, and the systems of intercropping and multiple cropping had also been practiced. In *Fan Sheng's Works on Agriculture*, a document composed in the Han Dynasty, it was pointed out that the system of one crop a year had been carried out in the Yellow River Valley, and that of two crops a year had been practiced in southern China (Shi, 1979; Wan, 1980; Lu, 1989).

The Period of the Wei and Jin Dynasties and the South and North Dynasty (200-580 A.D.)

Agriculture further developed in this period, especially in the Yellow River Valley. The great agricultural scientific work *Important Means of Subsistence for Common People* was written during this period as a result of agriculture development (Miao, 1981). In this book, the technical system of upland farming practices, with ploughing and harrowing as its core, to prevent drought and conserve soil moisture, was detailed. This book also described the technical measures concerning farming practices such as crop rotations, green manuring and seed-breeding. It also indicated that the technology of intensive cultivation in northern China was probably established in this period.

There are more than twenty kinds of crop rotations described in *Important Means of Subsistence for Common People*. Firstly, this book indicates that most crops need rotations. 'Paddy fields must be shifted each year' (The Chapter on Post-Planting); 'Paddy fields should be changed each year' (The Chapter on Rice Planting). Secondly, this book provides the theoretical and technical basis for the proper rotation of leguminous plants and cereal crops. It states that leguminous plants should be planted before cereal crops in crop rotations. Leguminous crops are considered as good pre-cropping for cereal crops. Thus the important position of leguminous plants in rotation is confirmed in this book. This kind of crop rotations, characterized by combining land use with land maintenance, is still widely applied. Thirdly, the importance of the rotation of grain and green manure plants is raised to a new level in this book. The chapter of farming in *Important Means of Subsistence for Common People* points out that multiple cropping of red bean, mung bean and flax can increase the soil fertility, and cites the "Way to Field Fertilization." In addition, multiple cropping of mung bean is also encouraged in vegetable cultivation.

In the South and North Dynasties, intercropping also developed. The first was the initiation of the intercropping of forests and grain crops. The chapter on planting mulberry in *Important Means of Subsistence for Common People* describes that the intercropping of mung bean or red bean with mulberry could lead to a bean harvest with good-quality and benefit both fields and mulberry. The second was the rudiments of under-crop sowing. In the chapter on planting fiber crops of *Important Means of Subsistence for Common People*, the experience of hemp under-crop sowing turnip is summarized as 'in June, spread turnip seeds into hemp fields, hoe the fields and plan to harvest its roots.' The third was the useful experience and lessons of the multiple cropping. The chapter on planting fiber crops of *Important Means of Subsistence for Common People* summarizes the lessons of the intercropping of hemp in the soybean fields: 'hemp can never be grown in soybean fields, otherwise the fields will be damaged and the harvests of the two crops will be poor.' In *Important Means of Subsistence for Common People*, the formation of upland farming practices are summarized as the system of 'ploughing and harrowing' conserving soil moisture and preventing drought (Miao, 1981).

In the period of the Wei and Jin Dynasties, south of the Yangtze River, double cropping of rice had become an important economic strategy. In this period, regenerate rice was widely introduced and grown. The system of the intercropping of grains with green manure plants developed (Tang, 1985; Guo, 1988).

The Period of the Sui, Tang, Song and Yuan Dynasties (580-907 A.D.)

The farming practices, as known, developed constantly in the period. The farming practices were generally a fixed model in northern China as early as in Han Dynasty, and the system of crop rotations was carried out widely. At that time, the area of the rotation of grain and green manure plants increased rapidly in northern China (Shi, 1979).

During the Sui and Tang Dynasties, the rotation of rice and wheat was developed in southern China (Li, 1980). In the South Song Dynasty, a system of three rice crops a year was initiated by way of intercropping, under-crop sowing and mixed cropping in southern China. In *Information on Southern China* written (probably by Zhou Qufei) in the 6th century, it is said: 'in the

Qin State, . . . paddy planted in January and February is called early rice, which will be harvested in April and May; that planted in March and April was called early-late rice, which will be harvested in June and July; that planted in May and June is called late rice, which will be harvested in August and September.' Obviously, three rice crops a year was conducted through intercropping, under-crop sowing and mixed cropping (Sang, 1982). At that time, while promoting the crop rotation and raising the efficiency of land use, the farmers also noticed the problem of soil fertilization. *Necessities of Farming* in the South Song Dynasty proposed the use of green manure, which was regarded as the best way to achieve soil fertilization (Jiao, 1984; Peng, 1985).

In the Yuan Dynasty, crop rotation and multiple cropping further developed in both southern and northern China. Because of much wetland in southern China, the key to the development of rice and wheat rotation was leveling land and irrigation (Peng, 1985; Tong, 1989).

In the Sui, Tang, Song and Yuan Dynasties, the rotation of grain and green manure plants spread to the Yangtze River Valley and the Huaihe River Valley in northern China. In short, application of green manure for soil fertilization, based on crop rotations, received increasing attention (Cao, 1984; Jiao, 1984).

The Period Since the Ming and Qing Dynasties (Since 1368 A.D.)

The Ming Dynasty was the period when systems of crop rotation and intercropping developed comprehensively, which improved the efficiency of land use to a new level. At that time, the systems of two crops a year and three crops a year based on the rotation of paddy field and upland concentrating on rice production were developed in southern China; however, the systems of three crops in two years and two crops a year with the rotation of grain between cotton crops were adopted in northern China. In addition, the crop rotation and intercropping of grains with green manure plants were adopted in both southern and northern China.

Song Yingxing, a great naturalist in the Ming Dynasty, detailed four systems for the rotation of grain and leguminous plants in southern China in his work of *God Creates Things*: "The first system is two crops a year with the rotation of rice and leguminous plants. Soybean is sown after the harvest of early rice in January and harvested in September or October. The second is two crops a year with the rotation of rice between wheat. Namely, late rice is cultivated after wheat harvest and wheat is cultivated again after the harvest of rice. The third is double cropping of rice. The fourth is the rotation of rice and mung bean"(Pan, 1988).

In the Ming Dynasty, the suitable time for ploughing in green manure was also studied in detail. The blossoming period of leguminous plants was considered the suitable stage for ploughing in because the plants at this time are tender and easy to decay after being buried, and the nutrients released are easily absorbed by crops.

Furthermore, the intercropping system developed greatly in the Ming Dynasty and several famous works concerning the intercropping system were composed. For example, *Complete Works of Agronomy* (republished in 1979) written by Xu Guangqi, a noted agronomist in the Ming Dynasty, summarized his experience on the intercropping of wheat and broad bean.

In the Qing Dynasty, the crop rotation, intercropping, and land use and land maintenance had developed more comprehensively.

Agriculture of Mashou County written in 1836 developed the experience of the crop rotations and shifts in northern China summarized by *Important Means of Subsistence for Common People*. It stressed that the rotation of grain and leguminous plants was the basic means to maintain soil fertility and increase yields in the grain production area of northern China (Lin, 1996).

Based on the experiences of past farming practice, the system of rotation between grain and green manure plants, and the system of intercropping, had been developed and applied much further in the Qing Dynasty. These developments were mirrored in many agricultural writings of that time.

In general, the basic principles of farming practices in the Ming and Qing Dynasties, as pointed out in the book of *Agricultural Theory* written by Ma Yilong in the Ming Dynasty, are that farming activities should be in accordance with seasons, climate, soil conditions such as soil moisture and soil fertility, and crop growth, i.e., farming practices are applied in light of local conditions (Song, 1990). Yang Shan, the Qing Dynasty, in particular, expounded that the basic task of farming was the adjustment of water, nutrients, atmosphere (Yang, 1984). Meanwhile, the new and comprehensive development had been made regarding the accumulation and creation of manure as well as the application of manure during this period.

CHARACTERISTICS OF THE CLASSICAL FARMING SYSTEMS IN CHINA

Making Full Use of Land

Shang-yang, an ancient Chinese politician in the Qin Dynasty (390-338 B.C.), advocated abandoning the Jing Fields and making footpaths between fields, reclaiming wasteland, and using fallow fields for the purpose of making full use of available land (Lin, 1996). Guanzhong (645 B.C.) also stressed natural systems and maintenance of the soil fertility (Ding, 1993). *Important Means of Subsistence for Common People* written by Jia Shixie (6th Century) in the Late Wei Dynasty emphasized to "be agreeable to natural conditions and measure soil fertility." In the South Song Dynasty, Chen Fu (1076-1154) in his work *Agricultural Treatise of Chen Fu* pointed out that planting should never be carried out on false days and harvests never in false months. Zhang Luxiang, in the Qing Dynasty, also proposed in his *Supplementary Agricultural Treatise* to use natural conditions, depend on manpower and promote soil fertility (Miao, 1982, Chen 1983).

Correctly Handling the Relationship Between Land Use and Land Maintenance

Only when the relationship between land use and land maintenance is handled in balance, can a high soil fertility level be maintained and a stable harvest and the product quality be guaranteed. *Agricultural Treatise of Chen Fu* proposed the concept of 'keeping the high fertility of land' (Miao, 1981). The reason why the soil fertility has not declined after thousands of years of agricultural activity is that Chinese farmers have combined the intensive land use with the positive land maintenance. The soil resources have been intensively managed and properly used with organic manure fertilization, proper farming practices and biological measures such as crop rotations and intercropping.

Intensive Farming

To improve the efficient use of land, Chinese farmers have developed many farming practices such as crop rotations, intercropping, and multiple cropping. In addition, engineering measures, such as building reserviors, leveling land, building terraces, and biological measures have also been adopted, which produced a significantly positive effect on agriculture. The intensive farming can be considered as a kind of agricultural technology to overcome adverse factors using proper and effective measures. Characterized by proper and intensive land uses combined with land maintenance, the classical farming systems have the embryonic forms of the modern ecological agricultural strategies and sustainable agricultural methods and technologies.

THE REALISTIC SIGNIFICANCE OF CLASSICAL FARMING SYSTEMS

Nowadays, the expanding population and the decreased cultivated area are putting more and more pressure on food production. Further development of agriculture is limited by worldwide soil degradation resulting from a variety of factors, among which the decreasing use of organic manure is of the most importance. In China, the applied amount of various chemical fertilizers is increasing year after year and now reaches 248 kg per hectare per year. In contrast, the applied amount of organic manure is becoming less and less.

In the recent decades, a variety of modern agricultural systems, including natural farming, organic farming, ecological agriculture and sustainable agriculture, have been proposed and put in practice not only for the sake of maintaining soil fertility and high production output of farmed land but for environmental conservation and human health. Among the modern agricultural systems mentioned above, actually the natural and the organic agricultural systems are not really new but are closely related to the traditional methods of farming systems practiced for thousands of years in China. Undoubtedly, the abundant and valuable experiences accumulated in the ancient Chinese farming practices, as discussed in the foregoing, can be very helpful to the modern forms of these agriculture systems.

The application of organic manure was the major way to fertilize soils in ancient China. It is estimated that there are approximately ten groups, more than a hundred kinds of organic manure documented in the ancient writings, including more than 10 kinds of nightsoil, dung and excrement of birds and animals, 12 kinds of cake manure, 12 kinds of residual manure, 9 kinds of bone manure, 22 kinds of green manure, 7 kinds of silt manure, 4 kinds of earth manure, 4 kinds of stalk manure and 4 kinds of mixed manure. In contrast, only seven kinds of inorganic fertilizers are recorded in the classical works, which are lime, gypsum, edible salt, halide water, sulfur, white arsenic, and sulfate salts of copper, zinc and iron. Inorganic manure is fewer than one-tenth of organic manure, indicating that in ancient China, people already knew that multiple manure could fertilize soil and raise soil productivity. Evidently, the knowledge on the resources and application of the organic manure, accumulated in the long history of classical farming systems, has realistic significance in the regeneration of organic agriculture in China and throughout the world.

For thousands of years, Chinese farmers have followed the principle of a close linkage of land use with land maintenance. They have not only made full use of land, but also positively maintained land. It is the quintessence of the experience of the classical farming practices in China that under the condition of making full use of land by the systems of crop rotation and intercropping. This treasure of the classic farming knowledge should be used as a database for the development of the ecological and sustainable agricultural systems.

REFERENCES

Cao, Longgong. (1984). *Talks on the History of Fertilizers* (in Chinese). Beijing: Agriculture Press, pp. 3-71.

Chen, Hengli. (1983). *The Collation and Interpretation of Supplementary Agricultural Treatise* (in Chinese). Beijing: Agriculture Press, pp. 29-177.

Ding, Peng. (1993). Study on the theories of agronomy from Guan-Zi (in Chinese). *Research on Agricultural History*, No. 1.

Guo, Wentao. (1981). *The Framing Methods and Cropping Systems in the Ancient China* (in Chinese). Beijing: Agriculture Press, pp. 8-202.

Guo, Wentao. (1988). *A Brief History of the Development of Agricultural Science and Technology in China* (in Chinese). Beijing: Agriculture Press, pp. 109-404.

Guo, Wentao. (1989). *Chinese Modern History of Agricultural Science and Technology* (in Chinese). Beijing: Agricultural Science and Technology Press, pp. 66-191.

Hu, Xiwen. (1959). *The Selections of Agronomic Heritages in China–Crops* (in Chinese). Beijing: Agriculture Express, pp. 106-108.

Jiao, Bin. (1984). The history and evolution of green manure in China (in Chinese). *Chinese History of Agriculture*. No. 2.

Li, Bozhong. (1980). *The Agricultural Developments in the South of the Lower Reaches of Yangtze River During the Tang Dynasty* (in Chinese). Beijing: Agriculture Press, pp.91-94.

Li, Changnian. (1981). *Talks on the History of Agriculture* (in Chinese). Shanghai: Shanghai Science and Technology Press, pp. 65-120.

Lin, Putian. (1996). *Soil Classification and Land Utilization in Ancient China* (in Chinese). Beijing: Science Press, pp. 100-178.

Lu, Huaxi. (1989). Study on the agricultural developments in Chinese Qin and Han Dynasties (in Chinese). *Agricultural Archaeology*. 1989 (2): 183-184.

Miao, Qiyu. (1981). *The Selecting Readings from "Agricultural Treatise of Chen Fu"* (in Chinese). Beijing: Agriculture Press, pp.1-10.

Miao, Qiyu. (1989). *The Collation and Interpretation of "Important Means of Subsistence for Common People"* (in Chinese). Beijing: Agriculture Press, pp. 16-272.

Pan, Qixing. (1988). *The Guides to "God Creates Things"* (in Chinese). Chengdu (China): Bashu Press, pp. 53-54.

Peng, Youliang. (1985). The Development of agricultural economy in Fujian in two Song Dynasties of China (in Chinese). *Agricultural Archaeology*. 1985 (1): 27-37.

Research Department of Agricultural Heritage of China. (1984a). *History of Chinese Agronomy I* (in Chinese). Beijing: Science Press, pp. 104-206.

Research Department of Agricultural Heritage of China. (1984b). *History of Chinese Agronomy II* (in Chinese). Beijing: Science Press, pp. 31-157.

Sang, Runsheng. (1982). Historic experiences of double cropping of rice in the Yangtze River valley (in Chinese). *Agricultural Archaeology*. 1982 (2): 52-64.

Shi, Shenghan. (1979). *The Selected Readings from the Agricultural Works of Two Han Dynasties* (in Chinese). Beijing: Agriculture Press, pp. 3-28.

Song, Zhanling. (1990). *Systematization and Research of "Agricultural Theory"* (in Chinese). Nanjing (China): South-East University Press, pp. 40-120.

Tang, Qiyu. (1985). *The Drafts of Agricultural History of China* (in Chinese). Beijing: Agriculture Press, pp. 206-566.

Tong, Enzheng. (1989). The origin and characteristics of agriculture in South China (in Chinese). *Agricultural Archaeology*. 1989 (2): 57-71.

Wan, Guoding. (1980). *The Collected Interpretations on "Fan Sheng" Works on Agriculture* (in Chinese). Beijing: Agricultural Press, pp. 23-150.

Wang, Yunsen. (1980). *Soil Science in Ancient China*. Beijing: Science Press, pp. 10-205.

Xu, Guangqi. (1879) (Ming Dynasty). *Complete Works of Agronomy* (Republished in 1879, in Chinese). Shanghai: Shanghai Press for Ancient Books.

Yang, Zhimin. (1984). About the series of agronomic books (Written in the late Qing Dynasty, in Chinese). *Agricultural Archaeology*. 1984 (1): 19-29.

Production and Application of Organic Materials as Fertilizers

Shuxun Mo

SUMMARY. The rapid social and economic changes in recent years have enhanced the development and availability of a large number of industrial and agricultural products. Such activities, in turn, have greatly increased the amounts and volumes of organic wastes and residues. It is important for Chinese scientists to know the types and amounts of waste materials produced each year and, if suitable for land application, utilize them as soil conditioners and fertilizers. Such information with respect to China is reviewed and discussed in this paper. *[Article copies available for a fee from The Haworth Document Delivery Service: 1-800-342-9678. E-mail address: getinfo@haworthpressinc.com <Website: http://www.HaworthPress.com>]*

KEYWORDS. Organic materials, organic wastes, urban refuse, compost

INTRODUCTION

Agriculture in China has a history of at least 3,000 years in applying organic fertilizers. Many ancient Chinese agronomists and agro-administrative officials had perceptual knowledge about the nature of many kinds of

---

Shuxun Mo is Research Soil Scientist, Institute of Soil Science, Chinese Academy of Sciences, Nanjing, China (E-mail: gbluo@issas.ac.cn).

[Haworth co-indexing entry note]: "Production and Application of Organic Materials as Fertilizers." Mo, Shuxun. Co-published simultaneously in *Journal of Crop Production* (Food Products Press, an imprint of The Haworth Press, Inc.) Vol. 3, No. 1 (#5), 2000, pp. 23-39; and: *Nature Farming and Microbial Applications* (ed: Hui-lian Xu, James F. Parr, and Hiroshi Umemura) Food Products Press, an imprint of The Haworth Press, Inc., 2000, pp. 23-39. Single or multiple copies of this article are available for a fee from The Haworth Document Delivery Service [1-800-342-9678, 9:00 a.m. - 5:00 p.m. (EST). E-mail address: getinfo@haworthpressinc.com].

organic manure. "To cultivate farmland well, fertilization and irrigation must be achieved." Such knowledge was recorded in the works by Han Feizi, called Yelao, in the last stage of the Warring States (475-221 B.C.). Beanstalk, silkworm manure, mud and a mixture of decomposed pig and human excreta were listed as fertilizers in Fan Shengzhi's Works on Agriculture in the Western Han Dynasty (206 B.C.). Ways and means to collect, manufacture and use various organic materials as fertilizers were described in great detail in *Important Means of Subsistence for Common People* written by Jia Simiao of the Northern Wei Dynasty (386-534 A.D.). These ancient writings have won high praise from many Western scientists (King, 1911; Allison, 1973; FAO, 1977; Liebig, translated text, 1983). They also aroused great interest to Roelcke, a student at the university of Munich in Germany, who came to China at his own expense to learn how to make silt grass pit manure, a waterlogged compost produced in the south of Jiangsu Province and reported the technology in his graduate thesis in 1987. Some of this knowledge is still handed down today, and is often the result of extensive research conducted in earlier times. In some areas inhabited by minority nationalities such as the Guangxi Autonomous Region, with a strong local economy and scientific support, organic manure is increasingly applied to designated "sanitation fields" (fields where fertilizers have never been used) along with increased applications of chemical fertilizers. According to statistics (Soil and Fertilizer Station of the Agricultural Department of Guangxi Province, 1997), 51.1 million tons of organic manure was used in the Region in 1997 alone, which was 94.6% more than in 1986. The average amount of organic manure applied increased from 2.89 to 5.79 t/ha. In the areas of Suzhou, Jiangsu Province, more straw was applied as manure to paddy fields in 1980s than animal excreta, which could help to maintain soil fertility and stabilize farm production (Huang and Yu, 1991).

Since the 1970s, China's chemical industry has developed rapidly. Large amounts of chemical fertilizer, especially nitrogen, are being applied to soil. With the improvement of other production inputs, agricultural yields have increased dramatically. The production and application of traditional organic manure, both requiring considerable hand labor gradually decreased, especially with the rapid development of commodity oriented economies in some coastal regions of China. Today, with increasing applications of chemical fertilizers, farmers need to resume the regular application of organic materials as manure on their fields for the following reasons.

Depleted Soil Fertility. Soil nutrients have been removed by intensive cropping and not replaced, also organic materials are applied sporadically, if at all. Consequently, soil fertility has declined. Organic materials provide a storehouse, reserve supply of plant nutrients (which chemical fertilizers can not do) that can maintain adequate soil fertility and guarantee sufficient long-term food production for consumers.

Low Soil Productivity. There are still many unproductive lands in China. They are infertile or have problems such as salinity, acidity or other poor chemical and physical properties for optimum crop production. The proper and regular application of organic materials is probably the most effective remedial measures to overcome these problems.

Environmental Pollution and Human Health. Suitable and non-toxic organic wastes should be recycled as organic fertilizers on land to avoid improper disposal resulting in environmental pollution and hazards to human health. This is an urgent problem for China now and in the future. To ensure environmental protection and a balanced agroecosystem it is very important to apply chemical fertilizers rationally, while utilizing organic materials as fertilizers as much as possible.

The purpose of this paper is: (1) to review the kinds and amounts of organic materials that can be used as organic fertilizers in China; (2) to consider some waste management practices and applications which improve on the traditional methods but with modern aspects; and (3) to identify problems in the disposal and application of organic materials, including technical, economic and administrative aspects, and to propose possible solutions.

PRODUCTION OF ORGANIC MATERIALS AS FERTILIZER

China has vast areas of land, a large population and a favorable climate. The production of organic materials is increasing at an average rate of 5 to 10% each year (Qi, 1998). Based on their possible use as organic fertilizers and soil conditioners for agricultural lands, organic materials are categorized into four groups.

Primary Organic Residues

This group includes mainly crop straws and other crop residues such as corn pith, pod vine and dropped vegetable leaves. In some regions these materials are burned as fuel sources. However, in areas where coal, gas, electricity and firewood are available considerable amounts of straw are conserved. Some basic data about the chemical composition of crop residues (i.e., straws) are shown in Tables 1a and 1b (Ma, 1995; Liu and Jin, 1991; Qian and Fang, 1997).

Straw is an important material for making compost and conditioning stable manure. For example, in north China, straw and grass are used in

TABLE 1a. Production and inorganic composition of major crop straws in China

Straw	Total yield (10^8 t/y)	Accounted for total bio-yield (%)	Inorganic composition of straws					
			N	P_2O_5	K_2O	Ca	S	Si
			%					
Rice	1.8	50	0.51-0.63	0.11-0.17	0.85-2.70	0.16-0.44	0.11-0.19	7.99
Wheat	1.1	52	0.50-0.67	0.20-0.34	0.53-0.64	0.16-0.38	0.12	3.95
Corn	2.0	60	0.48-0.50	0.38-0.40	1.67	0.39-0.80	0.20	--
Soybean	0.15	61	1.3	0.3	0.50	0.79-1.50	0.23	--
Rape	0.39	83	0.56	0.25	1.13	0.42	0.35	0.18

TABLE 1b. Organic composition of major crop straws in China

Straw	Organic composition of straws				
	Ash	Cellulose	Wax	Protein	Lignin
	%				
Rice	7.8	35.0	3.82	2.28	7.95
Wheat	4.3	34.3	0.67	3.00	21.2
Corn	6.2	30.6	0.77	3.50	14.8
Legume grass (dry)	0.1	28.5	2.00	9.31	28.3
Rape*	--	37.5	--	2.70	14.9

* Lin Xinxiong, "*Resources and Conservation*" Vol. 13, 150, 1987 (% of ash-free dry weight).

making compost and conditioning stable manure rather than traditionally-used mud or soil. Straw and grass will absorb greater amounts of feces; and the weight of compost made from these materials is much lower while its quality is higher (Table 2) (Lu, 1986).

The use of straw and grass as bulking materials in making compost results in a 67% saving in labor for transporting and applying the compost. A method of making stable manure with straw called "elevated stand for cattle bedding" (Mu and Zhao, 1996) is being widely adopted in Henan Province. A low, brick, enclosing wall is built around a shed with a center post for tethering the cattle. Then straw or other dry grass is spread for bedding at several day intervals. The stable manure is thereby accumulated each day and layer by layer, Thus producing greater amounts of manure than by traditional scattering, or from only one single layer of bedding material. This new method greatly reduces the potential for environmental pollution.

Secondary Organic Wastes

These include mainly livestock wastes. Poultry and animal excreta are the main resources of manure, which account for 67% of the total organic manure in the Chinese countryside. The estimated 1995 populations of livestock in China, (Agricultural Yearbook of China for 1996, Cai and Chen eds. 1996) and their annual excreta and nutrient production are shown in Table 3 (Lu and Shi, 1982).

In addition, other wing residues or by-products of agricultural, husbandry and industry, e.g., bran, chaff, husk; slaughterhouse wastes, fur, hooves and horns, residue of fish and shrimp in aquatic production; oil cake, dregs; wood shavings, sawdust, and by-products of paper mills. Almost all of them are very good organic materials for fertilizers if they could be collected and stored properly. But considerable amounts of them are discarded as wastes

TABLE 2. Composition of straw-manure compost and soil-manure compost (%)

Types of compost	Compost composition		
	OM	Total N	Total P
	%		
Straw-manure compost	15.2-27.8	0.30-0.89	1.82-3.15
Soil-manure compost	1.24-2.14	0.05-0.12	0.16-0.32

TABLE 3. Livestock population in China in 1995, and their annual production of excreta and nutrients

	Animal populations and excreta produced			Excreta produced (kg/head/y)			Total macronutrients produced (10^4 t/y)		
	(10^8 head)	(kg/head/y)	total(10^8 /y)	N	P_2O_5	K_2O	N	P_2O_5	K_2O
Pig	4.4	950	2.09	5.13	3.08	6.31	112.9	67.8	138.8
Cattle	1.3	7800	5.07	39.7	12.2	31.0	258.1	79.3	201.5
Horse	0.3	5300	0.69	33.9	13.0	26.1	44.1	16.9	33.9
Sheep	2.7	250	0.34	2.28	1.03	1.27	30.8	13.9	17.2
Chicken*	41	7.5	0.15	0.12	0.12	0.06	25.0	23.8	13.6
Total							470.9	201.7	405.0

* Chicken manure: N 1.63%, P_2O_5 1.54% and K_2O 0.85% from Fertilizer's Manual compiled by Beijing Agriculture University (in Chinese).
The totals show the annual production of macronutrients (N, P, K) × 10^4 (10,000 tons) from livestock in China.

without being recycled. "How wastes are managed has an effect on people everywhere." "Population and income levels are increasing worldwide. That means an increased demand for meat and poultry" (Bentley, 1975). It is clear that a large amount of poultry and animal manure is produced. Tremendous efforts have been made in many countries with regard to research and practice in managing livestock wastes. China has considerable experience in disposing of livestock wastes, but generally, it is done with hand labor on a small scale. With the rapid social and economic developments, the demand for ecological engineering technologies in managing livestock wastes properly and for the best advantage to society is increasing. There are many ecological engineering technologies, ranging from simple oxidation ponds to systematic engineering with equipment and machines for harmless disposal of poultry and animal excreta. Some advanced techniques of different scales have been introduced in recent years. Especially, the equipment and technique for disposal of chicken manure are now more advanced in China. For example, bio-organic manure was first developed in Hubei Province. Poultry or animal excreta was processed in only 5-7 days, i.e., fermented, deodorized, dried and stockpiled using a comprehensive bacterial technology and an advanced column-type fermentation tank. Then it is mixed with a certain amount of chemical fertilizers (N-P-K) and bacterial manure and finally granulated. The product contains 35 to 40% organic matter, 20 to 25% total N-P-K, and some microorganisms, including bacteria. Desirable results have been achieved with this product in a pilot scale experiment.

Compound granulated organic fertilizers are made by mixing some additives with the remaining sediment of methane production at Tian Hua Poultry and Animal Breeding Farm in Changsha, Hunan Province. This fertilizer contains humic acid at 26.0%, N 40.4%, P 0.64% and K 2.26% (Yang, 1996).

A systematic engineering approach for handling chicken manure has been installed at a pedigree farm near Harbin, Heilungjiang Province. Some 80% of its products are used as fertilizer and 10% as feed, while the remaining sediment might still be utilized to produce methane (Li and Wang, 1995).

The utilization ratio of livestock waste produced at the 729 large- or medium-size livestock farms in the Shanghai suburbs amounts to 87.4%. Of which 73.4% is fermented through high-temperature compost or septic tanks and then applied directly to soil, and 14% of which is manufactured as a granulated fertilizer after it is fermented for methane (Qi, 1998).

With a proper management of wastes, economic profit and ecological benefit can both be achieved.

Wastes from People's Daily Life

Human Excrement. The amount and composition of human excrement are shown in Tables 4 and 5 (Lu and Shi, 1982).

TABLE 4. Average composition of human excrement (% of fresh weight)

Item	Water	OM	Mineral	N	P_2O_5	K_2O	CaO	C:N
	(%)							
Feces	75	22.1	2.9	1.5	1.1	0.5	1.0	7.3
Urine	97	2.0	1.0	0.6	0.1	0.2	0.3	1.3

TABLE 5. Annual production and nutrient content of human excrement (fresh weight basis)

Item	Feces: urine	Amount	N	P_2O_5	K_2O	N	P_2O_5	K_2O
		kg/person/year				kg/t		
Human excrement	10:90	485	5.2	1.25	1.08	11.0	2.6	2.2

Generally, human excrement is a potential carrier of pathogenic bacteria. To use it directly as fertilizer or for general disposal is not safe. Since the 1970s many kinds of simple equipment, such as a vat or a small cement tank with cover, have been built and popularized. To conserve nitrogen in the excrement, straw, sod or peat, $CaSO_4$ or 3-5% calcium superphosphate are added. $FeSO_4$ is added to remove offensive odor and insecticide sprayed to kill insects and possible disease causing organisms. Several typical latrines are designed. For example, a latrine with a funnel-shaped pan and double urns has become very common in Henan Province (Mu and Zhao, 1996). A dry feces latrine was designed by Hui Yan and Dong Wan Counties, Guangdong Province (Liu and Jin, 1991) and a latrine with three squares joined together is used in the countryside of north China (Sanitation and Antiepidemic Stations of Shandong Province, 1977). All of these latrines are sealed to isolate them from the outside and to achieve safe disposal of human excrement. In most urban areas, flush toilets are used and the excrement is discharged to sewers.

Urban Refuse-Municipal Solid Waste and Garbage. There are almost 400 cities in China. With the increase in population, refuse production has increased at an annual rate of 10%. The current production is 55 to 65 million tons of refuse a year, much of which is now generally discharged to suburbs. In some areas, it is just simply dumped into rivers and lakes, which reflects a lack of recycling and disposal facilities. This has obviously caused serious environmental problems. Since the 1970s, studies and practices have been conducted to properly and safely dispose of refuse from daily life. It is mainly disposed of as compost. Because of their low investment, the composting facilities have developed rapidly in China (Chen and Zhang, 1990).

Composting of Refuse with Mechanized Technology. The modern compost production generally utilizes aerobic fermentation technology which includes processes of pre-disposal (crushing and sorting), later fermentation and disposal, and abating offensive odors. The required conditions are: organic matter 20 to 80%, water content 50 to 60%, temperature 50 to 65°C, C:N 30, P:N 75 to 150 and pH 5.8 to 8.5. Composting plants that can process 100 to 150 tons of refuse per day have been established in such cities as Wuxi and Suzhou of Jiangsu Province, Wuhan of Hubei Province, Anqing of Shanghai, Dagong of Tianjin, Xiamen of Fujian Province and Beijing (Chen and Zhang, 1990).

Simple and Easy High-Temperature Composting. Refuse may also be composted using manual operations, partly combined with machines and enclosed by a thin membrane. Compared with anaerobic processing, its composting period is shortened by almost 100%. Better results have been reported in Tianjin, Jixi of Shannxi Province, Baise of Guangxi Province, Nanping of Fujian Province, Shijingshan of Hebei Province and Beijing. Almost 4×10^4 tons of crude organic manure and 2×10^4 tons of refined manure were produced with this technology in 3 disposal yards within a period of 5 years in Anyang, Henan Province. They help a lot in supporting agriculture and removing refuse from the city. This technology is suitable for adoption by medium-and small-size cities or towns. More simple open-space composts are also tested in some regions. The composition of composts varies according to the technology used, feed stocks and raw materials, composting period and processes controlled. General requirements (dry weight basis) are: organic matter 35%, N 2%, P 0.8%, K 1.5%; disease causing organisms, phytopathogens, weed seeds and insect eggs are killed; no inert materials present such as glass and stone. Allowable concentrations of heavy metal are: As < 50 ppm, Cd < 5 ppm, Hg < 2 ppm, Pb < 3 ppm (wash and extract); C:N 20 and salt 1 to 2%, no foul smell, dark brown color and loose (Science and Tech. Dept. of Social Development of the National Scientific Commission et al., 1992). In the composting process, it is important that the easily biodegradable organic materials decompose rapidly and thoroughly, otherwise certain residual phytotoxic materials could injure growing plants. It is not enough to solely rely on the C:N ratio to judge the degree of maturity of urban refuse compost. CEC is a more reliable indicator for this purpose. The compost of urban refuse suitable for application as a fertilizer has a C:N of 20, total N of 2%, reducing sugar C: total C, 35 and CEC 60 meq/100 g (ash-free dry weight basis) (Harada and Inoko, 1989).

Natural Products

Peat and Peat-Like Materials. The supply of peat is abundant in China. Although most is of low quality, deposits are found in all provinces. Its general composition is shown in Table 6 (Mao, 1981).

Peat can be used to produce compost, compound fertilizer, granulated fertilizer, and bacterial manure. The insoluble humic acid or humate in peat could be made more soluble by ammonification, acidulation, salt double decomposition, extracted with alkali and nitric acid oxidation. The products of peat are: nitro-humic acid, humic acid and complex fertilizers of NH_4^+, Na, K, P and NP humic acid (Institute of Chemical Fertilizers and Pesticide of Shanxi Province, 1974). With a proper amount of humic acid as a binding agent, chemical fertilizer and micronutrient carrier, could be combined into compound fertilizers of varying composition (Mu and Zhao, 1996).

Sludge. The rain water or wind carries away the organic materials, inorganic nutrients and fertile sediment from the soil surface and they flow together into rivers and lakes and the planktons flourish, die and decompose on the river bed or lake bed. The amount of sediment deposited in Henan Province, a plain region, is estimated at 6.2 million tons (Mu and Zhao, 1996). Before the 1970s the sludge was collected and applied as fertilizer in Jiangsu Province, a region rich in rivers and lakes, accounting for 50 to 80% of the total "organic" or natural manure applied. Since the 1980s, because of rapid social and economic development, the application of this sludge has greatly decreased. According to a survey conducted in 1993, the sediment portion of the sludge accumulated only in the medium-size rivers and lakes of Jiangsu Province has reached 51×10^8 tons. Sludge rich in nutrients and organic matter is very good fertilizer (Yu, 1993), the composition of which is shown in Table 7.

TABLE 6. Composition of peat in China (6 regions)

Item	OM	Humic acid	C:N	pH	ash	N	P_2O_5	K_2O
	%	%	(%)					
Peat	40-70	20-40	10-20	4.5-6.5	40-60	1.2-1.9	0.02-0.3	0.24-0.5

TABLE 7. Comparison of nutrient content between soil and sludge in Jiangsu Province*

Name	OM (%)	Total N (%)	Available P (mg/kg)	Available K (mg/kg)
Soil	1.63	0.11	5.51	117.7
Sludge	3.08 ± 1.57	0.15 ± 0	40.3 ± 25.1	166.4 ± 49.4

* Analysed in 1989 and the results came from 12,764 times of analysis.

APPLICATION OF ORGANIC MATERIALS AS FERTILIZER

The application of organic manure differs from place to place because of such variables as natural conditions, soil types, cultivation systems, and even habits and customs of the Chinese people. Some farmers utilize limited organic materials they produce either entirely with traditional or with improved methods so as to maintain a reasonable level of soil fertility. Farmers who are engaged in other industries have no time to attend to their lands. Thus, most of the organic materials they produce are left there and go to waste. Some families of special husbandry or planting neglect their produced organic materials that they produce and pollute the environment. Some farmers produce commodity organic materials and sell them, hence the supply falls far short of the demands of the vast countryside. Most of them have very limited funds and the technology they adopt can not efficiently utilize organic materials as manure. The following methods are generally adopted by farmers or farm workers in China.

Direct Application

Straw, including chaff, husk, corn pith, pod and vine, etc., that are applied directly account for 15% of the total organic materials utilized as fertilizers. This approach is popular because it saves labor, reduces costs and is easily utilized. There are three methods for direct application: to spread it on the field evenly and then plow it down into the soil; to use it as mulch for controlling weeds, soil erosion, and soil moisture by covering all the field or only the interrow area; and to keep high stubble at harvest. The benefits of applying straw directly are as follows.

Supplementing Soil with Nutrients. Nutrients contained in the straw of several crops are reported in Figure 1 (Hidefuno, 1980).

The proportion of nutrients, especially K, Ca, Mg, B, and Mn, in the crop straws of Figure 1 are quite abundant. According to the study (Mo and Qian, 1981), more than 70% of the K taken up from soil by rice remains in straw. About 80% of this K may be extracted by cold water. Applying rice straw is very effective to supplement soil with K, especially for those fields lacking in K. Also, the silicon (Si) content of rice straw is generally high, especially for paddy rice. In one experiment, after three months of applying rice straw, the available Si content of soil was almost three times higher than without straw application (Zeng, 1995). Applying straw for longer periods would increase the soil fertility potential.

Protection and Improvement of Soil Fertility. Mulching soil with straw can reduce soil erosion, especially on sloping land with a loose surface soil and in the rainy seasons (Table 8) (Phillips and Young, 1973).

Straw rich in organic matter could begin to activate soil microorganisms,

FIGURE 1. Distribution of nutrients in straw and in economic products

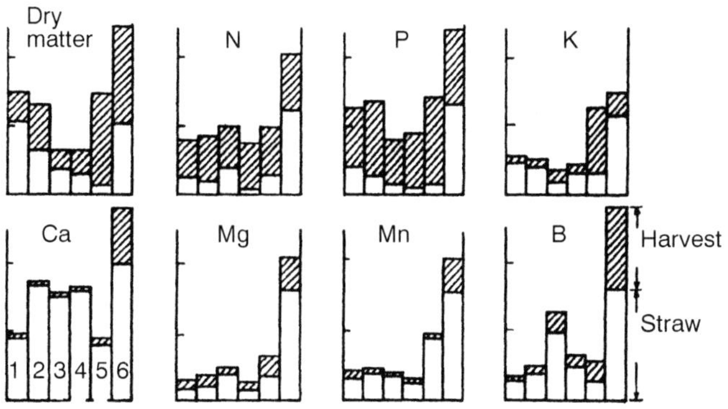

Crop: 1. Wheat, 2. Corn, 3. Soybean, 4. Kidney bean, 5. Barley, 6. Sugar beet

TABLE 8. Effect of mulch with straw on runoff, percolation and loss of soil

Amount of crop residue (t/ha)	Runoff (%)	Percolation or infiltration (%)	Amount of soil loss (t/ha)
0	45.3	54.7	15.4
0.31	40.0	60.0	4.0
0.62	24.3	74.7	1.75
1.24	0.5	99.5	0.37
2.47	0.1	99.5	0
4.94	0	100	0

renew soil organic matter and increase the level of physiologically-active materials (Table 9) (Xu et al., 1985).

The effect of straw on soil physical properties may require a longer time for expression.

Improvement of the Eco-Environment of Field. Mulching soil with straw could reduce evaporation, conserve soil moisture, and modify the temperature and humidity of fields (Zhao et al., 1996) and suppress weed growth (Tables 10 and 11) (Liu, 1997).

But disease-carrying straw must not be used. To achieve better results when applying straw as manure, technologies must be considered such as the amount of straw needed to return to soil, plowing time and depth, cutting length, soil moisture and inorganic fertilizer needed to promote its decom-

TABLE 9. Effect of mulching soil with wheat straw on numbers of N-fixing bacteria at different stages of soybean ($\times 10^4$ bacteria/g soil)

Treatment	Growing stages				
	Before sowing	Early blossoming	Podded stage	Milky grain	Ripe stage
Mulching (2.25 t/ha)	0.799	78.2	11.5	8.42	7.67
Control	0.799	7.43	10.6	1.12	6.15

TABLE 10. Effect of mulching with straw on conservation of soil moisture

Treatment	Crops			
	Winter wheat	Spring corn	Fallow in summer	Fallow in summer
	(mm)			
Mulching	153.4	207.0	389.9	374.9
Control	105.6	137.7	344.6	329.7

TABLE 11. Effect of mulching rape with straw on control of seeds

Treatment	Measuring date		
	Dec. 5, 1994	March 10, 1995	May 24, 1995
	(Weeds/m^2)		
Mulching	41	75	38
Control	109	131	96

position, etc. All of this should be done according to the agroecological conditions of different regions.

Besides straw, the sludge can also be applied directly. Pump boats equipped to remove and spray sludge have been used in some regions in place of partially manual labor operations. Sludge is very suitable for growing rice seedlings. An experiment that showed good results used a double plastic membrane, one with holes covering the seedbed on which sludge was spread for sowing, and another with the upper part covered to maintain temperature (Zhang and Ji, 1998).

Application of Compost After Decomposition

Composts which are wet (in arid regions) or waterlogged (in humid regions) are popular fertilizers in the countryside of China. It has no fixed model but is acceptable to local conditions and works simply and cheaply. Most organic wastes in the village can be used for compost; however, its quality differs greatly from place to place. In the suburbs of Shanghai, fecal wastes are collected, decomposed, stored and transported to fields for application by spraying or trickling. This system has very high efficiency (Wang et al., 1993).

Organic Materials as Fertilizers After Initial Energy Recovery and Food Chain Use

In general, all organic materials contain energy and organic or inorganic nutrients. To use them first as fuel sources (not to burn but to make methane) or fodder and then use their residues as fertilizer are ways of increasing their net return. Only 10 to 20% of the total energy can be obtained from burning straw while most of the nutrient elements contained are lost, and the process may cause environmental pollution. A recent experiment showed that 1.5 kg of pig feces could produce 1 m^3 methane while 50% of its organic carbon and 90% of N-P-K nutrients were preserved in this sediment (Jin, 1989). Now 5.5 million biogas units have been built. It is estimated that 13×10^8 m^3 methane could be produced each year in China and 33 million tons of digested slurry could be used as fertilizer (Qi, 1998). The residual material could be used to feed animals and produce edible mushrooms (He et al., 1984), earthworms (Ge and Ji, 1986) and fish (Zheng and Deng, 1998) and, after multiple recycling, could still be used as fertilizer. All of these alternatives have been practiced in some regions of China. Whatever models are adopted, social and economic benefits will be derived and good organic fertilizer will be produced.

PROBLEMS AND SUGGESTIONS

China has made great achievements in application of organic materials as fertilizer along with its economic development and social progress. However, many of these changes came too quickly and created many problems. Organic materials have accumulated like mountains in many cities because there are insufficient means to process them and return them to the land. The utilization of human excreta still involves less than 1/3 of the total amount produced. Basic facilities for disposing the ever increasing volume of urban refuse are few in number and less than adequate. Most of poultry manure and animal excreta are wasted and quickly become sources of environmental pollution. A great deal of straw is still burned or discarded by the farmers. Many operations for composting or processing organic materials as fertilizers are performed by hand. Ironically, soil fertility is being depleted on the one hand and organic materials are being wasted on the other (Liu, 1996). In view of this current situation in China, the following suggestions are proposed:

- Renew the once intimate relationship between human beings, especially farmers, and the land. We should rely heavily on government policies and the establishment of a series of related laws and regulations.
- Farmers need to acquire training and experience in agricultural production, but also acquire knowledge and technology related to organic recycling of wastes and residues as soil conditioners and fertilizers.
- National and local governments need to provide adequate funds to establish permanent installations for collection, storage and fermentation of organic materials and to provide necessary machines or animals for collection, transportation and application of organic fertilizers to the land.
- Scientific and technical experts are needed to ensure that scientific knowledge is translated into practical applications and best management practices for farmers, especially on the role of organic fertilizers in sustainable agriculture.
- Research is needed to develop advanced technologies and build large-scale industries to produce organic commodity fertilizers continuously and to alleviate the problems that cities face in treating and/or disposing of organic wastes.

CONCLUSION

China has a longstanding history of applying organic materials as fertilizer. With the development of the chemical industry, chemical fertilizers, particularly nitrogen sources are applied increasingly at high rates and has largely supplanted the application of organic fertilizers which used to be handled by traditional methods. It is important to reemphasize the use of organic materials as fertilizers and the value of their contained nutrients and organic matter, in order to (a) supplement soil nutrients and organic matter depleted by intensive and improper farming practices; (b) improve the yield and productivity of infertile and degraded soils; (c) protect the environment from pollution, and (d) balance the agroecosystem. The production of organic materials that can be used as fertilizers is enormous in quantity and varied in methods. In general, crop residues (i.e., straw), livestock manure and human excreta are most abundant. Substantial economic profits, and ecological and social benefits could be achieved by collecting various organic materials painstakingly, disposing them in accordance with local conditions and applying them rationally to land. But there are some serious problems and constraints in this regard. In order to fully utilize the resources that are being wasted or discarded, scientific knowledge, technological applications and administrative support will be needed.

In the final analysis, and to the fullest extent possible, acceptable organic materials from both rural and urban sectors must be utilized as soil conditioners and fertilizers to achieve a more sustainable agriculture worldwide.

REFERENCES

Allison, F.E. (1973). Some agricultural practices over the centuries. In *Soil Organic Matter and Its Role in Crop Production: Development in Soil Science*, New York: Elevier Scientific Publishing company of Amsterdam, Vol. 3. pp. 4-6.

Bentley, O.G. (1975). The livestock waste program in agriculture colleges. In *Managing Livestock Wastes (Proceedings of the 3rd International Symposium on Livestock Wastes)*, ed. State University of Illinois, Illinois: American Society of Agricultural Engineers, p. 2.

Cai, S.L. and W.L. Chen. (1996). Amount of main livestock of the country at the end of the year. In *Agricultural Yearbook of China (1996)*, ed. Editorial Committee of Agricultural Yearbook of China (in Chinese), Beijing, China: Chinese Agriculture Press, 348 p.

Chen S.H. and S.M. Zhang. (1990). *Principle and Technology of City Refuse Compost* (in Chinese), Shanghai, China: Fudan University Publishing House, pp. 19-21.

FAO. (1977). *China: Recycling of Organic Wastes in Agriculture.* FAO Soil Bulletin. Rome, Italy: FAO.

Ge, L. and S.Z. Ji. (1986). Effect of earthworm culture in the favorable cycle of agriculture. *Agro-Environmental Protection* (in Chinese) 2: 35-36.

Harada, Y. and A. Inoko. (1989). Relationship between cation-exchange capacity and degree of maturity of city refuse composts. *Soil Science and Plant Nutrition* 26: 353-362.

He, J.X., S.L. Zheng, L.Q. Hong and P.Y. Yang. (1984). Study on methane production from residual of culturing fungi *Pleurotus sajor-caju* and nutrient availability of ultimate residual. *Soils and Fertilizers* (in Chinese) 1: 19-21.

Hidefuno, O. (1980). Quantitative investigation of material cycling in crop production. In *Study Report Collection of Environmental Science* B49-R 12-10 (in Japanese). p. 7.

Huang S.J. and B.G. Yu. (1991). Researches of input-output of the field energy in Suzhou. *System Science and Comprehensive Studies in Agriculture* (in Chinese) 7: 248-250.

Institute of Chemical Fertilizer and Pesticide of Shanxi Province (1974). *Production and Application of Humic Fertilizer* (in Chinese), Beijing, China: Fuel Chemical Industry Publishing House, p. 8.

Jin, Y.K. (1989). Study on efficiency of applying residual of methane production from straw fermentation. *Soils and Fertilizers* (in Chinese) 1: 23-26.

King, F.H. (1911). *Farmers of Forty Centuries: Permanent Agriculture in China, Korea and Japan*. Emmans, Pennsylvania: Rodale Press, Inc., 438 p.

Liebig, J.V. (1983). *Application of Chemistry on Agriculture and Physiology* (in Chinese). Translated by Liu Gengling. Beijing, China: China Agriculture Press, pp. 48-52.

Li, H.J. and Q.Q. Wang. (1995). Discussion on the systems engineering of chicken manure. *System Science and Comprehensive Studies in Agriculture* (in Chinese) 11: 122-125.

Liu, C.G. (1996). To implement sustaining development strategy of agriculture and to develop a new aspect of organic manure work. *Journal of Soil Agrochemistry* (in Chinese). 11: 1-9.

Liu, G.L. and W.X. Jin. (1991). *Organic Manure in China* (in Chinese), Beijing, China: China Agriculture Press, p. 112.

Liu, W.D. (1997). Effect of straw mulch on rape plant. *Science and Technology of Qinghai Agriculture and Forestry* (in Chinese) 4: 47-48.

Lu, R.K. and T.J. Shi. (1982). *Agrochemistry Manual* (in Chinese) Beijing, China: Science Press, p. 142.

Lu, Z.L. (1986). Comparative study on straw-manure compost and soil-manure compost. *Journal of Soils* (in Chinese) 18: 211-214.

Ma, X.L. (1995). Present situation and prospect of the high-efficient utilization of crops straw in China. *Research of Agricultural Modernization* (in Chinese) 16: 399-400.

Mao, D.R. (1981). *Organic Manure* (in Chinese), Beijing, China: Chinese Agriculture Press, p. 57.

Mo, S.X. and J.F. Qian. (1981). Effect of plowing in straw on increasing potassium nutrient for rice crop. *Journal of Soil Science* (in Chinese) 1: 20-21.

Mu, C.G. and M.X. Zhao. (1996). Elevated stand for cattle bedding. In *Resources of Organic Manure in Henan Province* (in Chinese). Beijing, China: China Agricultural Science and Technology Press, p. 202.

Phillips, S.H. and H.M. Young. (1973). *No-Tillage Farming*. Milwaukee, Wisconsin: Reiman Associates Inc., p. 59.

Qian, Z. and J. Fang. (1997). Countermeasures, problems and effects of straw plowed to soil. In *Sciences and Technology Soil and Fertilizer and Sustained Development of Agriculture in Jiangsu* (in Chinese), ed. Soil Science Society of Jiangsu Province. Nanjing, Jiangsu: River-Ocean University Publishing House, pp. 115-116.

Qi, X.I. (1998). Ecological engineering for utilization of agricultural organic wastes in China. *Eco-Agriculture Research* (in Chinese) 6: 73-76.

Sanitation and Antiepidemic Station of Shangdong Province. (1977). *Harmless Disposal of Dung* (in Chinese), Tong County, Beijing, China: People Sanitation Press, pp. 77-78.

Science and Technology Department of the National Scientific Committee. (1992). *Popularized Project of Disposal Technique of City Refuse* (in Chinese), Beijing, China: China Architecture Press, pp. 260-268.

Wang, Y.Q., E. Zhu and H.Z. Tian. (1993). Present situation and ways of developing and using municipal waste in suburbs of Shanghai. *Soils and Fertilizers* (in Chinese) 4: 17-20.

Xu, X.Y., Y.M. Zhang, H. Xiang, R.S. Li and J.S. Hu. (1985). Studies on the practical significance of plant residue mulch and its effect on soil organisms. *Scientia Agricultura Sinica* (in Chinese) 32: 42-48.

Yang, R.L. (1996). Practice of methane ecological engineering in livestock and poultry ranch. *Research on Agricultural Modernization* (in Chinese) 17: 180-182.

Yu, B.H. (1993). An investigation on resources of canal mud and their use in Jiangsu Province. *Soils and Fertilizers* (in Chinese) 3: 30-32.

Zeng, M.X. (1995). Study on drawing the technical regulation of straw directly return to field. *Soils and Fertilizers* (in Chinese) 4: 8-13.

Zhang, J.Q. and Z.B. Ji. (1998). Key points of culture on the double membraned rice seedling nursery. with river muck in early rice. *Journal of Zhejiang Agricultural Science* (in Chinese) 1: 13-15.

Zhao, Z.B., X.R. Mei, J.H. Xue, Z.Z. Zhong and T.Y. Zhang. (1996). Effect of straw mulch on crop water use efficiency in dryland. *Scientia Agricultura Sinica* (in Chinese) 29:59-66.

Zheng, Y.H. and G.B. Deng. (1998). Benefit analysis and comprehensive evaluation of rice-fish-duck symbiotic model. *Eco-Agriculture Research* (in Chinese) 6: 48-51.

Biological Practices and Soil Conservation in Southern China

Xuezheng Shi
Ying Liang
Zitong Gong

SUMMARY. Southern China has abundant natural resources and is a major region of agricultural production. However, soil erosion is increasingly threatening the agricultural productivity of this region. Increases in soil erosion are caused by improper land use practices with soils of high rainfall erosivity. Much of the area prone to accelerated erosion was previously forested and subsequently logged four times since the 1920s. This paper discusses the effects of biological control practices on soil erosion according to erosion types, soil properties, and terrain. *[Article copies available for a fee from The Haworth Document Delivery Service: 1-800-342-9678. E-mail address: getinfo@haworthpressinc.com <Website: http://www.HaworthPress.com>]*

KEYWORDS. Southern China, soil erodibility, biological practice

INTRODUCTION

Southern China covers an area of 2.18 million km^2, accounting for 22% of the total land area of the country. It includes provinces of Fujian, Guangdong,

---

Xuezheng Shi, Ying Liang and Zitong Gong are Research Soil Scientists, Institute of Soil Science, The Chinese Academy of Sciences, Nanjing, China. This research was supported by the National Natural Science Foundation of China (No. 49571045).

Address correspondence to: Xuezheng Shi at the above address (E-mail: xzshi@issas.ac.cn).

[Haworth co-indexing entry note]: "Biological Practices and Soil Conservation in Southern China." Shi, Xuezheng, Ying Liang, and Zitong Gong. Co-published simultaneously in *Journal of Crop Production* (Food Products Press, an imprint of The Haworth Press, Inc.) Vol. 3, No. 1 (#5), 2000, pp. 41-48; and: *Nature Farming and Microbial Applications* (ed: Hui-lian Xu, James F. Parr, and Hiroshi Umemura) Food Products Press, an imprint of The Haworth Press, Inc., 2000, pp. 41-48. Single or multiple copies of this article are available for a fee from The Haworth Document Delivery Service [1-800-342-9678, 9:00 a.m. - 5:00 p.m. (EST). E-mail address: getinfo@haworthpressinc.com].

Hainan, Guanxi, Guizhou, Yunnan, Sichuan, Hunan, Jiangxi and Zhejiang. The climate of southern China is characterized by abundant rainfall and high temperature. Mean annual rainfall ranges from 1200 to 2500 mm, the mean annual temperature from 14 to 28°C. Southern China is the major area for production of grain, edible oils, and sugar for the country. The cultivated land in this region comprises approximately 28% of the total cultivated land of China and accounts for 42.7% of the nation's total cereal crop. All of the rubber and tropical fruits, such as banana, pineapple, lichee and longan, as well as most of China's citrus, tea and tung oil are produced from this region. In addition to its agricultural importance, this region supports 43% of the Chinese population. Recent increases in population are beginning to threaten the agricultural production of the area and contribute to unsound land use practices.

The promotion of unsound land use practices has contributed to accelerated soil erosion. If current trends continue, it will threaten not only the development of agriculture, forestry and animal husbandry, but also degrade the environment upon which the population depends. In China the biological practices to control soil erosion dates from the first century B.C. (Han Dynasty). At that time China had a population of 40 million. Afforestation on low relief hills was adopted to expand agricultural output. In order to conserve soils, inter-planting between crops and fruit trees, vegetables or tea and economic trees was carried out. From the 14th to 19th century A.D. (Ming and Qing Dynasties) various biological measures of soil control were practiced in different places in China. According to textural research, there were more than 50 biological control measures on soil erosion in use. Because forests were still present in mountains and higher hills, the biological practices were used in sloping cultivated land. Since the 1920s the forests land rapidly decreased with the increase in population. Since the 1950s various biological practices were utilized on both sloping and barren land to control soil erosion. This paper presents the current distribution and characteristics of soil erosion and a summary of different biological practices to control soil erosion in southern China.

DISTRIBUTION AND CHARACTERISTICS OF SOIL EROSION

Distribution

In southern China the area affected by soil erosion increased from 600,000 km^2 in 1950s to 690,000 km^2 in 1990s (Gong and Shi, 1992). Table 1 shows that the area for soil erosion in 1993 occupied 248,000 km^2 in the eastern part of southern China, making up 21.47% of the total land area. The area with light, medium and serious soil erosion is 112,000, 78,000 and 58,000 km^2, respectively. A soil erosion survey of Jiangxi, Hunan, Fujian, Guangdong and Guangxi provinces revealed that in 1993 soil erosion area in the five provinces increased by 239.5% over 1950s and by 18.4% over 1980s. In 1993 the soil erosion area for three provinces, Jiangxi, Hunan and Fujian, increased by 48,800 km^2 over the 1980s. For other three provinces, Guangdong, Hainan and Guangxi, the soil erosion area decreased by 20,600 km^2 (Liang et al., 1995).

Higher Rainfall Erosivity and Soil Erodibility

This region has abundant rainfall. However the rainfall does not evenly distribute in a year. The rainfall is usually concentrated from March to June and the accumulated rainfall in these months comprises 50% of the total annual rainfall. The maximum diurnal rainfall amounted to 200-300 mm and the maximum monthly rainfall was over 800 mm (Shi and Shi, 1992). The rainfall erosivity of Yingtan in Jiangxi Province is approximately 400 (unit: 17 MJ mm ha^{-1} h^{-1}) (Shi et al., 1997). Even both in Guangzhou and Guilin, the rainfall erosivity exceeds 1000 (Wang et al., 1995). Therefore, it is obvious that this region has high rainfall erosivity.

Parent materials in this region consist mainly of granite, purple shale, limestone and red sandstone as shown in Table 2. The soils derived from these parent materials have high erodibility. Field plot experiments show that all soils derived from purple shale, red sandstone and granite have a high erodibility factor K. For example, erodibility factor K of cambisols on purple shale or red sandstone is approximately 0.4 (Shi et al., 1997). Thus, the potential internal factor exists to cause serious soil erosion.

TABLE 1. Distribution of different soil erosion types

Erosion type	Light	Medium	Serious	Total
Erosion area (km^2)	112,000	78,000	58,000	248,000
Percentage (%)	45.13	31.39	23.42	

TABLE 2. Distribution of the parent materials

Plant material types	Distribution	Area (km^2)
Granite	Jiangxi, Hunan, Fujian and Guangdong	100,000
Purple shale	Sichuan, Hunan, and Guangdong	220,000
Limestone	Guangxi, Guizhou, Yongnan and Sichuan	400,000
Red sand stone	Jiangxi, Fujian and Guangdon	---

Artificial Acceleration of Soil Erosion

Vegetation plays an important role in soil erosion control. Prior to the 1920s, severe soil erosion did not occur in most of this region because of the plant cover that preserved the soil and water resources of the forested areas. Soil erosion sharply increased from the 1920s to 1980s. During this period there were instances of large-scale deforestation, as described in Table 3.

BIOLOGICAL PRACTICES ON BARREN LANDS

For Light and Medium Eroded Areas

In light and medium eroded areas, the abundant rainfall and high temperature can aid the use of biological control measures on soil erosion. Because fertile soil with A horizon is not eroded, the vegetation on such eroded lands can be gradually restored by plant succession and hillside afforestation. Pioneer tree species and masson pines can be planted on the barren land underlain by red soils. Then broad-leaved tree species will grow under the shade of masson pines. Thus, under natural succession of plants, the coniferous forest can be gradually converted to a broad-leaved forest along with soil fertility

TABLE 3. The period and general information of large-scale deforestation

The period	General information
1920s-1940s	Because of the wars
1958	Constructing small steel-smelting furnaces and a short coke, destroying a lot of virgin and secondary forests. For example, prior to 1958 the Daming Mountain district of Guangxi Province has 59,070 ha of forest land, while 20,900 ha of forest was cut down in 1958, and accounting to 35.4% of total area for forest land (Xi, 1989)
1966-1976 (The Cultural Revolution)	Excavation, quarrying and expanding cultivated land to accomplish officially mandated grain production increases. According to statistics of Qingxian County in Hunan, the deforested area from 1966 to 1976 was as much as the total area afforested for 30 years since 1949
Late 1970s to early of 1980s	Rural land use management changed rapidly, but adoption of erosion mitigation measures did not keep pace. Land use practices resulted in serious and widespread forest cover loss

improved. It takes ten years for canopy closure after hillsides are designated as off limits. For the light eroded area with fertile soil, emphasis must be placed on the planting of mixed broadleaf-conifer forest. At present, more than 95% of an afforested area in this region consists of conifer forest, such as China fir and masson pine. But in practice, it has been found that the China fir grows much better in the first generation than in the second, because conifer forest returns less quantity of nutrients to soil than broad-leaf species. Therefore, sustainable soil utilization can be carried out by planting a mixed forest of broadleaf and conifer trees in the light and medium eroded area.

For Seriously Eroded Area

Soils of the A horizon and, even in some cases, the B horizon are washed away in seriously eroded areas. Therefore, it is difficult to control soil erosion with only biological practices. In seriously eroded areas, a combination of biological and engineering measures is utilized for soil erosion control. Seriously eroded areas are modified with earth moving equipment to create a hill-slope profile with terraces placed at given intervals in the natural slope. Horizontal ditches perpendicular to the slope are constructed in order to preserve soil moisture and fertility. For seriously eroded areas, the choices of approaches for planting are subject to two dilemmas. On one hand, the engineering approach requires a lot of manpower and the costs are high. However, if rapidly growing trees are planted, effects are soon evident. On the other hand, the rate of revegetation is low if "natural revegetation" is used to reforest an area. Therefore, we may plant pioneer trees, such as *Pinus massomana*, *Dalgergia hupeana*, *Liquidambar farmosana* and *Schima eonfertiflora* as well as shrubs such as lespidiza and false indigo. All the vegetations grow well in infertile soil as found in the seriously eroded areas. Trees with higher economic value will be planted, after the soil moisture and fertility under these vegetative covers are improved through several generations of pioneer trees. The Botany Institute of Southern China has done experiments in the Heshan area of Guangdong Province. It was found that some tree species, such as *Acacia mangium*, *A. holoserious* and *Albizzia falcata*, grow well in the seriously eroded area.

BIOLOGICAL PRACTICES ON UNIQUE PARENT MATERIALS AND SLOPING LANDS

For Purple Shale Area

Because purple shale is characterized by a loose and fragile materials and is easily weathered, soils developed from purple shale are easily eroded. In the seriously eroded areas, most of the soil profile is washed away and denuded. Rill erosion and shallow gully erosion dominate the erosion process. The loose weathering fragments are easily washed from surface of the shale bedrock and the sediments are rapidly transported from the upper hillside to the foot slope after a storm. The length of the gully mouth in the eroded area is 10-20 times longer and 3.5-4.0 times deeper than the gully bottom. The gully depth usually ranges from 50 to 200 cm. Because the soils are easily eroded, soils derived from purple shale show many disadvantages. Despite the fact that it is easily eroded, the soil from purple shale has its advantages when restorative measures are imposed. At first, even if all the soil is washed away in the seriously eroded area, barren land is easily reclaimed. The purple shale is loosened by demolition during warm weather, in the next year looser fragile material is weathered to soil by freezing in winter. Secondly, recently reclaimed soils even contain rich mineral nutrition, and are suitable for productions of tobacco, beans and peanuts. When these crops are planted in such soils, both high-yield and good quality can be achieved. Therefore, the properties of purple shale should be used in the erosion control process. After purple shale is loosened by demolition, terraces are constructed for crops. The soils are especially suitable for planting tobacco, beans and peanuts, and also for arbors, such as *Albizia durazz*, *Melia toosendan Siebet zucc* and *Dalbergia hupeana*, shrubs, such as *Nerium indicum Mill* and *lespidiza*, and others, such as *Eulaliopsis honda*, *Evolvulus alsinoides* and *Agave sisalana* Perr.ex Engelm.

For Eroded Area of Collapsing Hill

Chinese soil scientists call the collapsing hill erosion as "Benggong." Its form looks like a ladle-shaped gourd. A collapsing hill is about 30-60 m in length, 20-50 m in breadth and 10-50 m in depth. It is a unique deep-gully erosion feature that occurs in the granite area in southern China where a very deep weathering crust is distributed. The weathering crust has a thickness of 30-50 m and consists of loose materials. A particle composition reveals that the soils and weathering crust are predominantly silt. Laboratory analysis shows that soils in eroded areas of collapsing hill has only about 5% of clay, while silt amounts more than 60%. According to a field survey, soil loss of the eroded area from collapsing hill accounts for 900 ton ha^{-1}. The sediments from eroded areas of the collapsing hills submerge rice field and silt up rivers and reservoirs. For example, rice fields as large as 3.6 ha are submerged by sediments in the Xinyi village of Wuhua County in Guangdong Province.

To control erosion from collapsing hills, the measure of "up blocking, down stopping, middle terracing and revegetating" is adopted. Firstly, horizontal ditches are constructed in the hillside slope above the collapsing section of a hill. The ditches channel water away, slowing the downward cutting and expansion of the collapsing hill. Secondly, check dams are built in the ditch head of gourd shaped collapsing hills, in order to stabilize the basal slope and store water for improving soils and establishing favorable conditions for plant growth. The earth fern, stony fence and biological check dams are popularly in use. Thirdly, narrow bench terraces on steep slopes of collapsing hill are built for re-vegetation. *Melinis munutiflora* Beauv, *Acacia* spp., *Acacia holosericea* and *Agave sisalana* Perr. ex Engelm are planted on the narrow bench terraces and *Bambus textilis* McClure, *Eucalyptus*, *Albizia flcata* Back, *Pinus elliottii* Engelm, *Pterocarya kunth* and *Hovenia thunb* are planted in the outwash area on the base of collapsing hills. The outwash area also support fruit trees such as peaches, plums, banana, lichee, dark plum and kiwi. Sometimes soil may have to be added to thicken the available topsoil. Plants can stabilize check dams. Such species as *Acacia* spp., *Pinus elliottii*, and *Dicranopteris pedata*, Nakaike are used for dam stabilization. According to one survey, after a collapsing hill is controlled for two years, the annual sediment yield from the collapsed area decreased from 418 tons to 2 tons. The erosion from the collapsing hill was effectively controlled by biological practices (Zhong et al., 1991; Yu, 1990).

For Sloping Cultivated Lands

In areas where crops are cultivated on sloping lands, commercial plants are grown. For example, orchards, economic forests and living fences are effectively used to control soil erosion. In some places of this region, shrub and grass zones 2 meters in breadth are planted in the lower parts of sloping cultivated land to stop offsite sediment transport. The shrub or grass species include day lily and jute because these shrubs or grasses not only are effective for soil and water conservation but have an economic value. Sloping cultivated land underlain with fertile soil has developed well with agroforestry and interplanting. These strategies promote soil erosion control and allow farmers to use the land for economic purposes. There are many good examples of agroforestry or interplanting use in sloping cultivated land in this region. For example, Chinquapin-tea interplanted with multiple layers of plant cover is found in Yunnan Province. In the interplanting system, the first plant layer is chinquapin, the second one is camphor tree and the third layer is tea. Another example is rubber tree interplanted with tea. In this agroforestry system, rubber tree, the first layer, is interplanted with tea, pepper, coffee and cocoa as the second layer, and with peanut, soybean and kudzu vines as the third layer.

CONCLUSIONS

South China is characterized by abundant rainfall and high temperatures. At the same time, problems of high rainfall erosivity and soil erodibility exist in this region. In addition, mountains and hills with steep slopes dominate the area, and represent locations from which sediment can be transported. For sustainable uses of sloping land, policymakers and scientists should pay close attentions to soil erosion controls. Though building terraces with stone dams is effective to soil erosion control, dams are costly and time-consuming. Due to the ubiquitousness and types of soil erosion in southern China, biological practices should be integrated with engineering methods to control soil erosion. Selection of biological practices depends on the magnitude of soil erosion, the properties of the soil and weathering materials as well as soil conditions of the sloping cultivated land. In the lightly and medium eroded areas "closing hillsides to facilitate afforestation" is adopted, while biological practices are combined with field engineering in seriously eroded areas. For the sloping cultivated land, soil erosion is controlled by interplanting and agroforestation.

REFERENCES

Gong Z.T. and X.Z. Shi. (1992). Rational land use and soil degradation control in subtropical China. *Research on Red Soil Ecosystem* 1:14-21.

Liang Y., T.L Zhang and D.M. Shi. (1995). The evaluation of soil erosion on hills in Southern China. *Research on Red Soil Ecosystem* 3: 50-56.

Shi X.Z. and D.M. Shi. (1992). The integrated utilization of red soil resources and control of soil erosion in China. *Chinese Journal of Soil and Water Conservation* 6: 33-39.

Shi X.Z., D.S. Yu and T.Y. Xing. (1997). Soil erodibility factor K as studied using field plots in subtropical China. *Acta Pedologica Sinica* 34: 399-405.

Wang W.Z., J.Y. Jiao and S.P. Hao. (1995). The calculation and distribution rainfall erosivity R in China (I). *Chinese Journal of Soil and Water Conservation* 9: 5-18.

Xi C.F. (1989). *The Soil Erosion and Its Control in Hill and Mountain Areas of Subtropical China*. Beijing: Science Press.

Yu Z.Y. (1990). Rehabilitation for the degradation ecological system of hilly barren land in southern subtropical area of Guangdong Province. *Research on Forest Ecological System in Tropics and Subtropics* 7: 1-11.

Zhong J.H., S.Y. Tang and J. Tan. (1991). Collapsing hill erosion and its control of granite weathering crust of mountain area in South China. *China Bulletin of Soil and Water Conservation* 11: 25-28.

Ecosystem Immunity as a Strategy for Controlling Insect Pests in a Biotic Community

Fumiki Takahashi

SUMMARY. Populations of insect pests sometimes increase beyond the economic threshold to become serious problems in crop production. However, their populations are usually maintained below the economic injury level by natural predator insects and parasites. Living bodies often develop a degree of immunity against pest invasion, such as macrophages in the human body. In this case, macrophages reproduce in the human body at the expense of blood and body fluid, but are suppressed below the level of illness to the human body. This is analogous to crops, which have developed defense mechanism in the biotic community against pests by means of natural enemies and provided immunity in a total system. However, natural enemies, parasites and predators, must be maintained themselves with little compensation or support from the agroecosystem. Some years ago I proposed a model "reproduction curve with two equilibrium points" to describe the fluctuation of insect populations (Takahashi, 1964). The lower equilibrium point in the model can be regarded as a latent period and the higher one as the outbreak level. This model is based on the S-shaped functional response curve of predator to prey density and is applicable in a biotic community where polyphagous predators predominate. To ensure stable populations of natural enemies in the biotic community, it is necessary to maintain their food supply in the field. The populations of monophagous predators, such as parasitoids, fluctuate in response to prey or host

---

Fumiki Takahashi is Professor, Osaka University of Commerce, Higashi-Osaka, Osaka 577-8505, Japan.

[Haworth co-indexing entry note]: "Ecosystem Immunity as a Strategy for Controlling Insect Pests in a Biotic Community." Takahashi, Fumiki. Co-published simultaneously in *Journal of Crop Production* (Food Products Press, an imprint of The Haworth Press, Inc.) Vol. 3, No. 1 (#5), 2000, pp. 49-61; and: *Nature Farming and Microbial Applications* (ed: Hui-lian Xu, James F. Parr, and Hiroshi Umemura) Food Products Press, an imprint of The Haworth Press, Inc., 2000, pp. 49-61. Single or multiple copies of this article are available for a fee from The Haworth Document Delivery Service [1-800-342-9678, 9:00 a.m. - 5:00 p.m. (EST). E-mail address: getinfo@haworthpressinc.com].

populations sometimes inducing dramatic increases in pest populations above the economic injury level. On the other hand, polyphagous predators can utilize ordinary organisms as alternate food sources and keep their population stable even when a pest population (i.e., their target food) decreases to low level. This model will be discussed with respect to the diversity of a biotic community. *[Article copies available for a fee from The Haworth Document Delivery Service: 1-800-342-9678. E-mail address: getinfo@haworthpressinc.com <Website: http://www.HaworthPress.com>]*

KEYWORDS. Ecosystem, immunity, predator, parasite, biotic community, pest insects, biological control, biocontrol, organic farming

INTRODUCTION

Insect populations fluctuate both in space and time. One of the goals of population ecology is to clarify the mechanism of population fluctuations to obtain a better understanding of natural pest control. Many studies have been conducted on the factors that regulate insect pest populations in the field. However, few of these studies have actually contributed practical information to the principles of biological control. Schwerdtfeger (1958) attempted to identify the mechanisms that regulate outbreak of forest insect populations from their latent period. Some scientists analyzed this fluctuation by focusing on the monophagous predator-prey interaction such as in the Lotka-Volterra model. Schwerdtfeger himself, however, considered that the outbreak could be induced by a random chance. Nevertheless, the mechanisms that induce population outbreaks from the latent level are virtually unknown.

Studies on the life tables and the estimation of population density are usually conducted when pest insects are abundant. The mortality factors prevailing at a high level of population density must be different from those at a low level. It is likely that the mortality factors operating at high population densities are different from those that maintain the population at lower levels although they may tend to suppress the high population levels. Nevertheless, only a few studies have been conducted in Japan to determine the mechanisms that control the outbreak of populations from their latent period including Furuta (1968, 1972), Sakuratani (1977) and Hidaka (1989).

DIFFERENT APPROACHES TO CONTROL PEST POPULATIONS

Populations of insect pests fluctuate and sometimes increase to a high level of economic importance, but usually they are maintained below the economic injury level in nature, often through biocontrol measures. They become a pest when they exceed the economic injury level (EIL) or biological injury level (BIL) (Takahashi, 1971). There are different approaches to control pest populations as shown in Figure 1.

A. Basic pattern of population fluctuation of a pest. (Damages occur by pest above the level of EIL or BIL.)
B. Reduction of the average level of population to keep the upper limit of fluctuation below EIL or BIL: Common idea of pest control. (Conceptually no damages.)
C. Conventional methods of control, for example the applications of pesticides, amplify the population fluctuation over EIL or BIL, even though the average level is reduced. (Damages occur by pest above the level of EIL or BIL.)
D. Reduction of the amplitude of fluctuation, even though the average level is unchanged. (No damage: This is the most important approach to conserve biodiversity in the agroecosystem.)

MECHANISMS TO STABILIZE THE POPULATION DENSITY

Questions continue to arise on what strategies or approaches might be used to reduce the magnitude of population fluctuations and how to maintain a

FIGURE 1. Different approaches to control pest populations.

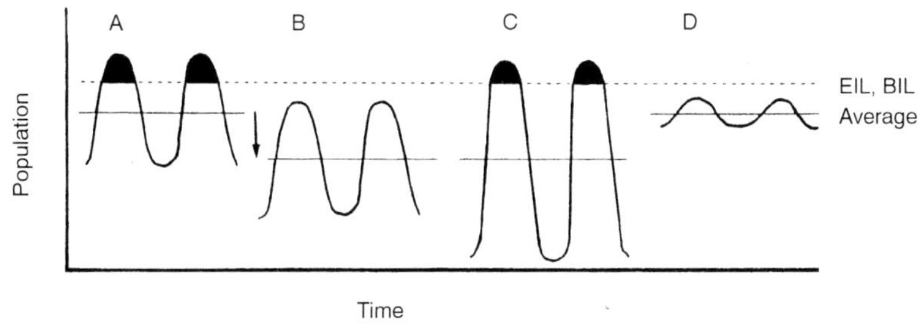

EIL: Economic injury level. Over this level economic costs surpass economic benefits.
BIL: Biological or ecological injury level. Over this level the biological resurgence of crops offset pest injury.
A: Basic pattern of population fluctuation of a pest (Damages occur by pest above the level of EIL or BIL).
B: Reduction of the average level of population to keep the upper limit of fluctuation below EIL or BIL: Common idea of pest control (conceptually no damages).
C: Conventional methods of control, for example the spray of pesticides, amplify the population fluctuation over EIL or BIL, even though the average level is reduced (damages occur by pest above the level of EIL or BIL).
D: Reduction of the amplitude of fluctuation, even though the average level is unchanged (no damage: This is the most important approach to conserve the biodiversity in the agroecosystem).

stable population. There are several hypotheses that may allow achieving a more stable population density which are summarized as follows:

- *Intra-Population Processes.* Density-dependent mortality processes need to be better understood, including the logistic growth equation and the reproduction curve with an equilibrium point when the environment is stable. In this case, food organisms (crops) are mostly consumed and heavily damaged because they tend to govern the mechanism that determines equilibrium population density.
- *Spatial-Temporal Relationships.* Space-time structure is especially important when the environment is unstable for insects. The mechanism is environmentally deterministic in the case of r-strategists. They have high productivity and high dispersal ability in space, and sometimes they carryover part of the population to future generations as in the case of cryptobiosis (Takahashi, 1977). However, damages can occur at some places of high density.
- *Predator-Prey Relationship.* The balance in the interaction between monophagous predator and prey is a basic principle of the biological control of pests by natural enemies. However, the populations of monophagous predators, such as parasitoids, tend to fluctuate in response to prey or host populations, often in a time-delayed manner and sometimes inducing dramatic fluctuations in pest populations above the EIL. They may be stabilized through the space-time structure (Pimentel et al., 1963). The r-strategic parasites can temporarily control pests to minimize crop damage. There is no stable relationship and, thus, a need for repeated introduction of the parasites into the field, analogous to an insecticide spray. If there are mutual relations between prey and predator through co-evolution or mutual evolution, r-strategic characters change to K-strategic ones, and food organisms (crops) will sustain damage as described in the next section on prudent predator-prey relationship.
- *Species Diversity and Population Stability.* The reproduction curve with two equilibrium points has been deduced from the S-shaped functional response curve of predator to prey density (Takahashi, 1964), though this response curve is a manifestation of the mutualism (Takahashi, 1976). This model is applicable in a biotic community where polyphagous predators predominate. They maintain their populations independently of prey pest population density.

PRUDENT PREDATORS-PREY RELATIONSHIP

Interaction between parasite population and host population is evolved to mutualism at certain population levels. This is shown in Figure 2 based on

FIGURE 2. Mutual relation of parasite and host on the population level (Adapted from Takahashi, 1985).

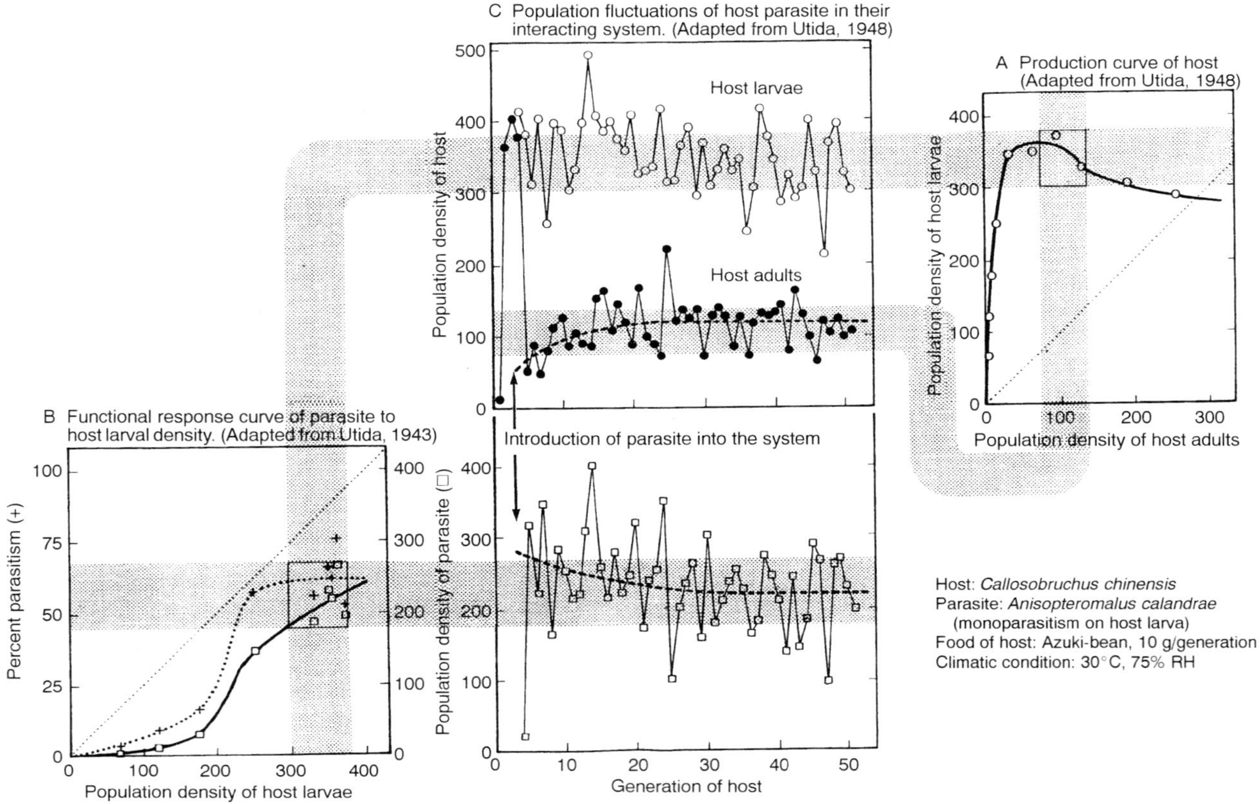

studies of Utida (1943, 1948) and Takahashi (1976) where both the parasite and host maintain their highest population productivity. Takahashi (1963) and Pimentel et al. (1963) also retested this tendency of mutual co-evolution. When the predator population consumes only part of the interest (i.e., surplus prey reproduction), thereby reserving part of the capital in the reproduction curve with one equilibrium point, the prey population can not only maintain its level but predator population can do likewise forever (Figure 3).

This mutual relationship between the prey (pests) and predator (natural enemies) is not an ideal economic situation. Therefore, on a practical basis we are using the r-strategic imprudent parasites for biological control. Sometimes we employ the parasite like pesticides, and utilize repeated introductions or releases as in the case of BT. Otherwise the situation evolves into the prudent parasite-host relation.

REPRODUCTION CURVE WITH TWO EQUILIBRIUM POINTS

The S-shaped functional response curve, which is observed in the prudent parasite-host relation, anticipates a promise in the case of polyphagous or euryphagous predators, because they can find foods to sustain their popula-

FIGURE 3. Reproduction curve: the relation between the density and the reproduction of the population.

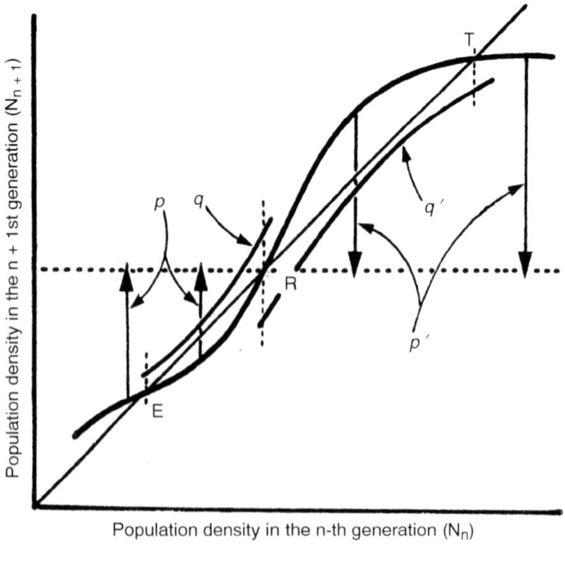

tions when the target pest populations decline. This relation can be explained by the model reproduction curve with two equilibrium points presented by Takahashi (1964, revised in 1976) in Figure 4. This S-shaped reproduction curve with two equilibrium points has been used by Morris (1963) in studying the dynamics of the spruce budworm, *Choristoneura fumiferana*. Sasaba and Kawahara (1970) also noted this relationship in studying the predation by spiders of rice leafhoppers, *Nephotettix cincticeps*. The lower equilibrium point in the model can be regarded as a latent period and the other higher one as a pest level. This model is deduced from the S-shaped functional response curve of predator to prey density. It is applicable in a biotic community where polyphagous predators predominate and have alternate food resources such as ordinary organisms. The factors and processes that tend to move the

FIGURE 4. Reproduction curve with two equilibrium points (Adapted from Takahashi, 1964, 1976).

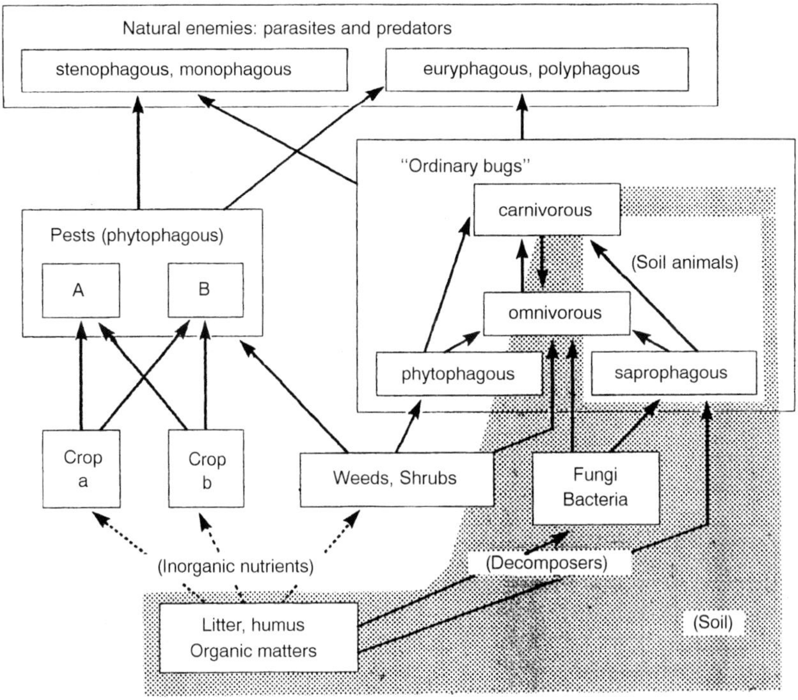

T: Equilibrium density at higher level
E: Equilibrium density at lower level
R: Release point

population densities between the two equilibrium points are shown in Figure 4 and Table 1.

ECOSYSTEM IMMUNITY

The living body has a system of immunity by means of antibodies against pest invasion, such as T-cells, B-cells and macrophages in the human body to keep the pest organisms below the level that causes a disease. In this case, the antibodies are self-supported by the human body itself at the expense of blood and body fluid. This is analogous to that of armaments supported by taxes to defend our nation (Table 2). Crop populations also have defense mechanisms in the biotic community against pests through natural enemies much like immunity in a total system. However, the population of natural

TABLE 1. The factors to move the population density between two equilibrium points (Adapted from Takahashi, 1964, 1976).

Mechanism of movement	Transfer to graduation level $E \rightarrow (R) \rightarrow T$ (p, q) (increase of reproductive rate)	Transfer to latent level $T \rightarrow (R) \rightarrow E$ (p′, q′) (decrease of reproductive rate)
	Simplify the biotic community	
Large change of reproductive rate in a short term (p, p′)	Creation of new farm	Eruption of parasites and disease after the pest outbreak
	Control by non-selective insecticides	
	Catastrophic change of climate	
	Change of cultivation system	
Small change of reproductive rate in a long term (q, q′)	Increased use of fertilizers	Introduction of resistant varieties of crop against pests
	Introduction of new crop varieties (high yield and high quality)	Recreation of biotic community
	Change in the pest-natural enemy relation	Use of real selective insecticides
	Extinction of a part of natural enemies	Improvement of climatic and culture conditions
	Change of ecological characters of pest and natural enemy	
	Time delayed fluctuation of population density	
	Long term change of climate	

TABLE 2. Comparison of system immunity between agroecosystem, human body, and nation (Adapted from Takahashi, 1991).

System	Target of defense	Items of defense mechanism	Foods or funds	Supporting system of foods or funds
Human body	Pathogens (fungi, bacteria)	Lymph, antibody (T-cell, B-cell, macrophages)	Haemolymph (a part of the body)	The system, itself (blood making organs)
Ecosystem	Pests (insects, pathogens)	Predators, Parasites	Preys, hosts, other organisms pollen, honey (only a component of the system)	No direct support of the system (self-perpetuation of food organisms)
Nation	Invaders from outside and rebellion inside	Armed forces and police	Taxes (expenses of the people)	The system, itself (of the people, by the people)

enemies, parasites and predators, must be maintained themselves with little (if any) compensation or support from the agroecosystem (Takahashi, 1991).

In forests, Furuta (1968, 1972) demonstrated that polyphagous predators are the main natural enemies controlling pine moth, *Dendrolimus spectabilis*, and gypsy moth, *Lymantria dispar*, at their latent state far below BIL. Sakuratani (1977) demonstrated that aphid populations are controlled at a low level by Chrysopidae (Neuroptera), Coccinellidae (Coleoptera), and Syrphidae (Diptera) in corn fields without insecticide treatment. In these studies the importance of polyphagous predators was indicated, while the action of monophagous parasitic species was not so effective. To maintain stable populations of natural enemies in the biotic community, it is necessary to maintain their food supply and stability in the field. Maksimovc (1978) showed that the gypsy moth was able to maintain its population at low levels by releasing an abundance of its eggs in the forests each year. It seems that gypsy moth eggs artificially deposited became the food of monophagous parasites, such as *Apanteles* (Braconidae) and *Hyposoter disparis* (Ichneumonidae) when pest populations were very low and parasite populations difficult to maintain. On the other hand, polyphagous predators can find ordinary organisms as alternate food sources to keep their population stable even when a pest population (i.e., target food), declines to a low level. This condition often occurs due to the diversity of a biotic community.

There are many types and numbers of non-pest insects in most fields (Figure 5), and as shown by Hidaka (1989) in organic rice fields (Figure 6) they support the polyphagous natural predators.

FIGURE 5. Structure of biotic community and insect pests (Adapted from Takahashi 1989).

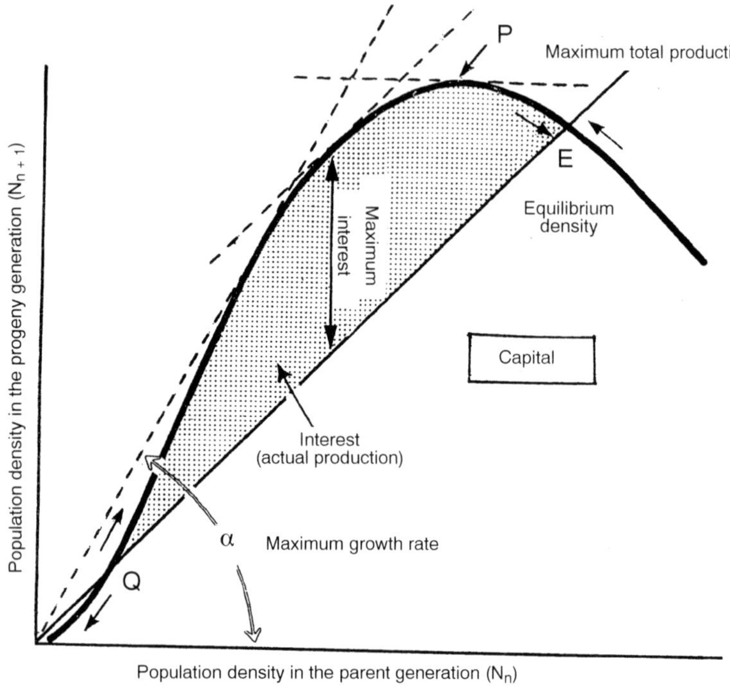

: Flow of materials and energies
: Flow of materials

EXTRAORDINARY ROLES OF ORDINARY INSECTS IN THE ECOSYSTEM

Ordinary insects in the field sometimes play an extraordinary role in the food chain and in the materials cycling in the ecosystem. For example, it is difficult to breed a spider, *Lycosa pseudoannulata* (Araneae; Lycosidae) with only the rice leaf hopper, *Nephotettix cincticeps*, as its food; but it becomes relatively easy when five species of prey collected from paddy fields are added as its food (Suzuki and Kiritani, 1974). This spider can not grow with only a single food source but needs variety of subsidiary foods. The larvae of *Dendrolimus spectabilis* excrete 100 mg of feces by eating 125 mg of pine leaves. But the soluble phosphorus content in the pine leaves (0.29 ppm) doubled to 0.56 ppm in the feces resulting in more efficient nutrient cycling (Washizuka, 1979). Soil animals *Sphaerillo dorsalis* and *S. russoi* (Isopoda;

FIGURE 6. Biotic community on rice stubbles (Adapted from Yoshikawa, 1985 and Hidaka, 1989).

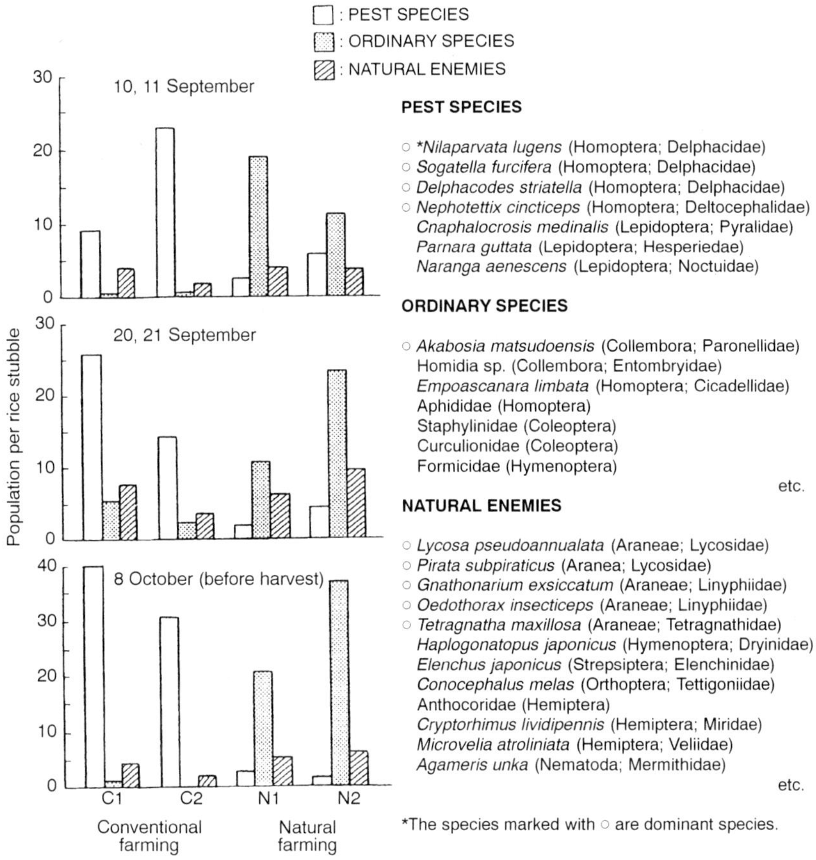

C_1: Sprayed selective insecticides
C_2: Sprayed non-selective insecticides
N_1 & N_2: Natural farming

Armadilidae) can significantly increase the soluble phosphorus content of the L-layer in a Japanese red pine forest in Hiroshima by their decomposition and feeding activities (Yamamoto et al. 1992). A collembolan species, *Akabosia matsudoensis*, feeds on fungi including the yellow stunt disease of rice plant (Hidaka, 1989).

Conventional agricultural practices destroy these ordinary but beneficial insects and their extraordinary roles in a biotic community, reducing the activity of polyphagous predators and making the system unstable. We need to promote the concept of a holistic and advanced agroecosystem that supports new agricultural practices for natural and organic farming, where the biotic diversity is still somewhat preserved.

CONCLUSIONS

It is more important to reduce the amplitude of fluctuation in pest populations than to reduce the average population level, thereby conserving biodiversity with many ordinary organisms in the field. Practices in conventional agriculture destroy these ordinary but beneficial insects and their extraordinary roles in a biotic community, thereby reducing the activity of polyphagous predators and making the system unstable. We need to promote the concept such as ecosystem immunity of a holistic and advanced agroecosystem that supports new agricultural practices for natural and organic farming, where the biotic diversity is still somewhat conserved.

REFERENCES

Furuta, K. (1968). The relationship between population density and mortality in the range of latency of *Dendrolimus spectabilis* Butler. *Japanese Journal of Applied Entomology and Zoology* 12: 129-136.**

Furuta, K. (1972). The relationship between population density and mortality in the range of latency of *Lymantria dispar* L. *Japanese Journal of Applied Entomology and Zoology* 16: 121-126.**

Hidaka, K. (1989). The ways for sustainable farming; a new ecological direction of neither intensive nor extensive. In *Natural, Organic Farming and Insect Pests*, ed. F., Nakasuji, Tokyo: Touzyusha, pp. 10-265. *

Maksimovic, M. (1978). Some research on the relation between the population densities of the gypsy moth and its natural enemies. *Plant Protection* (Beograd) 29: 127-139.

Morris, R.F. (1963). *The Dynamics of Endemic Spruce Budworm Populations*. Toronto: Entomology Society of Canada, 332 p.

Pimentel, D., W.P. Nagel and J.L. Madden. (1963). Space-time structure of the environment and the survival of parasite-host system. *American Naturalist* 97: 141-167.

Sakuratani, Y. (1977). Population fluctuations and spatial distributions of natural enemies of aphids in corn fields. *Japanese Journal of Ecology* 27: 291-300.**

Sasaba, T. and S. Kawahara. (1970). Roles of spiders as a predatious natural enemy in paddy fields. *Japanese Journal of Plant Protection* 24: 355-360.*

Schwerdtfeger, F. (1958). Is the density of animal populations regulated by mechanism or by chance? *Proceedings of the 10th International Congress of Entomology* (1956) 4: 115-122.

Suzuki, Y. and K. Kiritani. (1974). Reproduction of *Lycosa pseudoannulata* (Boesenberg et Strand) (Araneae : Lycosidae) under different feeding conditions. *Japanese Journal of Applied Entomology and Zoology* 18:166-170.**

Takahashi, F. (1963). Changes in some ecological characters of the almond moth caused by the selective action of an ichneumon wasp in their interacting system. *Researches on Population Ecology* 5: 117-129.

Takahashi, F. (1964). Reproduction curve with two equilibrium points: A consideration on the fluctuation of insect population. *Researches on Population Ecology* 6: 28-36.

Takahashi, F. (1971). Problems in the integrated control of pest insects. *Japanese Journal Plant Protection* 25: 259-266.

Takahashi, F. (1976). Natural enemies as a control agent of pests and the environmental complexity from the theoretical and experimental point of view. *Environmental Quality and Safety. Global Aspects of Chemistry and Technology as Applied to the Environment.* 5: 39-47.

Takahashi, F. (1977). Generation carryover of a fraction of population members as an animal adaptation to unstable environmental conditions. *Researches on Population Ecology* 18: 235-242.

Takahashi, F. (1985). The host-parasite relations, parasitism as a foraging strategy. In *Modern Biology 12b, Ecology B*, ed. M. Numata. Tokyo: Nakayama-shoten, pp. 10-13.

Takahashi, F. (1989). *Towards the Harmonized Pest Management from the Exclusive Pest Management.* Tpkyo: Nobunkyo Press. 462 p.

Takahashi, F. (1991). A concept of pest management in urban green zones. *Japanese Journal of Environmental Entomology and Zoology* 3: 210-216. *

Uchida, S. (1943). Host-parasite interaction in the experimental population of the azuki bean weevil, *Callosobruchus chinensis* (L.). 3. The effect of host density on the growth of host and parasite populations. *Ecological Review* (Sendai) 9: 40-45.**

Utida, S. (1948). Host-parasite interaction in the experimental population of the azuki bean weevil, *Callosobruchus chinensis* (L.). 5. Population fluctuations caused by host-parasite interaction. *Physiology and Ecology* 2: 1-11.**

Washizuka Y. (1979). Content of inorganic phosphorus in the larval dung of pine moth, *Dendrolimus spectabilis* Butler fed on needles of Japanese black pine, *Pinus thunbergii* Parl. *Japanese Journal of Applied Entomology and Zoology* 23: 117-118. **

Yamamoto, T., K. Nakane, and F. Takahashi. (1992). An experimental analysis of the effects of soil-macrofauna on the phosphorus cycling in a Japanese red pine forest. *Japanese Journal of Ecology* 42: 31-43.**

* In Japanese; ** in Japanese with English summary

Long-Term Changes in the Soil Properties and the Soil Macrofauna and Mesofauna of an Agricultural Field in Northern Japan During Transition from Chemical-Intensive Farming to Nature Farming

Yoshio Nakamura
Tokuko Fujikawa
Masao Fujita

SUMMARY. This paper reports the changes in soil chemical and physical properties, and in the populations and diversity of the soil macrofauna and mesofauna that occurred during the 10 years

---

Yoshio Nakamura is Senior Research Soil Scientist, Tohoku National Agricultural Experiment Station, Harajuku-Minami, Arai, Fukushima, Japan.

Tokuko Fujikawa is Soil Animal Scientist, Dannokoshi, Arai, Fukushima, Japan.

Masao Fujita is Research Soil Animal Scientist, International Nature Farming Research Center, Hata, Nagano, Japan.

The authors wish to express their sincere thanks and appreciation to Prof. Dr. J. Aoki of the Institute of Environmental Science and Technology of Yokohama National University and to the late Dr. Y. Kitazawa of Tokyo Metropolitan University for their advice and encouragement. The authors are greatly indebted to the Church of World Messianity and, especially to Mr. C. Sakakibara, whose kindness and valuable support allowed the survey to be conducted over the long term.

[Haworth co-indexing entry note]: "Long-Term Changes in Soil Properties and the Soil Macrofauna and Mesofauna of an Agricultural Field in Northern Japan During Transition from Chemical-Intensive Farming to Nature Farming." Nakamura, Yoshio, Tokuko Fujikawa, and Masao Fujita. Co-published simultaneously in *Journal of Crop Production* (Food Products Press, an imprint of The Haworth Press, Inc.) Vol. 3, No. 1 (#5), 2000, pp. 63-75; and: *Nature Farming and Microbial Applications* (ed: Hui-lian Xu, James F. Parr, and Hiroshi Umemura) Food Products Press, an imprint of The Haworth Press, Inc., 2000, pp. 63-75. Single or multiple copies of this article are available for a fee from The Haworth Document Delivery Service [1-800-342-9678, 9:00 a.m. - 5:00 p.m. (EST). E-mail address: getinfo@haworthpressinc.com].

(1976-1986) following the initiation of nature farming practices on this once chemically-managed field in Northern Japan. *[Article copies available for a fee from The Haworth Document Delivery Service: 1-800-342-9678. E-mail address: getinfo@haworthpressinc.com <Website: http://www.HaworthPress.com>]*

KEYWORDS. Nature farming, soil macrofauna, soil mesofauna, soil invertebrates, microarthropods, enchytraeids

INTRODUCTION

Soil animals or soil invertebrates (i.e., soil fauna) are classified into three groups, i.e., the microfauna (e.g., nematodes and protozoa), the mesofauna (e.g., mites and springtails), and the macrofauna (e.g., earthworms and termites) and together with soil microorganisms constitute the edaphon (France, 1921). The edaphon is one component of the pedosphere which has three functions, namely production, decomposition and self-purification (Mokan-joka-kenkyukai, 1983; Yahata, 1989; Nakamura, 1996). There are numerous studies on the relationships between the edaphon and these three functions. Since Charles Darwin's " The Vegetable Mould of Earthworm" published in 1881, it has been well known that earthworms, an important member of the soil fauna, contribute greatly to the improvement of the soil physical and chemical properties (Edwards, 1998). They have the function of "living hoes" and help the soil to form aggregates and humus. Recently, it was proved that the existence of earthworms results in less occurrence of club roots (Nakamura et al., 1995) and prevents the rotting of ginger (Enami et al., 1996). The vermicompost also tends to suppress club root (Szczech et al., 1993). The current research group has confirmed the possibility of controlling the infection of plant roots by hyphae of harmful fungi (i.e., *Rhizoctonia solani* and *Fusarium oxysporum* fsp. *Cucumerinum*) using fungivorous collembolans or oribatid mites (Nakamura et al., 1992; Enami and Nakamura 1996). A healthy and diverse edaphon enhances the productivity and fertility of agricultural soils (France, 1921) and ensures healthy soils (Nakamura, 1991). In modern intensive farming, however, large amounts of synthetic fertilizers and pesticides are applied and the soil is constantly tilled to obtain high yields from mono-crop culture. These practices have led inevitably to a reduction in the population and diversity of the soil animal community (Nakamura, 1991). The method of Nature Farming advocated in 1935 by the Founder of the Church of World Messianity, Mokichi Okada (1987), who emphasized the importance of soil health and microbial biodiversity is conceptually similar to the suggestions of France (1921). Actually, high population density and complex diversity of soil animals have been widely reported

(Fujikawa, 1976; Fujita, 1987; Nakamura, 1988, 1989; Nakamura and Fujita, 1988) in arable fields under a nature farming system. High stable yields of paddy rice were also reported by Horita (1981). In 1976, an abandoned crop field of the Nayoro Nature Farm in the northernmost part of Japan was converted to Nature Farming. Earlier it had been farmed by modern intensive methods which had led to crop yields too low to maintain cultivation. This paper reports on the changes in soil chemical and physical properties, and in the populations and diversity of the soil macrofauna that occurred during the 10 years (1976-1986) following the initiation of nature farming practices on this once chemically-managed field. The agricultural history, the field site characteristics and the experimental farming systems have been described elsewhere (Anonymous, 1987). The results of the first five years (Fujikawa et al., 1979 a,b; Nakamura et al., 1981) and evidence of new oribatid species were reported earlier (Fujikawa and Fujita, 1985).

MATERIALS AND METHODS

Site Description

The area studied (2000 m^2) at the Nayoro Nature Farm is situated in Nayoro (44°25′40″N, 142°26′30″E) of northern Hokkaido in the subarctic zone. The area was plowed or harrowed yearly immediately after disappearance of snow cover in the spring, and various farm machines were used five to twelve times during the growing season. In 1979, drainage was imposed, but it induced soil compactness. The next spring, rotary disking was conducted five times, and a pan breaker was used in 1982 and 1983. After spring plowing or harrowing, the crops were sown. The crop sequence of ten years was as follows: Azuki-bean in the first year and oats, oats, wheat, soybean, oats, oats, oats, kidney bean and oats in successive years. All seeds were obtained from the area surveyed except in 1976 and 1982. During the study period, no chemicals (synthetic fertilizers and pesticides) and no manure or excreta of livestock were applied. Instead, crop residues, Sasa manure (1 t/10a in 1976 and 1977) and litter (1.3 t/10a in 1983) gathered from the floor of broad-leaved forest (main tree species: *Ulmus davidiana* Planchon var. *japonica* (Rehder) Nakai, *Alnushirsuta* Turcz *Juglans ailanthifolia* Chrr.) near the study field were spread on the soil surface each year. Starting in 1980, clovers or oats were planted as a living mulch between the crop rows.

Soils and Vegetation

Soil at the study site is a heavy clay and the parent material is unconsolidated sedimentary rock. Gravel is scarce and viscosity is high (Nishimura et al., 1981). Soil samples from the 0-20 cm (plowed) layer were analyzed for macronutrients (Anonymous, 1970) in the spring seasons of 1976, 1980 and 1985.

Climate

During the survey period, the highest air temperature was 37.1°C (1978, Aug.) and the lowest was −37°C (1977, Feb.). Annual mean precipitation was about 1200 mm. Snowfall began in early December. Snow melting began at the end of March and the soil surface was exposed about the middle of April. Yearly average air temperature was 5.5°C, with an average high of 11.5°C and an average low of −0.4 (10-year mean). Among the seasons over ten years, air temperature was very low in the summer of 1980 and soil temperature at ground level was high in the spring and low in the summer of 1981.

Sampling Procedures

The study area was divided into 1,000 plots, each 2 m × 1 m, and four plots were randomly selected at each sampling time. To collect land snails, earthworms and all visible macroarthropods (which were called macro-animals) with the exception of ants, soil of a 50 cm × 50 cm quadrat was hand-sorted to a depth of 20 cm in each plot. Sampling for microarthropods (mainly Acari and Collembola) was conducted by first excavating a block of soil from each plot with 320 cm^2 surface area and 5 cm in depth. This sampling area was adequate for the determination of dominant oribatid species by analysis of the mean crowding-mean method by Iwao (1968, 1972). In the provisional survey on vertical distribution, the density of microarthropods became very low below 5 cm (Fujikawa et al., 1979a). The block was subdivided into sixteen adjoining samples each 4 cm × 5 cm in area using rectangular metal samplers. Microarthropods were extracted using a modified Tullgren apparatus (Fujikawa 1970) under a 20 W lamp for two days and a 40 W lamp for the third day. Soil sampling for enchytraeids was conducted for each plot by taking a block of soil with 80 cm^2 area and 20 cm depth. The block was subdivided horizontally into four 5-cm layers and into four adjoining samples each 4 cm × 5 cm in area using rectangular metal samplers. Enchytraeids were extracted using a modified O'Connor apparatus (Nakamura and Tanaka, 1979) under a 40 W lamp for three hours. The sampling schedule was based on the seasonal division by meteorological data after Allee et al. (1949): spring (s) (May 21 to July 10), summer (su) (July 11 to Aug. 20), autumn (a) (Sept. 21 to Oct. 31) and mid-winter (m) (Nov. 21 to March 30). Every year from 1976 to 1986 the sampling was done at each middle day of four seasons. The study period was divided into ten years. Each year comprised of four seasons, for example 1976 was designated as 1976s, 1976su, 1976a and 1977m. Sampling was conducted from 1976 to 1985 for macro-animals and microarthropods, and from 1981 to 1985 for enchytraeids.

RESULTS

Soils

Throughout the survey period, pH (1:1 water) was 5.3-5.4 (Table 1). Among the chemical properties, values for total P_2O_5 and exchangeable Ca and K in 1985 were significantly lower (P = 0.01) than in 1976. The humus content in 1985 was slightly higher than that in 1976. However, exchangeable Mg in 1985 was significantly (P = 0.01) higher than in 1976.

Microarthropods

As shown in Figure 1, the density of microarthropods varied seasonally and was high in the autumn of all years except 1982. It was significantly higher (P = 0.05) in the spring than in mid-winter of 1976 and 1979. The densities were very high in autumn of 1976 and 1979. Beginning in 1981 it decreased gradually from spring to mid-winter. Among the extracted micro-arthropods, Cryptostigmata (oribatid mites) was the most abundant animal,

TABLE 1. Change of soil properties* (mean ± s.e.)

Sampling	May 1976	May 1980	May 1985
pH (H_2O)	5.3 ± 0.1	5.3 ± 0.1	5.4 ± 0.1
Total nitrogen (%)	0.20 ± 0.06	0.24 ± 0.04	0.25 ± 0.10
Humus (%)	3.6 ± 0.8	4.5 ± 0.4	4.5 ± 0.2
CEC (meq/100 g)	21.6 ± 3.0	17.4 ± 1.2	16.8 ± 0.2
Exchangeable Mg (mg/100 g)	23.1 ± 4.4	39.4 ± 2.5	37.3 ± 5.7
Ca (mg/100 g)	263.0 ± 48.7	190.0 ± 26.9	197.0 ± 31.1
K (mg/100 g)	40.9 ± 11.3	14.8 ± 2.5	13.8 ± 7.7
Total P_2O_5 (mg/100 g)	227.0 ± 10.9	108.0 ± 24.6	118.0 ± 11.0
Water content (%, w/w)	19.1 ± 3.7		18.5 ± 1.0
Bulk density (g/100 cc)	89.5 ± 5.0		93.5 ± 6.2

*Soil samples were taken from the 0-20 cm depth after plowing.

FIGURE 1. Fluctuation in the density and biomass of microarthropods, enchytraeids and macro-animals.

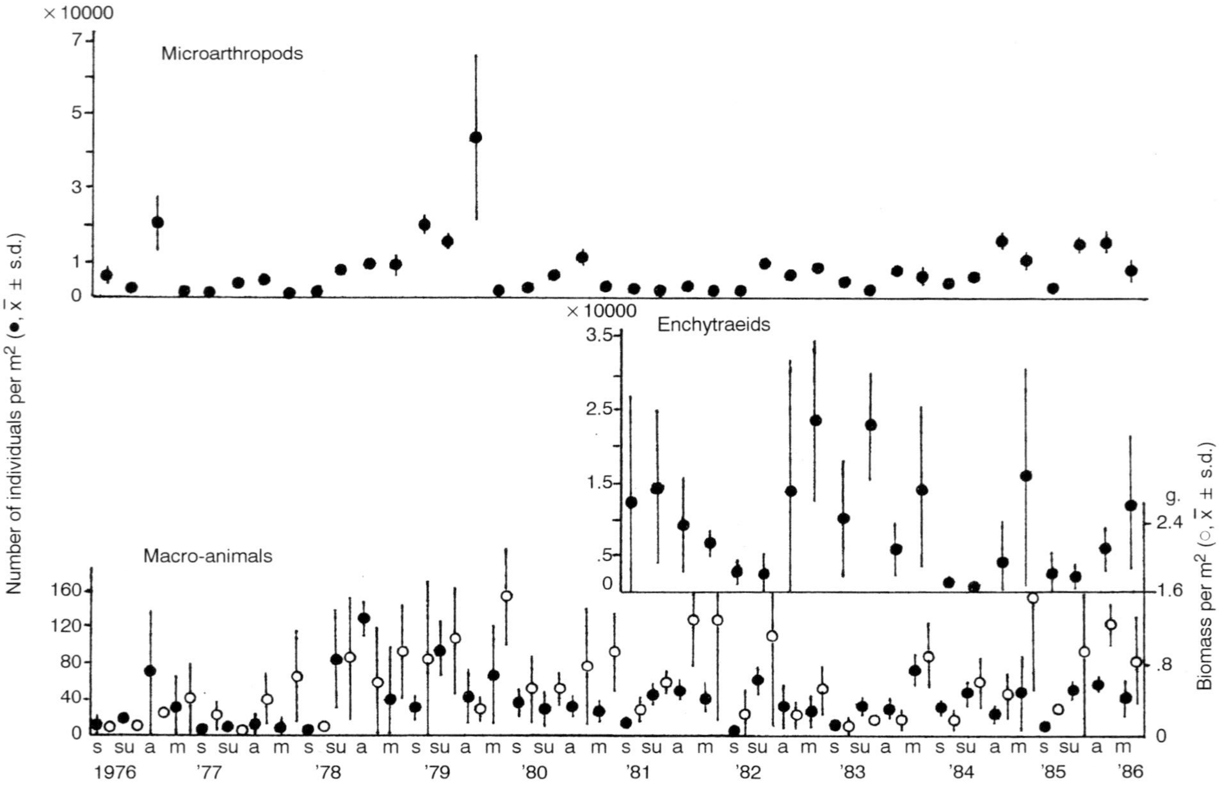

comprising 61.2% of the total count. Prostigmata and Collembola were also abundant and exceeded 10% of the total count (Table 2). Densities of these three animal groups varied annually. The densities of all these animals were the largest in the fourth year (1979), then decreased to a lower level in 1981 and 1982 and then again increased. For example, the densities of Prostigmata

TABLE 2. Relative abundance of macro-animals, enchytraeids and microarthropods collected from the soil during summer of 1976 to mid-winter of 1986.

Macro-animal	**(%)**	**Enchytraeida**	**(%)**
Gastropoda	0.4	*Achaeta*	1.3
Oligochaeta	3.7	*Enchytraeus*	12.7
Arachnida	16.4	*Fridericia*	23.2
Crustacea	3.3	*Henlea*	4.7
Diplopoda	0.1	*Lumbricillus*	26.1
Symphyla	< 0.1	*Sterucutus*	0.4
Chilopoda	12.1	Unidentified (small)	31.5
Insecta		**Microarthropods**	
Orthoptera	0.1	Pseudoscorpiones	< 0.1
Hemiptera	2.1	Acari	
Dermaptera	0.3	Mesostigmata	7.9
Lepidoptera	0.1	Prostigmata	12.7
Diptera	6.6	Astigmata	2.1
Coleoptera		Cryptostigmata	
Harpalidae	21.4	*Orbatula sakamorri* Aoki	12.7
Staphylinidae	14.0	*Punctoribates insignis* Berlese	7.0
Scarabaeidae	6.8	*Oppiella nova* (Oudemans)	18.4
Carabidae	0.1	*Tectocepheus velatus* (Michael)	17.4
Hydrophilidae	2.0	Other spp.	1.0
Silphidae	0.1	Immatures	4.7
Elateridae	2.1	Aranea <	0.1
Chrysomelidae	4.0	Diplopoda	< 0.1
Curculionidae	2.4	Chilopoda	< 0.1
Coccinelidae	0.6	Insecta	
Unknown larvae	0.5	Collembola	15.5
Hymenoptera	0.3	Others	< 0.1

and Cryptostigmata were higher in 1985 than in 1976. This trend in density variation was also observed for the total microarthropods which was higher in 1985 than in 1976. In the first year (1976), Mesostigmata comprised more than 42% of the total microarthropods, but its density was low in the remaining nine years. Among 31 oribatid species including 12 new species, the following four were abundant throughout the survey period: *Oribatula sakamorii* Aoki, *Punctoribates insignis* Berlese, *Oppiella nova* (Oudemans) and *Tectocepheus velatus* (Michael). These four species accounted for about 90% of the total number of oribatids identified during the ten-year survey (Table 2). The make-up of oribatid species in 1982 and 1984 was different than in other years. *P. insignis* was the most abundant in 1982 and *O. sakamorii* in 1984.

Enchytraeids

The density of total enchytraeids varied seasonally (Figure 1). It tended to increase from summer to mid-winter in three years except 1981 and 1983. It was lower in the spring than in the mid-winter for all five surveyed years. Among the six genera, *Lumbricillus*, *Fridericia* and *Enchytraeus* were most abundant. These three genera accounted for over 62% of the total (Table 2). *Lumbricillus* increased as times went on, and it occupied more than 63% of the total enchytraeids in 1984. The other two were most numerous in 1983 and decreased thereafter. The density was lower in 1984 and 1985 than in 1981. Such a trend was also observed for the total enchytraeids.

Macro-Animals

The density and biomass were lower in the spring than in the preceding mid-winter samplings throughout the survey period. Both increased gradually from spring to autumn in most years. The biomass was largest in the mid-winter of seven years except 1981, 1982 and 1985. Among the collected macro-animals, the following four carnivores were abundant: Arachnida, Chilopoda, Harpalidae and Staphylinidae (Table 2). These four accounted for more than 63% of the total. They were most abundant in the third year (1978) and thenceforth decreased. Among them, Staphylinidae and Chilopoda were more numerous in 1985 than in 1976, but Arachnida was less in 1985 than in 1976. Harpalidae was more abundant in 1985 than in 1976. Total number of macro-animals was also higher in 1978 and then decreased to the level of 1976 in 1980. The density changed little from 1980 to 1985, but was higher in 1985 than in 1976. On the other hand, each of the two groups of saprovores/microvores, Diptera and Scarabaeidae, comprised between 5 to 10% of the total macro-animals during the ten years. The former was most abundant in the first half of the survey period, while the latter was more abundant later on.

Composition of the macro-animals in 1985 was different than for the other nine years. In 1985, Lithobiidae spp. were most abundant and the following animals were collected for the first time: *Eisenia* (*Allolobophora*) *rosea* (Savigny) of Lumbricidae, *Lordithon* sp. of Staphylinidae and *Scelodonta lewisii* Baly of Chrysomelidae.

DISCUSSION

Throughout the ten-year survey period, the highest density of microarthropods was observed in 1979, and that of macro-animals in 1978. This highest density might have been induced by the application of Sasa manure in 1976 and 1977. It was followed by a rapid decrease, and might relate primarily to the disappearance of Sasa manure. In the summer of 1983, litter gathered from a forest near the study area was spread out between the crop rows, to supply humus and reproduce soil animals. The gathered litter materials contained various soil animals (Fujita and Fujikawa, 1986, 1987; Fujita et al., 1987). This practice increased the density of microarthropods, but the density of enchytraeids decreased. While the density of macro-animals did not change, that of Chilopoda, one of the abundant animal groups, increased. This practice also increased the crop yield of the year of application and introduced several soil animal species into the field soil (Fujikawa 1988d; Nakamura and Fujita 1988). Moreover, crop residues were returned annually to the field and intercrops were planted in 1980 and thereafter. These practices might have increased the humus content, the crop yield and the densities of microarthropods and macro-animals in the last surveyed year (1985). For example, two animal groups, Cryptostigmata and Prostigmata, were more abundant in 1985 than in 1976. The make-up of macro-animals in 1985 was different from that of the other nine years. In 1985, several soil animal species were newly observed.

Soil in the study area is heavy clay with low humus content and poor permeability. In spring, the soil has excessive water from melting of snow. Plowing and harrowing with large-sized machines under such conditions might create a plow sole (i.e., plow pan) that would enhance soil compaction upon drying (Bockel and Peerlkamp 1956; Nakano 1978). This plow sole pan could readily obstruct the migration of soil animals from a plow layer to deeper layers in a dry season. Its effect may be lethal to large-sized animals, e.g., macro-animals, or soft-bodied ones, e.g., enchytraeids. Moreover, various farm machines were used during the crop growing season and are thought to have contributed to a decrease in the densities of enchytraeids and macro-animals with time, though efforts were made to alleviate the conditions using soil amendments.

The densities of enchytraeids and macro-animals in all ten years and microarthropds in seven years were smaller in the spring than in the preceding mid-winter sampling time. The survey on soil animals carried out before and after the spring plowing and harrowing in 1980 and 1984 showed that the densities of microarthropods, enchytraeids or macro-animals decreased after these operations (Anonymous, 1987). Tillage (mold-board plowing, harrowing, etc.) generally decreases the abundance of soil animals (Nakamura, 1991; Edwards, 1998). Every spring sampling in this field study was conducted after plowing or harrowing; thus, the decreased density in spring is thought to be largely due to destruction of animals from spring tillage.

Among the ten survey years, the middle year, 1981, had many peculiar features (Anonymous, 1987). Some of these characteristics are described as follows. For example, there was no large seasonal change in the density of microarthropods. The density of macro-animals in summer was significantly larger than that in the first year, 1976. Among macro-animals, Diplopoda was observed only in 1976 and Gastropoda was not observed after 1976. Oligochaeta and Hydrophilidae showed the maximum densities during the ten years (Nakamura and Fujita, 1988). Four dominant oribatid species showed characteristics as follows (Fujikawa, 1987 a, b; 1988 a, b, c, d; 1995). There was no gravid-female of *O. sakamorii* in mid-winter and no female in spring. No individuals with any abnormal character were obtained from spring to summer for *O. sakamorii* and *P. insignis* and from autumn to mid-winter for *O. nova*. The average number of eggs per gravid female of *T. velatus* was large in summer. The value of Du/D1 (analysis for selecting optimum sampling unit by Iwao, 1972) in *T. velatus* was below 1.

Furthermore, the soil temperature at ground level was high in spring and low in summer, and the soil water content was high in summer and autumn. The gradual increase in crop yields stopped after 1980 (Anonymous 1987). The decrease in exchangeable Ca and K or total P_2O_5 content and increase in humus, exchangeable Mg or total N content were very sharp from 1976 to 1980, but gradual from 1980 to 1985. The time required for proper stabilization and equilibration of the soil and agroecosystem for transition from the modern chemical-intensive farming to the nature farming method is likely to be from 3 to 5 years. The middle year, 1981, in the present 10-year study with the above-mentioned characteristics, appears to be the turning point when the soil animal community became adjusted to the conditions of nature farming.

REFERENCES

Allee W.C., A.E. Emerson, O. Park, T. Park and K. Schmidt (1949). *Principles of Animal Ecology*. London: WB Saunders, Philadelphia, 837 p.

Anonymous (1970). *Methods of Analyzing Nutrients in Soils* (Committee of Methods Analyzing Nutrients in Soils). Tokyo, Japan: Yokendo (In Japanese), 430 p.

Anonymous (1987). Report on soil animals in Nayoro Nature Farm from 1976 to 1986. In *Nature Farming International Research Foundation*, eds. Y. Nakamura, T. Fujikawa and M. Fujita. Nagano (Japan): Internatinal Nature Farming Research Center, 230 p (In Japanese).

Bockel P. and P. Peerlkamp (1956). Soil consistency as a factor determining the soil structure of clay soil. *Netherlands Journal of Agricultural Sciences* 4: 122-125.

Darwin C. (1881). *The Formation of Vegetable Mould through the Action of Worms with Observation on Their Habits*. London: John Murray & Co. 298 p.

Edwards A.E. (1998). *Earthworm Ecology.* Boca Raton (Florida): St. Lucie Press, 389 p.

Enami Y., and Y. Nakamura (1996). Influence of *Scheloribates azumaensis* (Acari: Oribatida) on *Rhizoctonia solani*, the cause of radish root rot. *Pedobiologia* 40: 251-254.

Enami Y., Y. Nakamura, H. Shiraishi, H. and J. Tsukamoto (1996). Effect of the earthworm *Eisenia fetida* on protecting from rotting of stored ginger. *Abstract of Japanese Soil Zoological Society.* p. 15 (In Japanese).

France, R. 1921. *Das Edaphon. Untersuchungen Zur Okologie der bodenlewohnenden Mikroorganismen.* Stuttgartm, Germany, 99 p.

Fujikawa T. (1970). Notes on the efficiency of a modified Tullgren apparatus for extracting oribatid mites. *Applied Entomology and Zoology* 5: 42-44.

Fujikawa T. (1976). Oribatid mites in the natural farming and conventional fields. *Edaphologia* 15: 1-11 (In Japanese with English summary).

Fujikawa T. (1987a). Biology of *Oribatula sakamorii* AOKI, a dominant species of the oribatid fauna in nature farming field. In *Soil Fauna and Soil Fertility*, ed. B.R. Striganova. Moscow (Rusia): Nauka, pp. 544-552.

Fujikawa T. (1987b). Biological features of *Punctoribates insignis* Berlese in a nature farming field. *Edaphologia* 36: 13-20.

Fujikawa T. (1988a). Biological features of *Oppiella nova* (Oudemans) in a nature farming field. *Edaphologia* 38: 1-10.

Fujikawa T. (1988b). Two species belonging to the genus *Tectocepheus* from nature farm at Nayoro in Northern Japan (Acari; Oribartei) *Acarologia* 29: 205-213.

Fujikawa T. (1988c). Biology of *Tectocepheus velatus* (Michaels) and *T. cuspidendatus* Knule. *Acarologia* 29: 307-314.

Fujikawa T. (1988d). Fluctuation of oribatid mites in Nayoro nature farming filed during ten years. *Edaphologia* 39: 29-37.

Fujikawa T. (1995). Comparison among population of *Tectocepheus velatus* (Michaels) from forests, grasslands and cropfield. *Edaphologia* 55: 1-82

Fujikawa T. and M. Fujita. (1985). Five new species belonging to the genus *Epidamaeus* from Nayoro of Hokkaido, North Japan. *Edaphologia* 32: 19-28.

Fujikawa T., Y. Nakamura, M. Fujita, Y. Tsuzuki, Y. Hosono and K. Nishimura. (1979a). Effect of soil animals on heavy clay soil habitat modification. 1. Soil microarthropod community, especially oribatid mite community of crop field, after changing to nature farming. *Annual Reports of Research Institute of Environmental Science* (Kyoto, Japan) 7: 91-103 (In Japanese with English summary).

Fujikawa T., Y. Nakamura, M. Fujita, Y. Tsuzuki, Y. Hosono and K. Nishimura. (1979b). Ditto, 2 Ibid 7:105-109 (In Japanese with English summary).

Fujita M. (1987). A comparison of acari, collembola and enchytraeid numers in the soil of crop field under different soil managements. *Abstracts of Colloquium of the Japanese Soil Zoology* p. 24 (In Japanese).

Fujita M. and T. Fujikawa. (1986). List and description of oribatid mites in the forest litter as materials introducing soil animals into crop field of Nayoro (I). *Edaphologia* 35: 5-18 (In Japanese with English synopsis).

Fujita M and T. Fujikawa. (1987). Ditto, (II) Ibid 36:1-11 (In Japanese with English synopsis).

Fujita M., Y. Nakamura and T. Fujikawa. (1987). Soil fauna in the forest litter as materials introducing soil animals into crop field of Nayoro (Effects of soil animals on heavy clay soil habitat modification 7). *Edaphologia* 37: 9-16 (In Japanese with English synopsis).

Horita T. (1981). Investigations of profitability and extensibility of rice and citrus enterprises by natural farming. In *Studies on Natural Farming (Memorial Issue)*, Research Institute of Environmental Science (Kyoto). Kyoto: Research Institute of Environmental Science, pp. 412-419 (In Japanese with English summary).

Iwao S. (1968). A new regression method for analyzing the aggregation pattern of animal populations. *Research on Population Ecology* 10: 1-20.

Iwao S. (1972). Application of the m*-m method to the analysis of spatial patterns by changing quadrat sizes. *Research on Population Ecology* 14: 97-128.

Mokan-joka-kenkyukai. (1983). *Science of Pedosphere*. Tokyo: Mokan-joka-kenn-kyuukai, 352 p (In Japanese).

Nakamura Y. (1988). The effects of soil management on soil faunal makeup of a cropped andosol in central Japan. *Soil Tillage Research* 8: 177-186.

Nakamura Y. (1989). Oribatids and enchytraeids in ecofarmed and conventionally farmed dryland grain fields of central Japan. *Pedobiologia* 33: 389-398.

Nakamura Y. (1991). Utilization of soil animals for sustainable agriculture. *Tohoku Agricultural Research Reports* (Extra Issue) 4: 45-59 (In Japanese).

Nakamura Y. (1996). Interactions between earthworms and microorganisms in biological control of plant root pathogens. *Farming Japan* 30: 37-43.

Nakamura Y., T. Fujikawa and M. Fujita. (1981). Soil microarthropods and soil macroanimal communities in crop field of natural farming. Effects of soil animals on heavy clay soil habitat modification (5). *Annual Reports of the Research Institute of Environmental Science* (Kyoto) 7: 87-92 (In Japanese with English summary).

Nakamura Y. and M. Fujita. (1988). Abundance of lumbricids and enchytraeids in an organically farmed field in northern Hokkaido, Japan. *Pedobiologia* 32: 11-14.

Nakamura Y., J. Itakura, and I. Matsuzaki. (1995). Influence of the earthworm *Pheretima hilgendorfi* (Megascolecidae) on *Plasmodiophora brassicae* clubroot galls of cabbage seedling in pot. *Edaphologia* 54: 39-41.

Nakamura Y., I. Matsuzaki and J. Itakura. (1992). Effect of grazing by *Sinella curviseta* (Collembola) on *Fusarium oxysporum* f.sp. *cucumerinum* causing cucumber disease. *Pedobiologia* 36: 168-171.

Nakamura Y. and S. Tanaka (1979). Vertical distribution of Enchytraeidae in various habitats. *Edaphologia* 19: 1-12 (In Japanese with English synopsis).

Nakano K. (1978). Changes in soil physical properties of clayey soil by conservation from III-drained paddy field into upland field. *Bulletin of Hokuriku National Agricultural Experiment Station* 21: 63-94 (In Japanese with English synopsis).

Nishimura K., M. Fujita, Y. Nakamura and Y. Arakawa. (1981). Some effects of successive application of beet pulp on the growth of potato plant and on some chemical properties of the soil. In *Studies on Natural Farming* (Memorial Issue), Interdiscip Res. Inst. Envir. Sci. Kyoto pp. 207-215 (In Japanese with English summary).

Okada M. 1987. *True Health*. Church of World Messianity, Atami, Japan. 184 p.

Yahata T. 1989. *Glorious Pedosphere*, Ti-yuu-sha, (In Japanese), Tokyo, Japan. 167 p.

Szezech M., W. Rondomanskim, M.W. Brzeski, U. Smolinska and J.F. Kotowski (1993). Suppressive effect of a commercial earthworm compost on some root infecting pathogens of cabbage and tomato. *Biological Agriculture and Horticulture* 10: 47-52.

Phytophthora Resistance of Organically-Fertilized Tomato Plants

Ran Wang
Hui-lian Xu
Md. Amin U. Mridha

SUMMARY. Tomato plants (*Lycopersicon esculentum* L. cv. Momotaro T96) grown with organic fertilizer possessed higher resistance against phytophthora infection (*Phytophthora infestans*) in comparison with plants given chemical fertilization. However, the photosynthetic activity was not low in leaves of chemically-fertilized plants before phytophthora infection. There were no differences in leaf proteins between the two fertilization treatments, as revealed by electrophoresis. This suggested that high phytophthora infection was not related to physiological activity as indicated by photosynthetic rate and protein profile. Concentrations of leaf nitrogenous compounds, nitrate and amino acids were higher in chemically-fertilized plants. Nitrate concentration in the soil was also higher in chemical plots. On the other hand, leaf nitrate reductase activity and soil hydrogenase activity were lower in chemical plots. The integrated results suggested that nitrogen metabolism in organically-fertilized tomato plants accounted for high phytophthora re-

---

Ran Wang is Professor, Department of Horticulture, Laiyang Agricultural University, Shandong, China.

Hui-lian Xu is Senior Crop Scientist, International Nature Farming Research Center, 5632 Hata, Nagano 390-1401, Japan.

Md. Amin U. Mridha is Professor, Department of Botany, Chittagong University, Chittagong, Bangladesh. The authors thank Dr. S. Goyal, S. Kato, K. Katase and Dr. H. Umemura for their technical assistance and advice.

Address correspondence to: Hui-lian Xu at the above address (Email: huilian@janis.or.jp).

[Haworth co-indexing entry note]: "Phytophthora Resistance of Organically-Fertilized Tomato Plants." Wang, Ran, Hui-lian Xu, and Md. Amin U. Mridha. Co-published simultaneously in *Journal of Crop Production* (Food Products Press, an imprint of The Haworth Press, Inc.) Vol. 3, No. 1 (#5), 2000, pp. 77-84; and: *Nature Farming and Microbial Applications* (ed: Hui-lian Xu, James F. Parr, and Hiroshi Umemura) Food Products Press, an imprint of The Haworth Press, Inc., 2000, pp. 77-84. Single or multiple copies of this article are available for a fee from The Haworth Document Delivery Service [1-800-342-9678, 9:00 a.m. - 5:00 p.m. (EST). E-mail address: getinfo@haworthpressinc.com].

sistance. Therefore, organic farming can efficiently control this disease in tomato plants. *[Article copies available for a fee from The Haworth Document Delivery Service: 1-800-342-9678. E-mail address: getinfo@haworthpressinc.com <Website: http://www.HaworthPress.com>]*

KEYWORDS. Nature farming, organic fertilizer, chemical fertilizer, nitrogen metabolism, phytophthora, tomato, nitrogen metabolism

INTRODUCTION

About sixty years ago, chemical farming started to spread in Japan. Many farmers and agricultural scientists had a blind faith in chemical fertilizers and pesticides. Some people even believed that pests would be exterminated by the pesticides and starvation would end with the use of chemical fertilizers. Actually, the tragedy of agricultural chemicals started at this time. Mokichi Okada warned people that chemical fertilizers would foster pests and diseases and pesticides would lead to the appearance of resistant pests (Okada, 1941). Okada's predictions have now become well-established facts.

With concerns over environmental pollution from excessive use of chemicals, many farmers in Japan have adopted nature farming with little or no chemicals for disease control. However, the nature farming practitioners have to seek alternative methods of disease control. Farmers often use an organic-fertilizer produced by fermenting oil seed cake, rice bran and fish processing wastes. A microbial inoculant known as Effective Microorganisms or EM is utilized to enhance the fermentation process. This kind of organic fertilizer is also called "organic fertilizer" in Japanese. Reports have shown that this kind of organic fertilizer has good nutrient sustainability and improves crop quality (Xu et al., 1998; Wang et al., 1999). However, it is not clear whether this kind of organic fertilizer can increase resistance to crop disease resistance and what the mechanism might be. Therefore, this study was conducted to examine the effect of organic fertilization on phytophthora resistance of tomato plants compared with chemical fertilization.

MATERIALS AND METHODS

Plant Materials and Fertilization Treatments

Tomato seedlings (*L. esculentum* L. cv. Momotaro T96) with five expanded leaves were transplanted into an experimental field in a glasshouse. Four fertilization treatments were made as follows: (1) chemical fertilization (N-P-K: 13-16-17), (2) half chemicals and half organic fertilizer (100 g m^{-2}), (3) organic fertilizer (200 g m^{-2}), and (4) aerobically fermented organic fertilizer (200 g m^{-2}). Organic fertilizer was prepared using organic materials such as oil seed cake, rice bran and fish-processing wastes inoculated with EM (Effective Microorganisms, EM-1, International Nature Farming Research Center, Atami, Japan). This organic fertilizer so produced is also known as bokashi. The nitrogen level of the chemical fertilization treatment was adjusted to 70% of the total available fraction of the bokashi treatment. Photosynthesis, plant-soil nitrogen nutrition, and soil microbial activity were examined before phytophthora (*Phytophthora infestans*) infection occurred. Phytophthora was induced by increasing air humidity and temperature in the glasshouse. Phytophthora infection percentage and infection extent were recorded.

Phytophthora-Inducing Conditions and Examination

Three weeks after the seedlings were transplanted, plants were sufficiently watered and temperature was increased to 30°C and humidity to over 80%. Three days later the phytophthora had infected the tomato plants. The percentage of plants infected by phytophthora was defined as the infection rate. The ratio of number of infected leaflets to the infected plant number was defined as infection intensity.

Measurement of Photosynthesis

Photosynthetic rate (P_N) of the fifth leaf from the top was measured using a LI-6400 photosynthesis measurement system (LI-COR Inc., Lincoln, Nebraska, USA) under different photosynthetic photon flux (PPF) from 0 to 2000 μmol m^{-2} s^{-1}. Photosynthetic capacity (P_C), dark respiration (R_D) and the maximum quantum use efficiency ($Y_Q = KP_C$) were analyzed from the light-response curve modeled by an exponential equation as $P_N = P_C(1 - e^{-KI}) - R_D$, where K is a constant and I is PPF (Xu et al., 1995).

Biochemical Analysis and Enzyme Activity Assay

Electrophoresis was conducted for protein analyses using a double slab gel electrophoretic apparatus (NA-1121, Nihon Eido Co., Ltd., Tokyo, Japan). Nitrate reductase activity in the fifth leaf was assayed by the method of continuous spectrophotometric rate determination according to Gilliam et al. (1993). Nitrate concentration in the fifth leaf was measured using a reflectometer (RQflex, Merck, Kanto Kagaku Co., Ltd., Tokyo, Japan). Amino acids were measured by HPLC with UV-970 Detector and column of Shodex

RSPark KC-811 (Jasco, Tokyo, Japan). Dehydrogenase activity in soil was measured by the method of Casida et al. (1964).

RESULTS

High Phytophthora Resistance in Organically-Fertilized Tomato Plants

In this study, the phytophthora infection was shown by two indicators: (1) percentage infection, which showed the resistance of the plant to phytophthora, i.e., the ability not to be infected; (2) infection intensity, which showed tolerance to the phytophthora, i.e., the survival ability even if the plant was infected. Results in Table 1 show that tomato plants fertilized with organic fertilizer possessed higher resistance against phytophthora infection compared with plants fertilized with chemicals. Not only the percentage infection of the plants but also the infection intensity was lower for organic-fertilized plants. The fertilization treatment of half chemical fertilizer showed the infection percentage and infection intensity intermediate between the organically- and chemically-fertilized plants. Most of the plants in the chemical plots died completely two weeks after the phytophthora was induced. However, in the organic plots, some plants remained uninfected and some were still alive with partial green leaves. The results clearly indicate that organically-fertilized tomato plants showed both higher resistance and tolerance to phytophthora compared with the chemically-fertilized plants.

TABLE 1. Phytophthora infection of tomato plants on day 5, dead plants on day 12, and photosynthetic parameters before infection in plants fertilized with chemical and organic treatments.

Treatment	Infection on day 5		Dead on d 12	P_C	R_D	Y_Q
	%	Intensity	(%)	(μmol m^{-2} s^{-1})		(μmol mol^{-1})
Chemical	88.2 ±12.3	2.43 ±0.29	88.3 ±9.6	22.1 ±2.3	0.41 ±0.06	62.54 ±5.928
Half Chemical	85.7 ±7.6	1.83 ±0.13	71.4 ±6.7	23.2 ±1.9	0.46 ±0.05	61.48 ±5.23
Organic 1	60.5 ±7.2	1.57 ±0.17	63.7 ±8.2	19.2 ±1.4	0.47 ±0.03	60.48 ±4.93
Organic 2	48.6 ±5.3	1.61 ±0.15	59.2 ±5.4	21.1 ±1.6	0.41 ±0.06	59.71 ±4.16

Organic 1, anaerobically fermented organic fertilizer; Organic 2, aerobically treated organic fertilizer.

Physiological Activity Inconsistent

The question is what factors account for the higher phytophthora resistance in the organic-fertilized tomato plants. First, one may consider the physiological activity associated with plant growth, its common indication being the photosynthetic capacity. However, the photosynthetic activity was not lower but a little higher in leaves of chemical-fertilized plants before phytophthora infected the plants, compared with the organically-fertilized plants (Table 1). This suggested that high phytophthora infection was not related to physiological activity shown by photosynthetic activity. Growth and appearance of chemically-fertilized plant did not show any disadvantage compared with organically-fertilized plants. Hence, there must be other factors that account for the higher phytophthora resistance.

Protein Fractions

Electrophoresis determinations showed no difference in leaf proteins irrespective of the mode of fertilization treatment (electrophoretic gel plate not shown). If there are some proteins or enzymes missing in chemically-fertilized plants, the protein fraction would be expected to show some differences on the electrophoretic gel plate. Actually no differences were found. This suggested that the plants fertilized chemically or organically possess the same kinds of proteins. Apparently, no genetic change in relation to proteins was caused by either of the fertilization treatment.

Leaf Nitrate and Nitrogen Compounds

Table 2 shows that the nitrate concentration and the total concentration of amino acids were higher in the chemically-fertilized plants. However, the concentration of proline, an amino acid related with stress resistance, was higher in organically-fertilized plants than those treated with chemical fertilizer. Nitrate concentration in the soil was also higher in the chemical plots.

Leaf Nitrate Reductase and Soil Dehydrogenase Activities

Nitrate reductase activity (Table 2) was higher in organically-fertilized plants than in chemically-fertilized plants, as also confirmed by the results of leaf nitrate concentration mentioned before. Soil dehydrogenase activity is associated with nitrate reduction by microorganisms in soil. The activity of dehydrogenase was also higher in organically-fertilized plants in comparison with the chemically-fertilized plants. This was consistent with the nitrate concentration in the soil as shown in Table 2. The soil biomass carbon was

TABLE 2. Nitrate reductase (NRtase) activity (μmol kg^{-1}FM), nitrate concentration (g kg^{-1}DM), and total amino acids and proline concentrations (mmol kg^{-1}DM) in the fifth leaf as well as the soil biomass carbon (Bio-C, mg kg^{-1}), dehydrogenase (H-ase) activity (μg TPF g^{-1} d^{-1}) and ammonia and nitrate nitrogen concentration (mg kg^{-1}) in the soil fertilized with chemical and organic treatment.

Treatment	Leaf				Soil			
	NRtase Activity	NO_3^-	Amino acids	Proline	Bio-C	H-ase Activity	NH_4^+	NO_3^-
Chemical	5.47 ±0.51	2.7 ±0.21	2.33 ±0.19	0.43 ±0.037	263 ±16	21 ±1.2	55 ±4.4	224 ±21.5
Half Chemical	6.17 ±0.43	2.3 ±0.17	2.04 ±0.17	0.47 ±0.021	326 ±14	40 ±4.6	64 ±11.5	169 ±5.6
Organic 1	7.34 ±0.53	1.9 ±0.08	1.64 ±0.09	0.57 ±0.042	375 ±17	81 ±5.6	723 ±11.1	174 ±5.2
Organic 2	8.26 ±0.49	1.8 ±0.12	1.71 ±0.14	0.55 ±0.039	317 ±18	67 ±5.5	54 ±6.0	186 ±10.4

consistent with the activity of dehydrogenase in the soil. The integrated results suggested that nitrogen metabolism in organically-fertilized tomato plants accounted for the high phytophthora resistance.

DISCUSSION

Phytophthora is a severe disease of tomato caused by the pathogenic fungus (*Phytophthora infestans*). The stems, leaves, leaf petioles and even the root will rot very fast when tomato plants are infected by this fungus. Usually, scientists consider that the ease of infection is associated with genetic susceptibility of the cultivars (Yamada, 1997). Moreover, it is not conclusively known whether tomato phytophthora infection is associated with soil-plant nutrition or fertilization regimes. Sixty years ago, a Japanese philosopher, Mokichi Okada (1941) warned people that chemical fertilizer would foster diseases and insect pests. Now, scientists know that pesticides applications can lead to resistance by insects and disease pathogens.

This study found that the ease of tomato phytophthora infection is associated with nitrogen metabolism in the plant and the nitrogen status and metabolism in the soil. Tomato plants fertilized with chemical fertilizer contain more nitrate nitrogen and nitrogen compounds such as amino acids than plants grown with with organic fertilizer. The high concentration of nitrogen compounds might be favorable to the infection and development of the phytophthora pathogen. On the contrary, low nitrogen compounds in organically-fertilized plants might account for phytophthera resistance. The question is why the nitrate and nitrogen compounds were lower in concentration in organically-fertilized plants. Results showed that the high activity of nitrate reductase possibly accounted for their low concentration in organically-fertilized plants. This might be due to balance, even release and sustainability of nutrients in organic fertilizer. The activity of dehydrogenase in the soil was also higher in organic-fertilized plots than in chemical fertilized plots. This might enable nitrogen in soil to be supplied uniformly and in different types.

It is worth noting that the concentration of proline, an imino acid associated with stress resistance, was higher, although the total concentration of amino acids was lower, in organically-fertilized plants than in chemical-fertilized plants. This might also be associated with phytophthera resistance in organic-fertilized plots. However, the exact mechanism is not known.

Because of the high nutrient availability from the chemical fertilizer, the plant growth and photosynthetic activity at the early growth stage (before phytophthora infection) were high in leaves of chemically-fertilized plants. Results of electrophoresis also showed no difference in leaf proteins among the fertilization treatments. This suggested that high phytophthora infection in organically-fertilized plants was not related to physiological activity shown by photosynthetic activity and genetic characteristics of proteins. Therefore, the advantage of nitrogen metabolism in organically-fertilized tomato plants accounted for their high phytophthora resistance. It is concluded that organic farming can be an effective strategy to help control this disease in plant production systems.

REFERENCES

Casida, L.E. Jr., D.A. Klein and R. Santoro. (1964). Soil dehydrogenase activity. *Soil Science* 98: 371-378.

Gilliam, M.B., M.P. Sherman, J.M. Griscavage and L.J. Ignarro. (1993). A spectrophotometric assay for nitrate using NADPH oxidation by *Aspergillus* nitrate reductase. *Analytical Biochemistry* 212: 359-365.

Okada M. 1941. *The Basis of Paradise–Kyusei Nature Farming* (summarized and re-published in 1993). Press of Seikai Kyusei-Kyo, Atami (Japan), pp. 331-393.

Wang, R., H.L. Xu, M.A.U. Mridha and H. Umemura. (1999). Effects of organic fertilization and microbial inoculation on leaf photosynthesis and fruit yield of tomato plants. *Japanese Journal of Crop Science* 68 (Extra 1): 28-29.

Xu, H.L., L. Gauthier and A. Gosselin. (1995). Effects of fertigation management on growth and photosynthesis of tomato plants grown in peat, rockwool and MFT. *Scientia Horticulturae* 63: 11-20.

Xu, H.L., S. Kato, M. Fujita, K. Yamada, K. Katase, R. Wang, M.A.U. Mridha and H. Umemura. (1998). Sweet corn plant growth and physiological responses to organic fertilization. *Japanese Journal of Crop Science* 67 (Extra 2): 348-349.

Yamada, K. (1997). Fertilizations for high sugar concentration of tomato fruit. In *Agricultural Technology-Vegetables II: Tomato*, ed. Y. Kondo. Tokyo: Rural Culture Association, pp. 541-547.

Evaluating Soil Organic Matter Changes Induced by Reclamation

Xiaoju Wang
Hui-lian Xu
Zitong Gong

SUMMARY. Evaluating soil conservation and plant production on reclaimed lands requires an understanding of the rate of change in soil organic matter (SOM) under specific land use systems and management practices. By using a geographic information system (GIS), changes in the SOM status were assessed and mapped at an experimental station in subtropical China after eleven years of reclamation. Maps of soil survey, land use and topographic information of the experimental station were entered into the computer after digitalization, and a land unit map was produced by overlaying these maps. The SOM levels in different time periods were entered and combined with their corresponding land units in the land unit map, and their differences (ΔSOM) in time for each land unit were calculated. SOM changes at the experimental station were then assessed and mapped based on the ΔSOM values. The changing status in various land use systems, soil types and relief pat-

Xiaoju Wang is Research Soil Scientist, International Nature Farming Research Center, 5632 Hata, Nagano 390-1401, Japan.

Hui-lian Xu is Senior Crop Scientist, International Nature Farming Research Center, 5632 Hata, Nagano 390-1401, Japan.

Zitong Gong is Senior Soil Scientist, Institute of Soil Science, Academia Sinica, P. O. Box 821, Nanjing, China.

Address correspondence to: Xiaoju Wang at the above address.

This work was jointly supported by the National Natural Science Foundation of China and Chinese Academy of Sciences. The authors are grateful to Prof. J. Smith for his critical reading of this manuscript.

[Haworth co-indexing entry note]: "Evaluating Soil Organic Matter Changes Induced by Reclamation." Wang, Xiaoju, Hui-lian Xu, and Zitong Gong. Co-published simultaneously in *Journal of Crop Production* (Food Products Press, an imprint of The Haworth Press, Inc.) Vol. 3, No. 1 (#5), 2000, pp. 85-96; and: *Nature Farming and Microbial Applications* (ed: Hui-lian Xu, James F. Parr, and Hiroshi Umemura) Food Products Press, an imprint of The Haworth Press, Inc., 2000, pp. 85-96. Single or multiple copies of this article are available for a fee from The Haworth Document Delivery Service [1-800-342-9678, 9:00 a.m. - 5:00 p.m. (EST). E-mail address: getinfo@haworthpressinc.com].

terns could be shown in the computer. SOM changes varied with land use systems and the original SOM levels. After 11 years of reclamation, the soils with SOM levels of > 30 and < 10 g Kg^{-1} decreased 95.3% and 30.9% in area, respectively, while those with SOM levels of 20-30 and 10-20 g Kg^{-1} increased 44.3% and 8.4%, respectively. The SOM content in paddy fields, vegetable fields, pastures, and grasslands increased due to reclamation practices, but decreased in fuel woods, sparse weed land, and bare land. *[Article copies available for a fee from The Haworth Document Delivery Service: 1-800-342-9678. E-mail address: getinfo@haworthpressinc.com <Website: http://www.HaworthPress.com>]*

KEYWORDS. Geographic information system, reclamation, soil organic matter, land use

INTRODUCTION

The soil organic matter (SOM) content is a key component which affects many soil properties as well as overall soil quality (Schnitzer and Khan, 1978; Ven Der Linden et al., 1987; Gregorich et al., 1994; Franzluebbers and Arshad, 1996). The SOM is an important parameter when evaluating the effects of soil conservation practices. Maintenance of sufficiently high SOM levels is a prerequisite for sustaining soil quality and crop production (Ven Der Linden et al., 1987; Wang and Gong, 1995). Land uses and soil management practices play a vital role in SOM changes (Wang and Gong, 1996; 1998). The success of soil conservation efforts and management practices to maintain or improve SOM requires an understanding of how soils respond to agricultural use and management inputs over time (Parton et al., 1987; Jenkinson et al., 1987; Gregorich et al., 1994; Wang, 1995; Franzluebbers and Arshad, 1996). Thus, it is important to know the time-rate of change in SOM levels as affected by management practices in different land use systems.

SOM change is a temporal and spatial process (Jenny, 1980; Hoosbeek and Bryant, 1992; Wang and Gong, 1996). Therefore, only by comparing and analyzing SOM changes during two or more time periods, can the nature and mechanism of SOM changes be determined and the role of land use systems and management practices be assessed. However, in the current studies, monitoring and modeling work is usually confined to small plots or soil profiles (Jenkinson et al., 1987; Harding and Jokela, 1994; Gregorich et al., 1996; Franzluebbers and Arshad, 1996). Studies at the regional level on mapping the SOM content are usually based on the current SOM status and therefore do not directly reflect SOM changes over time or provide a comprehensive assessment and analysis of the dynamic nature of SOM changes (Wang, 1995).

The objectives of this research were to demonstrate a methodology for assessing SOM changes on an experimental station scale, and to analyze and discuss the dynamics of SOM changes in different land use systems and soil types in south China after reclamation.

MATERIALS AND METHODS

Site Description and Soil Collection

The study was conducted at the Qianyanzhou Experimental Station of the Chinese Academy of Sciences. This station is in Jiangxi Province, south China, at 115°57′ E, 27°42′ N, with a subtropical climate, hill relief, elevations from 66 to 131.5 m and an area of 204.2 ha. The mean annual temperature is 18.6°C and the annual precipitation is about 1360 mm. According to the Chinese soil taxonomic classification (Gong, 1991), the soils are classified into four soil groups and fifty species. The four soil groups are Red soils, Paddy soils (PS), Meadow soils (MS), and Umbrihumus Meadow soils (US), which approximately correspond to the Haplic Acrisols, Aric Anthrosols, Eutric Fluvisols and Umbric Fluvisols, respectively, in the FAO-UNESCO World Map Revised Legend. Based on the soil parent materials, the Red soils were further differentiated as argillaceous Red soils (AS) and sandstone Red soils (SS).

Prior to 1983, the station was a very poor village with poor soils, low crop yields, and much wasteland. That year, scientists of the Chinese Academy of Sciences conducted a detailed survey of soils, vegetation, water resources, geology and geomorphology, which yielded a great deal of important background data (Cheng, 1993; Wang, 1995). Reclamation efforts were then initiated, and various land use systems were adopted. There were more than twenty land use patterns including cropland, orchard land, forest land, shrubby land, pasture land, grassland, and bare land, which had been managed using typically local (i.e., indigenous) methods and practices (Cheng, 1993; Wang and Gong, 1995). This provided unique conditions for the present study.

Based on the detailed soil survey of 1983, surface soil samples (0-20 cm) were collected in 1994 using a grid method (75 m × 75 m). Additional samples were taken according to specific conditions, so that every land unit was sampled. The vegetation, management practices, land cover and erosion conditions of each sample point were recorded. SOM was determined using the potassium dichromate-sulfuric acid oxidation method (Institute of Soil Science, Academia Sinica, 1978), which was the same method as that used in 1983.

Development of Soil Information System and Soil Change Database

A soil information system, QYZSIS, was developed by using PC-ARC/INFO and FOXBASE software, and was used for quantitative evaluation and mapping of SOM changes. The steps in developing this system are summarized as follows.

Related maps of relief, soils and land use were entered into the computer database. This required digitizing, editing and updating these maps, developing a DEM (digital elevation map), and producing a land unit map by editing an overlay of soil, relief and land use maps. SOM contents in 1983 and 1994 were then added to the database and combined with their corresponding land units in the land unit map to calculate the differences between them (ΔSOM). The SOM changes were then evaluated and mapped together with a statistical summary of the results to show the SOM changes under various land uses, soils and relief types. Mean values of the SOM contents in 1983 and 1994 under different land use systems and soil types were calculated, and the difference between 1983 and 1994 was compared using Tukey's test and the least significant difference (LSD) method.

Evaluation of Soil Organic Matter

Changes in the SOM content for each land unit were obtained by subtracting the corresponding SOM values of 1983 from those of 1994. This provides the net change in SOM, designated as ΔSOM. The resulting ΔSOM (g kg^{-1}) values were evaluated quantitatively by dividing them into six changed classes as follows: great increase, > 10; moderate increase, 5 to 10; slight increase, 0 to 5; slight decrease, -5 to 0; moderate decrease, -10 to -5; and great decrease, < -10.

RESULTS AND DISCUSSION

General Changes in Soil Organic Matter

Large changes in the SOM content were found after eleven years of reclamation practices (Table 1). Generally, the areas for soils with very high (> 30 g kg^{-1}) and very low (< 10 g kg^{-1}) SOM levels decreased, whereas those for soils with medium SOM (10-30 g kg^{-1}) increased. In 1983, the areas for soils with SOM levels of > 30, 20-30, 10-20, and < 10 g kg^{-1} accounted for 4.4, 9.5, 67.8 and 18.3% of the total soil area, respectively. After 11 years of reclamation, the areas for soils with SOM levels of > 30 and < 10 g kg^{-1}

TABLE 1. Area (ha) of soils with different soil organic matter (SOM) after eleven years of reclamation.

SOM ($g\ kg^{-1}$)	Year	Area (ha) of different soils					
		AS[1]	SS	PS	MS	US	Total
> 30	1983	5.18[2]	2.23	0.00	0.00	1.08	8.49
	1994	0.00	0.00	0.00	0.00	0.40	0.40
20-30	1983	9.44	5.39	0.00	0.00	3.36	18.19
	1994	12.15	10.26	0.45	1.85	1.53	26.24
10-20	1983	48.75	41.05	11.39	24.15	5.13	130.47
	1994	45.03	47.03	14.07	28.60	6.65	141.38
0-10	1983	0.00	23.50	3.28	8.43	0.00	35.21
	1994	6.19	14.88	0.15	2.13	0.99	24.34

[1] AS: argillaceous Red soils; SS: sandstone Red soils; PS: paddy soils; MS: meadow soils and US: umbrihumus meadow soils.
[2] The soil area for a specific SOM range and soil type is the sum of all map units whose SOM and soil type belong to this condition.

decreased to 0.2 and 12.7%, respectively, while those with SOM levels of 20-30 and 10-20 $g\ kg^{-1}$ increased to 13.6 and 73.5%, respectively (Table 1). This showed that the SOM content for some low SOM soils had increased after reclamation, but high SOM soils had not.

The SOM dynamics varied according to soil type. Argillaceous Red soils and Umbrihumus Meadow soils generally showed a significant loss or decline in SOM. These two soil types had higher SOM levels in 1983, with 8.2 and 11.3% above 30 $g\ kg^{-1}$ and no soil below 10 $g\ kg^{-1}$. However, those areas with a SOM level above 30 $g\ kg^{-1}$ virtually disappeared, while the areas of these two soil types with SOM < 10 $g\ kg^{-1}$ increased to 6.19 ha (9.8%) and 0.99 ha (10.3%), respectively. The sandstone Red soils with both high (> 30 $g\ kg^{-1}$) and low (< 10 $g\ kg^{-1}$) SOM levels decreased in area. Contrarily, the SOM status for the Meadow soils and paddy soils had generally increased. They were initially low in their SOM content with no SOM > 20 $g\ kg^{-1}$ in 1983. By 1994, the areas of soils with SOM > 20 $g\ kg^{-1}$ had increased by 3.8 and 5.7%, and those with SOM < 10 $g\ kg^{-1}$ had decreased by 74.7 and 95.4%, respectively, showing that the SOM levels had improved substantially. The apparent improvement of low SOM soils and degradation of high SOM soils in this study strongly suggest that research should focus on developing best management practices for improving marginal soils and also maintaining the quality of highly productive soils.

Characteristics of Soil Organic Matter Changes Under Different Land Use Systems

Paddy fields, vegetable fields and uplands showed distinct differences in SOM changes. SOM was generally improved in paddy fields because of better fertilizer management and rotations with green manure crops or legumes after reclamation. Figure 1 shows that the mean SOM levels in 1994 for the five soil types increased or remained unchanged compared with those in 1983, especially for meadow and paddy soils with increases of 83.2 and 52.6%, respectively. The vegetable field also significantly increased its SOM level in the meadow soils, and did not decrease in the other soils. The upland

FIGURE 1. Changes in average contents of soil organic matter under different land use systems and soil types. ns = no significant differences by t-test; * = significant differences at 5% probability level; ** = significant differences at 1% probability level.

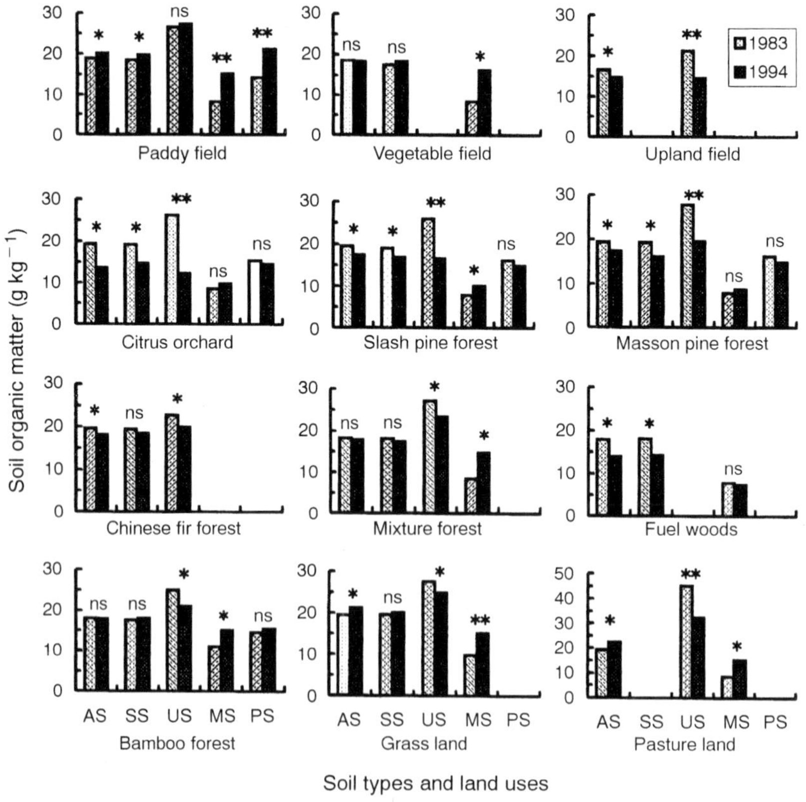

field decreased its SOM level for the two soil types tested. Changing status of SOM in paddy and vegetable fields was therefore better than that in the upland field.

Citrus is one of the most important cash crops in south China. The citrus orchards, with a total area of 36.69 ha, are mostly distributed on Meadow soils at river terraces. The density is about 1900 tree/ha with an orchard age of 8-10 years. Chemical and organic fertilizers were applied once a year at higher rates than to farmland, and equivalent to 173 kg N, 70 kg P_2O_5, and 80 kg K_2O per ha. With the exception of Meadow soils and Paddy soils, the SOM levels decreased significantly in the other three soils compared with those found in 1983. Umbrihumus Meadow soils had the largest decrease in SOM from 26.23 g kg^{-1} in 1983 to 11.13 g kg^{-1} in 1994. Argillaceous Red soils and sandstone Red soils also declined in their original SOM by 3-6 g kg^{-1}. This indicates that the current inputs and management practices neither maintained nor enhanced the SOM level (Wang, 1995).

Coniferous trees are usually planted extensively in hilly regions in subtropical China as the pioneer (i.e., early established) tree species, because of their fast growth even on poor soils. Coniferous forests and mixed forests of coniferous and broadleaf trees occupied the largest area of all the land use patterns at the experimental station (Table 2). After 11 years of growth, slash pine, Masson pine, Chinese fir and the mixed forests on Umbrihumus meadow soils, argillaceous Red soils and sandstone Red soils reduced the SOM contents, which was similar to the study reported by Chen (1995). However, some improvement in SOM was observed in the mixed forests. The coniferous forest on the sandstone Red soils had less SOM decline than on the Umbrihumus soils. SOM decrease was not found in the coniferous forest on the Meadow soils (Figure 1). This suggests that planting coniferous trees in poor soils can improve the SOM content, but may decrease the SOM in better soils. The fertility status of soils in south China should always be considered before extensive planting of coniferous trees.

In bamboo forests, SOM increased for the Meadow soils and decreased for Umbrihumus Meadow soils. Fuel woods decreased the SOM content, especially for argillaceous Red soils and sandstone Red soils which showed a decrease of 4.28 and 3.80 g Kg^{-1}, respectively (Figure 1).

Natural grassland and pasture fields showed the greatest improvement in SOM because of their vast underground biomass (Wang and Gong, 1996). SOM content in Meadow soil and two kinds of Red soils were 5 to 8 g kg^{-1} higher than the original levels. However, SOM in the Umbrihumus Meadow soils showed a decrease due to the higher SOM content before reclamation, and rapid decomposition of SOM after reclamation (Figure 1). The study also showed that SOM accumulation significantly decreased as the extent of cover

TABLE 2. Evaluation results of SOM changes (ΔSOM) under various land use systems

Land use	Total area (ha)	Area (Ha) with different SOM in g kg^{-1}					
		> 10	5-10	0-5	−5-0	−10-−5	<−10
Paddy field	8.24[1]	0.48	1.07	4.10	2.59	0.00	0.00
Upland	0.79	0.00	0.00	0.02	0.49	0.00	0.28
Vegetable field	0.63	0.10	0.12	0.33	0.08	0.00	0.00
Citrus orchard	36.69	0.00	1.33	25.33	5.23	2.55	2.25
Hawthorn orchard	0.19	0.00	0.00	0.00	0.19	0.00	0.00
Chestnut orchard	0.23	0.00	0.00	0.12	0.11	0.00	0.00
Masson pine forest	47.81	0.00	2.26	27.06	8.99	6.39	3.11
Slash pine forest	41.16	0.00	4.67	23.40	3.53	7.70	1.86
Fir forest	5.38	0.38	0.44	1.29	0.49	1.06	1.72
Mixed forest	12.18	0.95	1.76	3.42	4.79	1.26	0.00
Broadleaf forest	5.91	0.45	1.79	1.71	1.19	0.00	0.77
Fuel woods	2.45	0.00	0	0.20	2.09	0.00	0.16
Bamboo forest	12.38	0.00	3.30	4.08	2.38	2.12	0.50
Bush	0.43	0.00	0.23	0.20	0.00	0.00	0.00
Paulownia forest	0.24	0.00	0.00	0.24	0.00	0.00	0.00
Oil-tea	1.67	0.00	0.00	0.14	1.03	0.50	0.00
Tung oil trees	0.61	0.00	0.00	0.21	0.40	0.00	0.00
Pasture land	2.78	1.44	0.77	0.47	0.00	0.10	0.00
Grass land	2.33	0.52	0.32	1.21	0.06	0.22	0.00
Spares weed land	9.96	0.00	0.00	2.18	3.66	3.28	0.84
Bare land	0.30	0.00	0.00	0.08	0.03	0.04	0.15
Total (ha)	192.36	4.32	18.06	95.83	37.29	25.22	11.64

[1] The soil area for a specific ΔSOM and land use is the sum of all map units whose ΔSOM and land use belong to this condition.

from herb vegetation decreased, suggesting that herbs play an important role in conserving SOM in south China.

In summary, directions and intensity of SOM changes were different according to the land use system. The direction and intensity were also greatly influenced by the original SOM levels and soil types. Therefore, one can not simply conclude that a specific land use system will cause a constant change in SOM. The study showed that the SOM in Meadow soils under most land use systems increased after reclamation, but the degree of increase varied with land use systems. Umbrihumus Meadow soils usually had the highest SOM content in the station before 1983, but after 11 years of reclamation, the SOM content either declined or maintained its previous level under a specific land use system. SOM in the Red and Paddy soils increased after being utilized as paddy fields, vegetable fields or grasslands. However, they decreased or were unchanged after being utilized as upland, fuel woods, or coniferous forests. This indicates that paddy fields, vegetable fields, and grasslands may help to maintain or increase SOM.

Quantitative Assessment of Soil Organic Matter Changes

SOM changes under various land use systems and soil types were evaluated quantitatively and mapped using a GIS and computer database according to the changed values of SOM (ΔSOM) from 1983 to 1994 (Figure 2). In the computer database, this map and its connected database can display the detailed SOM changes for various land use systems, soil types and relief conditions in the study area. Only the results of changes under the specific land use systems in this paper are discussed (Table 2).

The results (Figure 2 and Table 2) indicate that 69.3% (5.17 ha) of the total area of paddy field improved the SOM status with ΔSOM above 0, and 18.8% (1.55 ha) above 5. About 87.3% of the vegetable field had a ΔSOM above 0, and 34.9% above 5. This showed that the SOM status in most of the paddy field and vegetable field was increased. Results also indicate that SOM increased in the fields where more organic fertilizers were applied or green manure crops were part of the crop rotations. However, SOM in 62.0% of the upland showed a slight decrease, which likely reflected the lower fertilizer input and inadequate management.

SOM in citrus orchards showed a small improvement wherein 72.7% (26.66 ha) of the soil had a ΔSOM above 0, but only 3.6% (1.33 ha) was above 5, and no ΔSOM was above 10. Citrus soils with ΔSOM above 0 were mainly Meadow soils with the initial organic matter content less than 10 g kg^{-1} and distributed on a flat stream terrace, while those with ΔSOM below 0 were usually located on slopes with Red soils of higher initial SOM, which proved difficult to maintain. Moreover, these orchards were usually given lower inputs and less intensive management than those on terraces because unterraced slopes are more difficult to manage.

SOM in the coniferous forests and mixed forests of coniferous and broadleaf trees did not change greatly, as the ΔSOM values mostly ranged from −5 to +5. However, where low SOM soils, such as the eroded Red soil, were planted with coniferous trees, its SOM content was increased. For the high SOM soil, such as the thick Red soil, a considerable decrease in SOM was observed (Wang, 1995). This indicated that for the soil with a poor SOM status, coniferous trees were good pioneer (i.e., early established) tree species

FIGURE 2. Evaluation map of changes in soil organic matter after 11 years of reclamation in Qianyanzhou Experimental Station.

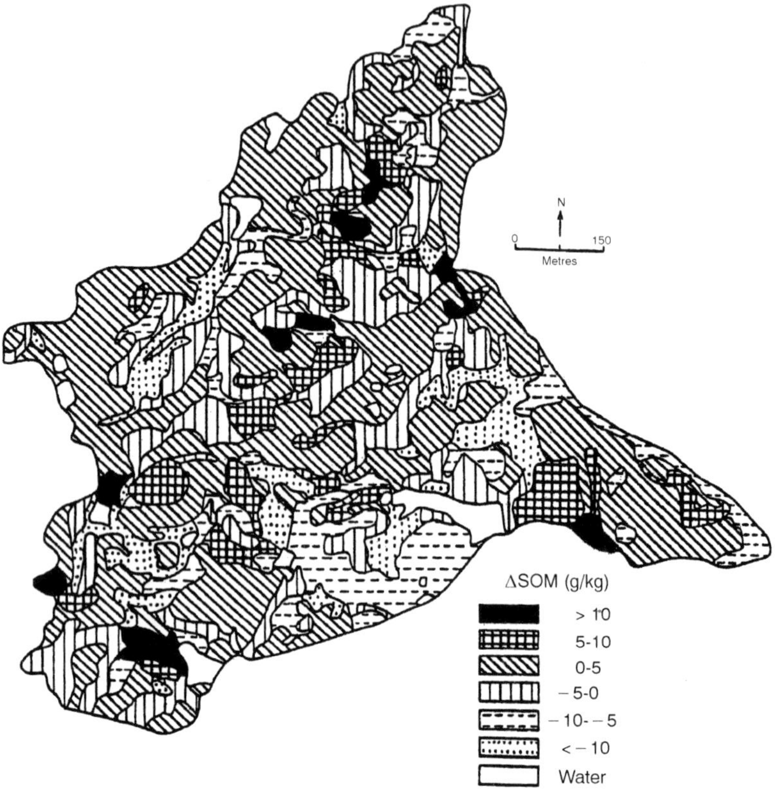

to improve SOM and to produce wood. However, for high SOM soils, coniferous forests could not maintain the original SOM level.

For pasture land, SOM increased significantly. About 79.5% of the total area had a ΔSOM above 5, and 51.8% above 10. Only a small part had a ΔSOM between −5 and 0. This occurred mainly on the Umbrihumus Meadow soils with a high initial SOM level and a rapid decomposition loss of SOM after transition to pasture. Organic matter in soils of natural grassland was also characterized by an increase because of the higher grass density, but it was less significant than that of pastureland. Most of the sparse grassland and all of the bare land continued to degrade with a concomitant decrease in the SOM content (Table 2).

CONCLUSIONS

The combination of a soil database with a GIS was effective in evaluating and mapping changes in soil organic matter in time and space. This study provided basic data as well as a method for monitoring and assessing SOM changes as affected by soil type, land use, topography and management inputs. All of these resulted in degrees and intensities of soil organic matter changes. Results indicated that the SOM levels of pasture land, grass land, paddy fields, vegetable fields and bamboo forest were generally increased with time, whereas the SOM levels of citrus orchards, coniferous forest, and fuel woods, spare grass land and bare land decreased. Herb vegetation played a very important role in conservation and improvement of the soil organic matter content in subtropical China.

SOM changes were greatly influenced by the original SOM levels. Soils with higher SOM contents showed a tendency to be degraded (loss of SOM) during reclamation, while soils with very low SOM were usually more amenable to improvement (i.e., to increasing the SOM level). Research needs to focus on management practices that will increase the SOM levels of both marginal or poor soils and high fertility soils during reclamation efforts.

REFERENCES

Chen, Y.R. (1995). Study on changes in soil fertility under different artificial forests. *Natural Resources* 2: 46-51 (in Chinese).

Cheng, T. (1993). *Studies of Ecological System Restoration and Agricultural Sustainable Development of Red Earth Hills*. Beijing: Earthquake Press (in Chinese).

Frarnzluebbers, A.J. and M.A. Arshad. (1996). Soil organic matter pools during early adoption of conservation tillage. *Soil Science Society of America Journal* 60: 1422-1467.

Gong, Z.T. (1991). *Chinese Soil Taxonomic Classification*. Beijing: Science Press.

Gregorich, E.G., M.R. Carter, A. Angers, C.M. Monreal and B.H. Ellert. (1994). Towards a minimum data set to assess soil organic matter quality in agricultural soils. *Canadian Journal of Soil Science* 74: 367-385.

Gregorich, E.G., B.H. Ellert, C.F. Druiy and B.C. Liang. (1996). Fertilization effects on soil organic matter turnover and corn residue C storage. *Soil Science Society of America Journal* 60: 472-476.

Harding, R.B. and E.J. Jokela. (1994). Long-term effects of forest fertilization on site organic matter and nutrients. *Soil Science Society of America Journal* 58: 216-221.

Hoosberk, M.R. and R.B. Bryant. (1992). Towards the quantitative modelling of pedogenesis: a review. *Geoderma* 55: 183-210.

Institute of Soil Science, Academia Sinica. (1978). *The Physical and Chemical Analyses of Soil*. Shanghai: Shanghai Science and Technology Press (in Chinese).

Jenkinson, D.S., P.B.S. Hart, J.H. Rayner and L.C. Parry. (1987). Modeling the turnover of organic matter in long-term experiments at Rothamsted. *INTECOL Bulletin* 15:1-8.

Jenny, H. (1980). *The Soil Resource: Original and Behavior*. New York: Springer-Verlag.

Parton, W.J., D.S. Schimet, C.V. Cole, and D.S. Ojima. (1987). Analysis of factors controlling soilorganic matter levels in Great Plain grasslands. *Soil Science Society of America Journal* 51: 1173-1179.

Schnizer, M. and S.U. Khan. (1978). *Soil Organic Matter Development in Soil Science*. Amsterdam: Elsevier.

Ven Der Linden, A.M.A., J.A. Van Veen, and M.J. Frisset. (1987). Modeling soil organic matter levels after long-term applications of crop residues, and farmyard and green manures. *Plant and Soil* 101: 21-28.

Wang, X. and Z. Gong. (1998). Assessment and analysis of soil quality changes after eleven years of reclamation in Subtropical China. *Geoderma* 81: 339-355.

Wang, X. and Z. Gong. (1995). Ecological effects of different land use patterns in red soil hilly region. *Pedosphere* 5: 163-170.

Wang, X. (1995). Monitoring and evaluation of soil changes under different land use systems in red soil hilly region of China. Ph.D. Dissertation, Institute of Soil Science, Chinese Academy of Sciences, Nanjing, China (in Chinese with English summary).

Wang, X. and Z. Gong. (1996). Monitoring and evaluation of soil changes under different land use systems. *Pedosphere* 6:373-378.

Organic Wastes for Improving Soil Physical Properties and Enhancing Plant Growth in Container Substrates

Nsalambi V. Nkongolo
Jean Caron
Fabienne Gauthier
Mitate Yamada

SUMMARY. Increasing rates (5, 10, 25 and 40% v/v) of 6 sources of organic wastes were substituted for peat to assess changes in the physical properties of peat-perlite substrates and investigate the relationship

---

Nsalambi V. Nkongolo is Soil Scientist, Vegetable Cultivation Technology System Laboratory, National Agriculture Research Center, 3-1-1 Kannondai, Tsukuba, Ibaraki 305-8666 Japan; present address: Laboratory of Soil Science, Hokkaido University, Sapporo 060-8589, Japan.

Jean Caron is Associate Professor, Soil Science and Agrifood Engineering Department, Faculty of Agricultural and Food Sciences, Pavillon Paul-Comtois, Laval University, Ste-Foy, Quebec, Canada G1K 7P4.

Fabienne Gauthier is Graduate Student, The Horticultural Research Center, Faculty of Agricultural and Food Sciences, Pavillon de l'Envirotron, Laval University, Ste-Foy, Quebec, Canada G1K 7P4.

Mitate Yamada is Senior Agronomist, Vegetable Cultivation Technology System Laboratory, Agriculture Research Center, 3-1-1 Kannondai, Tsukuba, Ibaraki 305-8666 Japan.

Address correspondence to: Nsalambi V. Nkongolo at his present address as above (E-mail: Felly_Nkongolo@yahoo.com).

[Haworth co-indexing entry note]: "Organic Wastes for Improving Soil Physical Properties and Enhancing Plant Growth in Container Substrates." Nkongolo, Nsalambi V. et al. Co-published simultaneously in *Journal of Crop Production* (Food Products Press, an imprint of The Haworth Press, Inc.) Vol. 3, No. 1 (#5), 2000, pp. 97-112; and: *Nature Farming and Microbial Applications* (ed: Hui-lian Xu, James F. Parr, and Hiroshi Umemura) Food Products Press, an imprint of The Haworth Press, Inc., 2000, pp. 97-112. Single or multiple copies of this article are available for a fee from The Haworth Document Delivery Service [1-800-342-9678, 9:00 a.m. - 5:00 p.m. (EST). E-mail address: getinfo@haworthpressinc.com].

between plant response and these properties. Wastes were either fresh or composted bio-filter sludge (FBF and CBF), sewage sludge (FSS and CSS), and de-inked paper sludge (FDP and CDP). Geranium plants (*Pelagornium* × *hortorum 'Orbit Hot Pink'*) were grown in the substrates. Growing substrates' saturated hydraulic conductivity (K_S), air-filled porosity (f_a), pore tortuosity (τ), and relative gas diffusivity (D_S/D_O) all increased linearly ($p = 0.0001$) as the rate of organic wastes increased. *Geranium* plant height (PHT), shoot dry mass (SDM) and root dry mass (RDM) were either linearly or quadratically decreased ($p = 0.0001$) as the amount of waste increased in the substrates. During both growing seasons, Geranium SDM and RDM were either linearly or quadratically correlated with D_S/D_O and τ. Organic waste types and their rate of application strongly affected the aeration status of the substrates. D_S/D_O and τ better expressed the relationship between plant growth and the physical conditions of the root zone.

KEYWORDS. Aeration criteria, plant response, peat substrate, organic waste, gas diffusivity, pore tortuosity, geranium

INTRODUCTION

Environmental concerns are urging us to recycle as much waste as possible. In fact, current waste management practices such as ocean dumping, landfills and incineration are increasingly viewed as unsuitable because of their effects on ocean pollution, ground water contamination and air pollution. The use of organic wastes in container grown plant substrates (Bugbee et al., 1991) represents, therefore, one of the best alternatives from both environmental and energetic considerations.

There is a general agreement on the beneficial effects of organic wastes as substitute components in container grown plants (Cline and Chong, 1991; Chong and Cline, 1993). However, conflicting results have also been reported on effects of organic waste on physical properties of horticultural plant substrates. Tyler et al. (1993) reported that the addition of composted turkey litter to pine bark did not affect total porosity, but container capacity, available and unavailable water and bulk density increased with increasing compost rate. Bugbee et al. (1991) found a decrease in available water and bulk density of pine bark amended with municipal leaf, sewage sludge and street sand compost. Warren and Fonteno (1993) obtained a linear increase of total porosity and unavailable water, but a decrease in air-filled porosity of a loamy sand amended with composted poultry litter. Bowman et al. (1994) found that amending sand-sawdust substrates with ground automobile tires did alter (though not appreciably) total porosity, container capacity and bulk density but increased air-filled porosity.

In the assessment of the physical properties of horticultural plants' substrates, most of research work has been devoted to the routine monitoring of the aeration (water and air) status of substrates. While substrates with improved water holding capacity and air-filled porosity are desirable, a good growing substrate should also contain a significant number of large pores to facilitate the rapid exchange of gases and water (Allaire et al., 1994; 1996 a, b; Caron and Nkongolo, 1999) to and from the root zone. In addition, a reliable measure of the aeration status, which is related to plant response is also needed (Paul and Lee, 1976). The objective of this study was to assess changes on substrates storage and exchange properties following amendment with 6 different sources of organic wastes and to relate plant response to these properties.

MATERIAL AND METHODS

Substrates Formulation

The experiment was conducted at the Horticultural Research Center, Laval University, Quebec, Canada in the summer season of 1994 and winter season of 1995. Four rates (5, 10, 25 and 40% by volume) of each of 6 types of organic wastes were combined to four rates (25, 40, 55 and 60% by volume) of peat and to a single rate (35% by volume) of perlite (Table 1). Organic wastes were both fresh (FBF) and composted (CBF) bio-filter sludge (Les Tourbières Premier, Rivière du Loup, Quebec), fresh (FSS) and composted (CSS) sewage sludge (Fafard & Frères, Bonaventure, Quebec) and fresh (FDP) and composted (CDP) de-inked paper sludge (Les Compost du Quebec, Saint Henri, Quebec). Some of the chemical properties of waste materials are given in Table 2. After formulation, the substrates were placed in regular greenhouse containers of 11.5 cm in diameter by 20 cm deep. The containers were arranged in a randomized complete block design with 5 replications. A total of 120 (4 rates of organic wastes × 6 types of organic wastes × 5 replications) containers were used in each growing season's study. At the beginning of each growing season, containers with their substrates were immersed in degassed tap water and allowed to saturate. After physical parameters were measured, Geranium plants were potted in.

Measurement of the Physical Properties of the Substrates

Volumetric Water Content (θ). We first of all measured the dielectric constant (k_a) by inserting time domain reflectometry (TDR) probes (13.5 cm long) into each 20-cm long cylinder. The measured k_a was converted into the volumetric water content at saturation (total porosity) using:

$$(\theta = -0.0055 + 0.0425\,k_a - 0.000975\,k^2{}_a + 0.00000907\,k^3{}_a \qquad [1]$$

for k_a between 5 and 58 (Paquet et al., 1993). The volumetric water content at container capacity (θ_c) was measured after saturation and drainage for two hours.

Air Filled Porosity (f_a). Air-filled porosity was calculated as the difference between the volumetric water content (estimated by TDR) at saturation (θ_{sup}) and after saturation and drainage (θ_c):

TABLE 1. Volumetric composition ($cm^3\ cm^{-3}$) of experimental substrates (for each of the 6 types of organic wastes)

Substrates	Peat	Perlite	Organic waste
S1	60	35	5
S2	55	35	10
S3	40	35	25
S4	25	35	40

TABLE 2. Some physical and chemical properties of organic wastes used in this experiment

Organic wastes (OW)	Rate of application	EAW ($cm^3\ cm^{-3}$)	f_a ($cm^{-3}\ cm^{-3}$)	CEC (meq/100 g)	Salinity (mS)	C/N ratio	pH
Fresh bio-filter (FBF)	5%	0.43	0.13	96.4	1.26	52.6	5.27
	40%	0.52	0.17				
Composted bio-filter (CBF)	5%	0.30	0.13	83.1	6.67	45.8	4.38
	40%	0.08	0.13				
Fresh sewage sludge (FSS)	5%	0.23	0.13	58.1	3.75	43.1	6.42
	40%	0.08	0.22				
Composted sewage sludge (CSS)	5%	0.43	0.13	68.1	3.82	51.5	6.09
	40%	0.29	0.17				
Fresh de-inked paper sludge (FDP)	5%	0.15	0.11	8.5	0.17	41.7	9.11
	40%	0.31	0.20				
Composted de-inked paper sludge (CDP)	5%	0.33	0.18	21.1	1.73	40.7	7.42
	40%	0.30	0.16				

EAW = Easily available water, f_a = Air-filled porosity.

$$f_a = \theta_{sup} - \theta_c \qquad [2]$$

where f_a is the air-filled porosity ($m^3\ m^{-3}$), θ_{sup} the volumetric water content at saturation ($cm^3\ cm^{-3}$) and θ^c the volumetric water content at container capacity ($m^3\ m^{-3}$).

Saturated Hydraulic Conductivity (K_S). Before assessing volumetric water contents, cylinders with their substrates were subjected to K_s measurements. A Côté infiltrometer was used to measure the water flux density in each container under a constant (3 cm of water) pressure head (Allaire et al., 1994). A nylon mesh (0.15 cm) was placed over the surface of each substrate to prevent light particles from floating and plugging the opening of the Mariotte bottle during the flow. The water flux was measured after a steady state had been reached, and the final height of the substrate determined thereafter. Saturated hydraulic conductivity (K_S) was finally calculated using Darcy's law as follows:

$$q = Q/A = K_S\,(h + Z)/L \qquad [3]$$

where q = Flux density ($m^3\,m^{-2}\,s^{-1}$), A = area normal to the flux (m^2), K_S = saturated hydraulic conductivity ($m\ s^{-1}$), h = hydraulic head (kPa), L = length of the cylinder filled with substrate (m):

$$Z = \text{Gravity head (kPa) and } (h + Z)/L = \text{hydraulic gradient } (m\ m^{-1}). \qquad [4]$$

Pore Tortuosity Factor (τ). The Pore tortuosity factor was calculated with the two-point method (Nkongolo, 1996). Estimates of pore tortuosity with the two-point method are obtained solely on the basis of four unknowns: α, θ_{sup}, θ_c and K_S. K_S is the saturated hydraulic conductivity and is measured from a water flow experiment as explained earlier. θ_{sup} and θ_c are the water contents measured successively at saturation and container capacity. α is the slope of the linear regression line between Ψ and θ (early phase of the water desorption curve). The value of α is obtained from two points: θ and ψ at the point of air entry (θ^a, Ψ^a), and θ and Ψ at container capacity or after saturation and drainage (θ_c, Ψ_c). In fact, in the case of peat substrates, the early phase (saturation to -1.0 kPa approximately) of the water desorption curve can be well approximated by the following linear equation:

$$\theta = a\Psi + b \qquad [5]$$

where Ψ is the soil water potential (kPa) that can be transformed into an equivalent mean pore radius using Jurin's law (Musy and Soutter, 1991). By developing Equation 5 as shown in Nkongolo (1996), the pore tortuosity factor can be calculated as follows:

$$\tau = \frac{10\rho_g}{8\eta K_s} * 0.000225 * \frac{\theta_c - \theta_{sup}}{(\text{half of the height of substrate in pot (kPa)} - \Psi_a)} \left[\frac{\Psi_{sup} - \Psi_{inf}}{\Psi_{sup} * \Psi_{inf}} \right] \quad [6]$$

where τ is the pore tortuosity factor (cm cm^{-1}), g is gravity (m s^{-2}), ρ is the water density (g cm^{-3}), and η is the water viscosity (Pa s^{-1}), K_s is the saturated hydraulic conductivity (cm s^{-1}), Ψ_{inf} is the water potential at residual water content (θ_{inf}). The value of (α) is calculated from:

$$a = \frac{\theta_{inf} - \theta_{sup}}{\Psi_{inf} - \Psi_a} = \frac{\theta_{inf} - \theta_{sup}}{(\text{half the height of the substrate (kPa)} - \Psi_a)} \quad [7]$$

Relative Gas Diffusivity (D_s/D_o). Relative gas diffusivity was estimated according to the relationship proposed by King and Smith (1987) as follows:

$$D_s/D_o = (f_a)/\tau \quad [8]$$

where D_s = gas diffusion coefficient within the substrate ($m^2 s^{-1}$) and D_o = gas diffusion coefficient in the free air ($m^2 s^{-1}$).

Plant Propagation

Geranium (*Pelagornium × hortorum 'Orbit Hot Pink'*) seeds were sown in multi-cellular plates of 200 units, filled with commercial potting mix (PRO-MIX 'PG®' Les Tourbières Premier, Rivière du Loup, Quebec). Substrate temperature was maintained at 22 ± 2°C with heating cables during the first two weeks after sowing. Seedlings were irrigated and fertilized as needed with a solution of 150, 36, 168, 120, and 28 ppm of N-P_2O_5-K_2O, Ca, and Mg, respectively. They were also supplied with a solution of micro-nutrients containing 0.75, 0.64, 0.22, 0.20, 0.07 and 0.03 ppm of Fe, Mn, Zn, B, Mo and Cu, respectively. A month after sowing, seedlings of about the same height, size and appearance were selected and transferred into 11.5 cm diameter pot (type Ultra®, Kord Products Limited, Ontario) filled with substrate made of peat, perlite and organic wastes as described earlier. Seedlings were sub-irrigated and waste water was recycled. At intervals of two to three days, plants were provided with a nutritiet solution containing 100, 30, 106, 80 and 35 ppm of N-P_2O_5-K_2O, Ca and Mg, respectively, and a micronutrient solution containing 0.2, 0.75, 0.30, 0.25, 0.46 and 0.13 ppm of Fe, Mn, Zn, B, Mo and Cu, respectively. Greenhouse conditions were standardized for commercial production. Air temperature was maintained at 22 ± 2/18 ± 2°C day/night. Mist fans were used to maintain relative humidity of 90%. A 16-hour photoperiod was maintained in the greenhouse. A thermal shading tissue (LS

16®), giving 65% shading was used between 1100 a.m. and 1500 p.m. during sunny days. These conditions were maintained throughout the experiment. Data on plant height and width were regularly collected and served to calculate the growth index. After four months of growth, plant height was measured and plant shoots were removed, oven-dried and weighed. Substrates were again subjected to *in-situ* measurements of physical properties. The same experimental procedure was used in both winter 1995 and summer 1994 growing seasons.

Statistical Analysis

Statistical analyses were conducted using SAS/STAT Release package version 6.03 (Statistical Analysis System Institute, 1988). A 4 $\times$ 6 factorial in a randomized complete block design was used. The two factors were six sources of organic wastes and four organic waste application rates. The experiment was replicated five times for a total of 120 observations in each growing season (6 organic wastes $\times$ 4 rates of application $\times$ 5 replications). Least square methods were used to fit Equation [6] to measured data using Mathcad software package version 4.0 (Mathsoft, Cambridge, MA). Computations of Equation [7] were performed using the same software.

RESULTS AND DISCUSSION

Initial Physical and Chemical Properties of Organic Wastes

Analyses were conducted to characterize the initial physical and chemical properties of the organic wastes in this study (Table 1). These properties were very important in determining the optimum rates of organic wastes to be incorporated into the substrates. The rate of perlite (35% volume per volume) remained constant while that of peat varied with the amount of wastes incorporated (Table 2). The proportion of easily available water (EAW) in substrates formulated with 40% of FSS and CBF is very low (0.08 cm^3 cm^{-3}). This suggests frequent irrigation compared with substrates having a higher proportion of EAW such as FBF. The air-filled porosity (f_a) of organic wastes was within the range (0.13-0.22 cm^3 cm^{-3}) of acceptable values. In fact, many authors have suggested that at least 10% of the pore space should be filled with air in order to sustain plant growth in the field as well as containers (Paul and Lee, 1976). Substrates made of 5% FDP and CDP had lower CEC values compared with others, suggesting lower ability to retain nutrient elements. However, the sub-irrigation technique used in this study allows nutrients to be supplemented at intervals of two to three days during the experiment. FBF followed by CBF had the highest CEC. The levels of salinity in CBF, FSS and CSS seemed too high for growing plants in a controlled environment with artificial light and fertilization. Acceptable levels often range between 0.5 and 1.0 mS. In fact, contrary to surface irrigation, the sub-irrigation technique used in this study does not enhance percolation so that high levels of salinity could be decreased. Fortunately, no leaf damage was observed as a result of high salinity of the wastes. The nutrient solutions used had a lower elemental concentration compared with commercial solutions. However, the wastes in this study already had a high nutrient content (Bugbee et al., 1991). In addition, sub-irrigation is a system requiring half of the recommended fertilization (Biernbaum, 1991). It was not necessary to correct organic wastes pH, but peat was corrected from pH 5.0 to 7.0 by addition of 5.8 kg of $CaMg(CO_3)_2$ per m^3 of peat. Finally, FDP had a very high C/N ratio (C/N = 173.75) implying low release and availability of nitrogen for plant growth. The correction was made by adding 0.015 g of N-P_2O_5-K_2O 20-20-20 per g of humid waste to reach a C/N ratio of 50.

Changes in the Physical Properties of Substrates

During both growing seasons, the substrate f_a, saturated hydraulic conductivity (K_s), pore tortuosity (τ) and relative gas diffusivity (D_s/D_o) were all significantly affected by both organic waste type (OW) and their rate of application (RA). In summer 1994 (Table 3), K_s ($p = 0.0001$ and $p = 0.0001$), f_a ($p = 0.02$ and $p = 0.0001$), ($p = 0.0001$ and $p = 0.0003$) and D_s/D_o ($p = 0.05$ and $p = 0.004$) all very significantly responded to OW and RA with very significant OW $\times$ RA interactions for K_s ($p = 0.0025$), f_a ($p = 0.0003$), τ ($p = 0.003$) and D_s/D_o ($p = 0.05$ and $p = 0.017$). Close examination of these OW $\times$ RA interactions showed that only sewage sludge and de-inked paper sludge had an effect on substrate K_s. K_s increased linearly with increasing RA. However, the rate of increase was faster for FSS and CDP as compared with CSS and FDP. Substrates formulated with composted forms of both bio-filter sludge (CBF) and de-inked paper sludge (CDP) had the highest linear increases in f_a. The OW $\times$ RA interaction for τ was only significant for FDP and CDP. τ increased linearly with increasing RA for both of these organic wastes. However, CDP provided a more tortuous path for gas exchange than FDP. Substrates made from CDP had both the highest air storage (f_a) and the longest path for the exchange τ. This is proof that in container substrates, air content may be adequate but its exchanges may be restricted. It confirms our hypothesis that routine assessment of air storage alone is not adequate in evaluating the physical properties of substrates. Even though an overall significant OW $\times$ RA interaction was observed for D_s/D_o, the statistical test used was unable to detect the differences among organic wastes. This is also true for the significant OW effect on this physical property. It was, however,

TABLE 3. Substrates' physical properties after the first season of geranium production (summer 1994)

Organic waste (OW)	K_s (cm s^{-1})	f_a (cm^3 cm^{-3})	τ (cm cm^{-1})	D_s/D_o (cm^2 s^{-1} cm^{-2} s)
FBF	0.066	0.216	30.6	71.10^{-4}
CBF	0.081	0.228	35.5	64.10^{-4}
FSS	0.081	0.210	54.5	39.10^{-4}
CSS	0.064	0.204	43.7	47.10^{-4}
FDP	0.085	0.200	50.5	40.10^{-4}
CDP	0.115	0.237	56.6	42.10^{-4}
Rate of application (RA)				
5%	0.061	0.208	46.7	45.10^{-4}
10%	0.067	0.188	35.3	53.10^{-4}
25%	0.076	0.215	42.4	51.10^{-4}
40%	0.124	0.253	50.2	50.10^{-4}

Analysis of variance

		K_s		f_a		(τ)		D_s/D_o	
Sources of variation	dl	F	Prob.	F	Prob.	F	Prob.	F	Prob.
Organic waste (OW)	5	6.41	0.0001	2.81	0.0205	12.87	0.0001	2.29	0.0518
(FBF vs. CBF)		1.98	0.1630	0.98	0.3236	3.57	0.0617	0.58	0.4488
(FSS vs. CSS)		2.87	0.0935	0.180	0.6733	3.86	0.0524	1.02	0.3147
(FDP vs. CDP)		8.58	0.0043	9.45	0.0027	0.81	0.3713	0.80	0.3738
Rate of application (RA)	3	23.70	0.0001	14.72	0.0001	6.93	0.0003	4.77	0.0038
Rate linear		62.57	0.0001	33.88	0.0001	5.35	0.0228	12.29	0.0006
Rate quadratic		7.35	0.0072	6.64	0.0115	5.96	0.0165	1.03	0.3116
Interaction (OW × RA)	15	2.61	0.0025	3.20	0.0003	2.54	0.0033	2.08	0.0174
(FBR vs. CBF) × Rate Linear		0.04	0.8393	22.90	0.0001	2.08	0.1523	1.29	0.2590
(FSS vs. CSS) × Rate Linear		7.28	0.0082	0.61	0.4360	0.56	0.4569	0.14	0.7048
(FDP vs. CDP) × Rate Linear		3.98	0.0513	6.09	0.0150	4.13	0.0156	0.15	0.6987
Error	92		0.0010		0.0015		.5523		0.0110

apparent that D_s/D_o decreased linearly with increasing RA and that FBF was the best organic waste in improving substrate gas diffusion properties with a D_s/D_o value of 0.0071 cm^2 s^{-1} cm^{-2} s. This gas diffusivity coefficient value corresponded to the lowest pore tortuosity factor of 30.5 cm cm^{-1}, confirming that gas exchange was to some extent, not restricted in substrates made of this form of organic waste. This trend is in agreement with results previously found by Allaire et al. (1996 a, b).

Results obtained in summer 1994 were confirmed in winter 1995 (Table 4) experiment. In fact, as in summer 1994, K_S (p = 0.0167 and p = 0.0001), f_a (p = 0.0001 and p = 0.0001), τ (p = 0.0001 and p = 0.0001) and D_s/D_o (p =

TABLE 4. Substrates' physical properties after the second season of geranium production (winter 1995)

Organic Waste (OW)	K_S ($cm\ s^{-1}$)	f_a ($cm^3\ cm^{-3}$)	τ ($cm\ cm^{-1}$)	D_s/D_o ($cm^2\ s^{-1}\ cm^{-2}\ s$)
FBF	0.093	0.212	31.2	68.10^{-4}
CBF	0.096	0.231	33.9	68.10^{-4}
FSS	0.100	0.214	56.7	38.10^{-4}
CSS	0.066	0.207	42.9	48.10^{-4}
FDP	0.091	0.207	50.9	41.10^{-4}
CDP	0.124	0.238	59.5	40.10^{-4}
Rate of application (RA)				
5%	0.820	0.211	49.5	43.10^{-4}
10%	0.680	0.197	36.3	54.10^{-4}
25%	0.810	0.216	41.5	52.10^{-4}
40%	0.1480	0.248	48.2	51.10^{-4}

Analysis of variance

		K_S		f_a		(τ)		D_s/D_o	
Sources of variation	dl	F	Prob.	F	Prob.	F	Prob.	F	Prob.
Organic waste (OW)	5	2.93	0.0167	5.83	0.0001	49.19	0.0001	7.06	0.0001
(FBF vs. CBF)		0.03	0.8581	6.43	0.0128	3.84	0.0530	2.23	0.1307
(FSS vs. CSS)		4.89	0.0294	0.90	0.3465	19.69	0.0001	1.25	0.2666
(FDP vs. CDP)		4.42	0.0382	16.09	0.0001	4.79	0.0310	3.97	0.0492
Rate of application (RA)	3	16.37	0.0001	25.26	0.0001	21.25	0.0001	9.98	0.0001
Rate linear		35.78	0.0001	59.79	0.0001	4.67	0.0332	25.27	0.0001
Rate quadratic		13.23	0.0004	11.49	0.0001	26.48	0.0001	2.86	0.0942
Interaction (OW × RA)		16.37	0.0001	10.59	0.0001	13.31	0.0001	3.93	0.0001
(FBR vs. CBF) × Rate Linear		3.84	0.0528	51.14	0.0001	38.05	0.0001	0.01	0.9242
(FSS vs. CSS) × Rate Linear		7.59	0.0070	5.07	0.0266	0.010	0.9033	0.52	0.4728
(FDP vs. CDP) × Rate Linear		5.28	0.0232	9.15	0.0032	19.42	0.0028	1.60	0.2095
Error	92		.0024		.0006		.1644		.0049

0.0001 and p = 0.0001) were all very significantly affected by OW and RA with significant OW × RA interactions for K_S (p = 0.0001), f_a (p = 0.0001), τ (p = 0.0001) and D_s/D_o (p = 0.0001). We examined the (FBF vs. CBF) × R linear and (FDP vs. CDP) × R linear for f_a in Figure 1. The figure shows strong linear and quadratic trends between f_a and the rate of application of each of the four types of organic wastes. However, regardless of these geometric trends, the figure also shows that f_a increased faster with CBF and CDP than it did with FBF and FDP. Similarly (graph not shown), K_S increased faster when FSS and CDP were substituted for peat in the substrates than when the substitution was made with CSS and FDP. τ increased faster in CBF and CDP compared with FBF and FDP. This tendency was observed during the first season experiment. Except for FSS which had a high value compared with CSS, τ was generally higher in the composted than the fresh form of organic wastes studied. Despite an overall significant OW × RA interaction, there was no significant difference between wastes for D_S/D_O. Finally, the composted form of de-inked paper sludge (CDP) and fresh form of sewage sludge (FSS) were most effective in changing the physical properties of substrates.

Plant Growth Parameters

In magnitude, geranium plant height (PHT), shoot dry mass (SDM) and root dry mass (RDM) were higher in summer 1994 than in winter 1995. In

FIGURE 1. Substrates' air-filled porosity versus rate of application of wastes for four sources of organic wastes: ◆ Fresh bio-filter, ■ Composted bio-filter, ▲ Fresh de-inked paper and ● Composted de-inked paper.

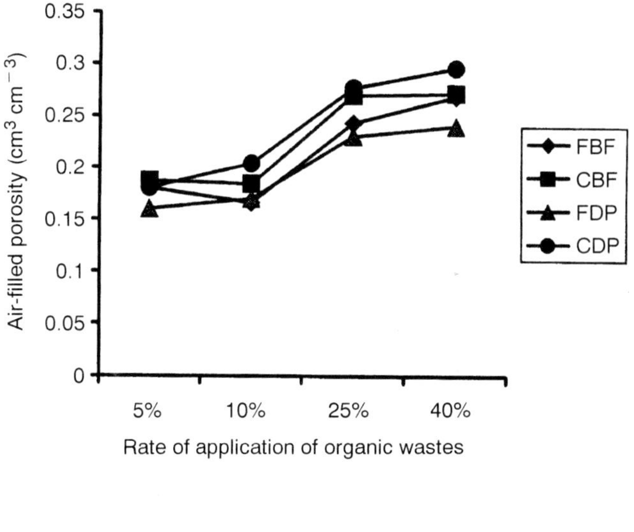

summer 1994 (Table 5), strong OW × RA interactions were observed for PHT (p = 0.0066), SDM (p = 0.0011) and RWD (p = 0.0001). In addition, PHT responded equally to OW, but was linearly reduced by RA (p = 0.0115). RDM responded to both OW (p = 0.0001) and RA (p = 0.0001) and SDM was only significantly affected by OW (p = 0.0096). SDM linearly decreased

TABLE 5. Geranium plant height (PHT), shoot dry mass (SDM) and root dry mass (RDM) after the first season of production (summer 1994)

Organic wastes (OW)	PHT (cm)	SDM (g)	RDM (g)
FBF	26.95	15.53	8.757
CBF	27.90	14.98	8.955
FSS	27.95	14.31	11.554
CSS	28.35	13.64	6.980
FDP	27.78	14.26	10.761
CDP	27.75	16.37	11.712
Rate of application (RA)			
5%	28.70	15.34	9.903
10%	28.67	15.44	8.883
25%	28.23	14.68	10.290
40%	26.50	13.93	10.162

Analysis of variance

		PHT		SDM		RDM	
Sources of variation	dl	F	Prob.	F	Prob.	F	Prob.
Organic waste (OW)	5	.61	0.6940	3.24	0.0096	72.03	0.0001
(FBF vs. CBF)		1.30	0.2577	0.50	0.4810	0.41	0.5219
(FSS vs. CSS)		0.23	0.6327	0.73	0.3944	216.5	0.0001
(FDP vs. CDP)		0.00	0.9762	7.38	0.0078	9.32	0.0029
Rate of application (RA)	3	3.88	0.0115	2.40	0.0724	11.95	0.0001
Rate linear		7.52	0.0073	6.94	0.0098	9.61	0.0025
Rate quadratic		1.03	0.3117	0.12	0.7323	0.14	0.7095
Interaction (OW × RA)	15	2.35	0.0066	5.00	0.0011	23.30	0.0001
(FBR vs. CBF) × Rate Linear		1.40	0.2403	3.99	0.0486	1.48	0.2271
(FSS vs. CSS) × Rate Linear		0.39	0.5343	0.42	0.5197	24.29	0.0001
(FDP vs. CDP) × Rate Linear		1.00	0.9641	6.02	0.0081	0.16	0.6920
Error	92		6.961		6.043		0.0605

with increasing organic wastes RA while RDM increased. We closely examined the (FBF vs. CBF) × Rate linear and (FDP vs. CDP) × Rate linear interactions for both growth parameters and found that SDM decreased faster with FDP and CBF than CDP and FBF. RDM decreased faster with CSS than FSS. In the winter 1995 experiment (Table 6), very significant OW × RA

TABLE 6. Geranium plant height (PHT), shoot dry mass (SDM) and root dry mass (RDM) after the second season of production (winter 1995)

Organic wastes (OW)	PHT (cm)	SDM (g)	RDM (g)
FBF	28.88	10.14	8.176
CBF	24.15	9.13	6.804
FSS	24.10.	8.86	8.815
CSS	24.03	9.59	6.316
FDP	21.45	8.99	7.632
CDP	24.18	9.88	8.683
Rate of application (RA)			
5%	24.45	10.36	8.473
10%	23.90	9.99	7.667
25%	24.13	9.53	9.068
40%	22.70	8.76	5.751

Analysis of variance

		PHT		SDM		RDM	
Sources of variation	dl	F	Prob.	F	Prob.	F	Prob.
Organic waste (OW)	5	4.25	0.0016	2.51	0.0350	10.96	0.0001
(FBF vs. CBF)		0.79	0.3769	4.72	0.0323	9.93	0.0022
(FSS vs. CSS)		0.01	0.9270	2.51	0.1166	33.45	0.0001
(FDP vs. CDP)		11.13	0.0012	3.71	0.0571	5.84	0.0176
Rate of application (RA)	3	2.63	0.0543	17.25	0.0001	32.99	0.0001
Rate linear		5.79	0.0180	47.98	0.0001	39.61	0.0001
Rate quadratic		0.99	0.3216	2.93	0.0902	38.47	0.0001
Interaction (OW × RA)	15	1.52	0.1149	4.57	0.0001	14.01	0.0001
(FBR vs. CBF) × Rate Linear		0.06	0.8149	0.62	0.4320	1.71	0.1945
(FSS vs. CSS) × Rate Linear		1.43	0.2347	2.11	0.1494	0.25	0.6160
(FDP vs. CDP) × Rate Linear		5.31	0.0310	2.15	0.1463	9.53	0.0026
Error	92		6.673		2.111		0.1185

interaction were observed for SDM (p = 0.0001) and RDM (p = 0.0001). In addition, PHT (p = 0.0016 and p = 0.0543), SDM (p = 0.035 and p = 0.0001) and RDM (p = 0.0001 and p = 0.0001) were significantly affected by both OW and RA. Figure 2 examines the relationship between geranium SDM and RA for FSS, CSS, FDP and CDP in winter 1995. As for physical parameters, strong linear and quadratic trends were observed between SDM and RA for each of the 4 sources of organic wastes. However, the slopes of these trends are different. SDM decreased faster with CDP than FDP while it increased faster with FSS than CSS for the first 2 rates of application of organic wastes (5 and 10%). SDM maintained a constant growth rate when the rate of application changed from 25 to 40%, for all the four sources of wastes. Similarly (graph not shown), in summer 1994, FSS improved RDM faster than the 5 other sources of organic wastes. SDM responded quicker to CDP and FBF than the four other types of wastes. Overall, the best responses of SDM and RDM were achieved in substrates containing CDP in summer 1994. This substrate had also the highest value of f_a. Substrates made of FBF provided the highest growth in PHT and SDM in winter 1995. In relation to substrates physical properties, FBF containing substrates had an acceptable level of f_a, the lowest t and the highest gas diffusion (Ds/Do). Both CDP and FSS seem to have a greater effect on both physical properties and plant growth parameters than the four other types of organic wastes studied.

FIGURE 2. Geranium shoot dry mass versus waste application rates for four sources of organic wastes: ◆ Fresh de-inked paper, ■ Composted de-inked paper, ▲ Fresh sewage sludge and ● Composted sewage sludge.

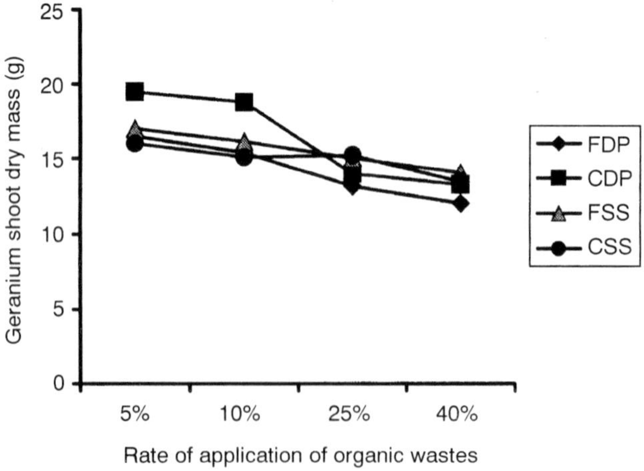

Relationship Between Plant Growth Parameters and the Physical Conditions of Substrates

The relationships between geranium growth parameters and the physical conditions of the root zone are summarized in Table 7 for both growing seasons. In winter 1995, the pore tortuosity factor (τ) was quadratically correlated with RDM and linearly with SDM. The relative gas diffusivity coefficient (D_S/D_O) showed a quadratic relationship with RDM and a linear one with SDM. During the first growing season (summer 1994), τ was again found to quadratically correlate with RDM and linearly with SDM. D_S/D_0 linearly correlated with both RDM and SDM of geranium. Allaire et al. (1996 a, b) and Caron and Nkongolo (1998) and Nkongolo et al. (1995) have reported similar results. Both τ and D_S/D_0 better expressed the relationship between plant growth and the physical conditions of substrates.

CONCLUSION

The physical properties of container substrates are often reported to be a problem. Amendment of substrates with organic wastes is one of the ways to improve these properties. Unfortunately, depending on the initial chemical and physical properties and the rates of application of organic wastes used, the physical properties of the resulting substrates may be improved or impeded. In addition, routine assessment of the physical properties of the organically-amended substrates is restricted to their air and water storage status. However, a better understanding of the relationship between plant growth and the physical conditions of the root zone requires an assessment of both storage and exchange properties of substrates. This investigation has found that overall, organic wastes used did not impair the physical conditions of the root zone. It suggests that gas diffusivity (D_S/D_O) and pore tortuosity (τ) could be used as additional indices when assessing the aeration status of substrates. These two gas exchange indices were highly correlated with geranium growth parameters.

TABLE 7. Correlation coefficient between geranium shoot dry mass (SDM) and root dry mass (RDM) with some physical properties of substrates after 2 seasons of production.

Physical parameters	Summer 1994		Winter 1995	
	SDM	RDM	SDM	RDM
τ	0.55***	0.92****	0.46****	0.83***
D_S/D_O	0.52****	0.85****	0.44****	0.75****

τ = pore tortuosity factor, D_S/Do = gas relative diffusivity coefficient.

, * Significant at P = 0.001 and 0.0001, respectively.

REFERENCES

Allaire, S.E., J. Caron, L.E. Parent, I. Duchesne, and J.A. Rioux. (1996a). Air-filled porosity, gas relative diffusivity, and tortuosity: indices of *Prunus* × *cistena* Growth in Peat substrates. *Journal of American Society for Horticultural Science* 121: 236-242.

Allaire, S.E., J. Caron, L.E. Parent, I. Duchesne, and J.A. Rioux. (1996b). Corrigenda. *Journal of American Society for Horticultural Scienc* 121: 592.

Allaire, S.E., J. Caron, and J. Gallichand. 1994. Measuring the saturated hydraulic conductivity of peat substrates in nursery containers. *Canadian Journal of Soil Science* 74: 431-437.

Bowman, C.D., R.Y. Evans, and L. Dodge. (1994). Growth of *Chrysanthemum morifolium* with ground automobile tires used as a container soil amendment. *HortScience* 29: 774-776.

Bugbee, G.J., C.R. Frink, and D. Migneault. (1991). Growth of perennials and leaching of heavy metals in substrates amended with a municipal leaf, sewage sludge and street sand compost. *Journal of Environmental Horticulture* 9: 47-50.

Caron, J. and N.V. Nkongolo. (1999). Aeration in growing media: recents devlopments. *Acta Horticulture* 481: 541-551.

Chong, C. and Cline, R.A. (1993). Response of four ornamental shrubs to container substrate amended with two sources of raw paper mill sludge. *HortScience* 28: 807-809.

Cline, R.A. and C. Chong. (1991). Putting paper mill waste to use in agriculture. *Highlights Research* (Ontario). 14: 16-19.

King, J.A. and K.A. Smith. (1987). Gaseous diffusion through peat. *Journal of Soil Science* 38: 173-177.

Musy, M. and S. Soutter. (1991). Physique du sol. Lausanne: Press polytech et Université Romandes.

Nkongolo, N.V. (1996). Tortuosité de l'espace poral: Importance, évaluation et effet sur la croissance végétale. Ph.D. thesis, Univ. Laval, Quebec, Canada. 133 p.

Paquet, J.M., J. Caron, and O. Banton. (1993). In situ determination of the water desorption characteristics of peat substrates. *Canadian Journal of Soil Science* 73: 329-339

Paul, J.L. and C.I. Lee. (1976). Relation between growth of chrysantenums and aeration of various container substrates. *Journal of American Society for Horticultural Science* 115: 500-503.

Statistical Analysis System Institute, Inc. (1988). SAS/STAT user's guide, release 6.03. SAS Institute, Inc., Cary, NC. 1028 p.

Tyler, H.H., S.L. Warren, T.E. Bilderback, and W.C. Fonteno. (1993). Composted turkey litter: I. Effect on chemical and physical properties of a pine bark substrate. *Journal of Environmental Horticulture* 11: 131-136.

Warren, S.L. and W.C. Fonteno. (1993). Changes in physical and chemical properties of a loamy sand soil when amended with composted poultry litter. *Journal of Environmental Horticulture* 11: 186-190.

'Kachiwari'–
A Disease Resistant and Nature-Farming Adaptable Pumpkin Variety

Toshio Nakagawara

SUMMARY. A new disease resistant and delicious pumpkin variety, 'Kachiwari', has been released from the Agricultural Experiment Station, International Nature Farming Research Center. The growth characteristics of 'Kachiwari' pumpkin are rapid and competitive. Fruit setting occurs one week earlier than other early varieties. The vines are robust large and resistant to diseases, especially powdery mildew. Even if stored over the entire winter, the fruit retains excellent flavor with high starch and sugar concentrations. *[Article copies available for a fee from The Haworth Document Delivery Service: 1-800-342-9678. E-mail address: getinfo@haworthpressinc.com <Website: http://www.HaworthPress.com>]*

KEYWORDS. Nature farming, organic farming, pumpkin, storability

INTRODUCTION

Many new vegetable varieties with high yield potentials and improved quality have been developed in recent years (Aoba, 1982). Some of these varieties have been grown successfully in nature farming and/or organic

---

Toshio Nakagawara is Plant Breeder, International Nature Farming Research Center, Hata, Nagano 390-1401, Japan (E-mail: huilian@janis.or.jp).

[Haworth co-indexing entry note]: " 'Kachiwari'–A Disease Resistant and Nature-Farming Adaptable Pumpkin Variety." Nakagawara, Toshio. Co-published simultaneously in *Journal of Crop Production* (Food Products Press, an imprint of The Haworth Press, Inc.) Vol. 3, No. 1 (#5), 2000, pp. 113-118; and: *Nature Farming and Microbial Applications* (ed: Hui-lian Xu, James F. Parr, and Hiroshi Umemura) Food Products Press, an imprint of The Haworth Press, Inc., 2000, pp. 113-118. Single or multiple copies of this article are available for a fee from The Haworth Document Delivery Service [1-800-342-9678, 9:00 a.m. - 5:00 p.m. (EST). E-mail address: getinfo@haworthpressinc.com].

farming systems with only organic materials as nutrient sources. However, the application rates of compost and organic materials must often be increased to keep pace with the yield increases. Moreover, the problems of pests and diseases that are not generally found with the nature farming adaptable varieties often occur with these high yielding varieties. This is because the new varieties are often selected for their response to large applications and high rates of chemical fertilizers. A proverb says "Eat more and live shorter; eat less and live longer." Excessive rates of chemical fertilizers are a principal cause of pest problems. When little or no pesticides are used, it is important to know that low fertilizer applications are sufficient for good growth. However, the focus of today's vegetable breeding is to develop varieties that give high yield with high rates of chemical fertilizers. Thus, there are very few varieties available that are entirely suitable for nature farming and organic farming systems. Typical crop varieties with low fertilizer requirements are millet and sorghum which have a well-developed fibrous root system with a high nutrient scavenging and uptake ability. These crops grow and yield well in poor infertile soils but are characterized by a high nutritional quality (Kudo, 1997). Scientists at the International Nature Farming Research Center have conducted research studies on soil improvement and plant breeding by adopting rotation-cropping and inter-cropping with green manure crops. A promising new vegetable variety resulting from this research is 'Kachiwari' pumpkin, which shows high disease resistance (especially to powdery mildew) and can be stored over winter without losing food and nutritional quality.

SCHEDULES OF THE BREEDING

The 'Kachiwari' pumpkin was bred using varieties from yard gardens of the countryside residents (Figures 1 and 2). These garden varieties, such as 'Masakari' and 'Tochi', showed high disease resistance but short storability and quality declines when stored for more than three months. Moreover, these varieties are locally cultivated and mainly adaptable to specific climate and agroecological conditions of a particular area. These varieties were used in the breeding program at the International Nature Farming Research Center with the goals of disease resistance, long-term storability and retention of high quality, especially when stored over winter.

CHARACTERISTICS OF THE 'KACHIWARI' PUMPKIN

High Disease Resistance

The vegetation is large and robust and the competitive ability is excellent compared with other varieties planted in close proximity (Table 1). Fruit

FIGURE 1. Appearance of 'Kachiwari' pumpkin fruit.

setting is one week earlier than other early varieties (Figure 2). The vines are vigorous with long life and resistant to powdery mildew.

Hard Fruit with High Dry Mass Density

The fruit is pear-shaped (Figure 1). The fruit skin color is green-gray with cream fancies on the surface (Figure 1). Although the size is slightly smaller than that of other varieties, the dry mass per fruit is comparable because the fruit has especially high density (Table 1).

Long-Period Storability and Flavorful and Resistant to Rotting

It is known that this delicious pumpkin contains more starch and less water. 'Kachiwari' can retain a high starch content, a low water content and remain delicious for five months (Table 2). The non-reducing sugars that contribute to its taste and flavors are in the highest concentration three months after harvest. It tastes floury two months after harvest, and soft and floury three months after harvest like the taste of a walnut. Usually, rotting occurs during storage because the fruit is infected by anthracnose and other fungal pathogens. A carbon to nitrogen ratio (C:N) below 2.5 is favorable for infections of these pathogens. The C:N ratio in 'Kachiwari' is higher than in

FIGURE 2. Comparison of the vegetative appearances between 'Kachiwari' (left) and the common variety 'Miyako' (right).

other early ripening varieties, and changes very little during storage (Table 2). In the cool area, the fruit can be stored for five months without air conditioning.

Cultivation Under Low Fertilization and Non-Pesticides

'Kachiwari' pumpkin is suitable for low fertilizer cultivation. With high levels of chemical fertilizer, problems with aphids will occur, the vines will be overgrown, and accumulation of starch in fruit will be depressed resulting in low fruit quality. Adequate growth and yield can be achieved by application of organic fertilizer at 30% of the usual amount. If the soil is very fertile, 'Kachiwari' pumpkin can be cultivated without fertilizers. Following a low fertilizer and non-pesticide approach, the seeds should be sown directly in the field without pruning and intercropped with green manure crops. The interspace of the crop lines was mulched with rye or clover hay produced the previous year. When the main vines mature, the green manure crops are cut and placed under the main vines, which are oriented in the same direction. The small branch vines should be allowed to grow freely. The merits of the direct-sowing, non-pruning and green manure intercropping practices are robust and highly developed root systems with large populations of the natural predators including insects such as ladybird, beetles and frogs. Similar results are also found in the case of apple orchards (Fujita, 1995).

TABLE 1. Comparison of 'Kachiwari' with other varieties.

Variety	Vegetation	Leaf color	Leaf size (L × W cm)	Fruit node	Av. internode (cm)	Flower date (d after sow.)	Av. Fruit weight (g)	Horizon/ Vertical	Shape	Fruit color	Flesh color	Fruit density
Kachiwari	Strong	Middle	27.6 × 37.8	15.7	11.2	56	1440	1.3	Pear	Gray-green	Orange	1.0335
Miyako	Weak	Deep	20.0 × 30.8	14.4	8.1	49	1287	1.6	Flat round	Deep green	D. yellow	0.966
Ebisu	Middle	Deep	22.4 × 32.6	13.3	8.1	49	1870	1.9	Flat round	Deep green	Yellow	0.944

Sown on 10 May and harvested 13-21 August.

TABLE 2. Fruit quality and rotting ratio during storage.

	Kachiwari					Variety from market				
Storage stage	Starch (%)	Sugars (%)	Water (%)	C/N (%)	Rot (%)	Starch (%)	Sugars (%)	Water (%)	C/N (%)	Rot (%)
Harvest	6.1	4.48	70.9	32.8	0.0	5.9	3.49	77.7	31.9	0.0
1 month	6.4	4.14	73.2	28.2	0.0	3.6	3.39	82.7	25.7	14.3
2 months	6.4	5.39	75.2	26.8	0.8	3.2	5.49	82.1	21.9	66.7
3 months	5.8	11.06	73.1	25.9	3.3	2.5	6.69	85.0	22.0	95.2
4 months	5.9	8.47	72.4	23.9	5.0	0.4	9.07	85.9	22.2	100.0
5 months	5.8	7.47	78.4	23.1	19.2	--	--	--	--	--

The fruits were stored at room temperature. Starch and sugars (non-reducing sugars) are based on dry mass.

REFERENCES

Aoba, T. (1982). *Vegetables in Japan*. Tokyo: Yatsusaka-Books, 162 pp.

Fujita, M. (1995). An approach of apple cultivation towards nature farming. The 4th Conference on Technology of Effective Microorganisms, Saraburi, Thailand.

Kudo, K. (1997). What is the future for millets? Millet Newsletter 10: 1-2.

Nature Farming Practices for Apple Production in Japan

Masao Fujita

SUMMARY. Apples were grown in a study using nature farming practices, including organic fertilizer to supply nutrients, microbial inoculants for biocontrol of pest insects, and low- or non-pesticides measures. When low- or non-pesticide measures were adopted, the numbers of pest insects increased accordingly; however, the populations of beneficial organisms and natural predators of pest insects also increased. Even though the reduction in pesticide use caused some fruit damage, the enhanced biocontrol of pest insects by increased numbers of natural predator insects more than compensated for the damage. Thus, low- or non-pesticide nature farming practices were cost-effective because they increased biocontrol of pest insects and decreased the amount of pesticide needed and the cost of application. *[Article copies available for a fee from The Haworth Document Delivery Service: 1-800-342-9678. E-mail address: getinfo@haworthpressinc.com <Website: http://www.HaworthPress.com>]*

KEYWORDS. Apple, microbial inoculant, nature farming, pest control, organic fertilization

INTRODUCTION

Apple fruit production in Japan decreases readily with the aging of apple growers increases by the Statistics Bureau, Management and Coordination

---

Masao Fujita is Research Soil Animal Scientist, International Nature Farming Research Center, Hata, Nagano 390-1401, Japan (E-mail:fujita@janis.or.jp).

[Haworth co-indexing entry note]: "Nature Farming Practices for Apple Production in Japan." Fujita, Masao. Co-published simultaneously in *Journal of Crop Production* (Food Products Press, an imprint of The Haworth Press, Inc.) Vol. 3, No. 1 (#5), 2000, pp. 119-125; and: *Nature Farming and Microbial Applications* (ed: Hui-lian Xu, James F. Parr, and Hiroshi Umemura) Food Products Press, an imprint of The Haworth Press, Inc., 2000, pp. 119-125. Single or multiple copies of this article are available for a fee from The Haworth Document Delivery Service [1-800-342-9678, 9:00 a.m. - 5:00 p.m. (EST). E-mail address: getinfo@haworthpressinc.com].

Agency (1995). Urban people often believe that apples are produced in a beautiful green environment. Actually, apples are produced in Japan with more than fifteen pesticide applications per season. Consequently, apple trees may often appear white with pesticides (rather than green) during the fruiting season. Because a larger apple fruit sells at a higher price in the Japanese marketplace, growers make every effort, including the use of chemical fertilizers, to increase the fruit size larger than 300 g (Fukuda, 1994). Consumers demand and select apple fruit by size and appearance. This makes it difficult for apples produced by nature farming to compete with chemically produced apples. Nevertheless, non-pesticide or low-pesticide apple production has been studied and practiced successfully despite the many difficulties (Anonymous, 1986). In 1993 the Nagano Orchard Experimental Station studied low-pesticide apple production (6 pesticide applications per season), and non-pesticide production (one application per season of lime sulfur and machine oil) compared with chemical production (12 pesticide applications per season). Fruit grown with low- and non-pesticide methods were infested with codling moth caterpillars, and some fruit were inedible. This suggested that successful non-pesticide and low-pesticide cultivation of apples is very difficult in Japan. Thus, Japan started to import apples from New Zealand and the United States in 1994 (Nakata, 1994). Since then, post-harvest and residual pesticides have become a concern with imported apples. At the same time, associated chemical pollution and production costs of domestic apples were being taken seriously. Most farmers, however, have not changed their way of thinking, and are still using chemical inputs in apple production. There are a few apple growers in Nagano Prefecture who have been practicing low-pesticide apple cultivation for many years. Their current chemical usage as well as their low-pesticide apple cultivation practices in this area were investigated. It was decided to study apple production by nature farming practices. Thus, in 1993 a study began with an apple orchard of about 0.1 hectare using a low-pesticide approach began in 1993. At the same time, some apple growers in Nagano and Aomori Prefectures expanded their low-pesticide apple production area to include organic fertilizers and a microbial inoculant. This paper reports the results of this nature farming low-pesticide apple production study compared with conventional chemical-intensive apple production in Japan.

MATERIALS AND METHODS

An organic amendment was prepared by fermenting a mixture of organic materials (i.e., rape oilseed meal, rice bran and fish processing by-product) with a microbial inoculant called EM (produced by the International Nature Farming Research Center, Atami, Japan). EM is widely known as "Effective

Microorganism" and consists of a mixed culture of naturally-occurring, beneficial microorganisms, predominantly lactic acid bacteria, photosynthetic bacteria, yeasts, actinomycetes and certain fungi that is reported to enhance soil quality and growth, yield and quality of crops (Higa and Parr, 1994). The organic amendment serves as an organic fertilizer, soil conditioner and source of beneficial microorganisms, and is often referred to as EM-Bokashi. It was applied to soil three times per season, in March, September and November (with total nitrogen at 160 kg ha^{-1}) either by spreading on the soil surface or burying in holes because the orchard was managed as a no-till system. Green manure was utilized as a supplemental source of plant nutrients. Cover crops were grown to provide organic matter and ensure biological diversity. Herbicides were not used because of their adverse effects on habitats and food chains of soil biota near the surface that often leads to disease and insect infection. Foliar applications of another microbial inoculant, EM-5, were made at a dilution of 1:1000 to control some diseases (Anonymous, 1994).

RESULTS AND DISCUSSION

Pest Control

Table 1 shows the record of pest control at the nature farming orchard in 1994. Spray application with the microbial inoculant did not show any direct effect on controlling disease and insects. However, microbes and the sugars from EM-5 applied to soil might help to diversify the soil biota. To control apple scab, twice a season fungicides were sprayed in a concentration of half the normal by mixing in wood vinegar (1:1000): Trifumin wettable powder in early May, at the beginning of the bloom period; and Diebolt wettable powder in middle May, at the time of blossom drop. The disease appeared on leaves but hardly on fruit. Although *Alternaria* blotch appeared in July, it was controlled by spraying Bordeaux mixture at the initial stage (early July). In 1995, excessive rainfall provided a favorable environment for disease infestation. Since we did not spray fungicides in July as had been done up until 1994, *Alternaria* blotch appeared on fruit of some trees of the 'Tsugaru' variety, which is more vulnerable to disease than the 'Fuji' variety.

Since insecticides were not used in the spring of either 1994 or 1995, some young fruit were eaten by larvae of moths from late May to early June. Although we once suspected that no fruit would be left undamaged by larvae, all the fruit of each cluster were hardly eaten. Actually, there were enough fruit to select and many undamaged fruit were thinned out. In case of chemical cultivation, all the work of fruit thinning, including chemical thinning, must be done manually, which requires much labor. While spider mites

TABLE 1. Record of pest control at the apple orchard in 1994

Pesticide	date		Amount of spraying per 10 a
	Apr. 11 (Gemination period)	Water	400 L
		EM-5	2 L
	Apr. 19	Water	500 L
		EM-5	1 L
	Apr. 27	Water	500 L
		EM-5	1 L
⊙	May 6	Water	400 L
	(Beginning of the blooming period)	Trifumin wettable powder (fungicide)	0.1 L
		Wood vinegar	0.8 L
⊙	May 12	Water	400 L
	(Blossom drop period)	Diebolt wettable powder (fungicide)	0.2 L
		Wood vinegar	0.8 L
	May 20	Water	500 L
		EM-5	1 L
	May 30	Water	500 L
		EM-5	1 L
⊙	June 7	Water	500 L
		Jimandaisen wettable powder (fungicide)	0.5 L
		Dasuban wettable powder (insecticide)	0.25 L
		Wood vinegar	
	June 8	Water	500 L
		EM-5	1 L
⊙	July 2	Water	500*
		Spreader	
		Bordeaux mixture (fungicide) Quick lime	6 kg
		Copper sulfate	2 kg
		Nicotinic sulfate (insecticide)	0.5 L
		Zinc sulfate (fungicide)	1 kg
	Aug. 19	Water	500 L
		EM-5	1 L
		EM-3	0.25 L
	Sept. 13	Water	500 L
		EM-5	1 L
		EM-3	0.25 L
	Nov. 10	Water	500 L
		EM-5	1 L
		Kyusei EM-1	1 L

(Tetranychidae) were not observed in the spring, they were observed in August, both in 1993 and 1994, when pesticides were sprayed in July. They were found most on dwarf-cultured trees of the 'Fuji' variety, and then on those of normal cultivation of the same variety. They were hardly found on trees of the 'Tsugaru' variety for both cultivation types. Ladybird beetles were found, which are the natural enemy of aphids, together with predatory spiders, larvae of syrphid flies, and predatory mites (Phytoseiidae). These results show that many natural enemies of insect pests arise during low-pesticide cultivation using a microbial inoculant. This indicates that the diversification of the biota creates an environment which prevents a large infestation of insect pests. Furthermore, we found that the hardness of leaves was different with different methods of cultivation (Table 2). The apple leaves in the experimental orchard with nature farming had lighter colors, a smaller total nitrogen content, and a higher ratio of carbon to nitrogen than those grown by chemical farming. This means that the apple leaves are hard and resistant to insects.

Frequency of Pesticide Application and Their Cost

The frequency and the cost of pesticide applications are compared for the two cultivation methods in Table 3. According to the chemical pest control schedule in the Nagano area, the frequency of pesticide spraying was 15 times a year, including two special applications (J. A. Matsumoto Highland, 1994). The cost of pesticides per hectare was ¥814,780 in 1993 and ¥797,420 in 1994. Meanwhile the frequency of pesticide spraying at the orchard was 6 times a year in 1993, and its cost per hectare was ¥243,430. In 1994, sprays made were four times a year and cost ¥129,030 (16.2% of the cost in chemical method). In 1995, the frequency of spraying was 2 times a year, and the cost was expected to be reduced to less than 10% of the chemical method. The cost reduction was achieved by reducing the frequency of pesticide applications through use of the microbial inoculant and reducing the concentration of pesticides by mixing with wood vinegar.

TABLE 2. Carbon and nitrogen content and C:N ratio of apple leaves as affected by chemical farming compared with nature farming's low-pesticide practice (Nagano Orchard Experimental Station, 1994).

Farming	Total C ($g\ kg^{-1}$)	Total N ($g\ kg^{-1}$)	C/N
Chemical	488	28	17.4
Nature	499	21	23.8

TABLE 3. Frequency and cost of pesticide applications for apple production using chemical farming compared with nature farming's low pesticide practice (Nagano Orchard Experimental Station, 1994).

	Cost of pest control					
	Frequency (Times/y)			Cost (1000¥)		
	1993	1994	1995	1993	1994	1995
Chemical	15	15	15	814.78	797.42	791.00
Nature	6	4	2	243.43	129.03	89.00

Sales and Distribution

Although some trees of the 'Tsugaru' variety were infested by *Alternaria* blotch in 1995, the fruit were sold to consumers after explaining that the flesh or freshness was not different from that of ordinary apples. Growers need to consider using pesticides according to the weather and try to inform consumers how apples are cultivated. Before shipping to the local agricultural cooperatives, apples are sorted into three groups by the grower: two groups to the fruit market and one group for juice processing. Usually, marketable apples are further sorted into 4 classes by color and shape and eight classes by size, a total of 32 classes in all, before they are shipped to the market (personal communication from J. A. Matsumoto Highland, 1994). Since growers spend considerable time on harvesting and sorting, and a lot of labor is required for further sorting at the agricultural cooperative, a grower receives about 70% of the market price. The nature farming apples were sold directly to consumers without sorting, and the consumers were informed that the value of apples produced by nature farming without pesticides and chemical fertilizers can not be estimated from their appearance. The time for harvesting and sorting was greatly reduced.

Since there was less rainfall and higher temperatures than usual during the growing period in 1994, the appearance of diseases was very low, and the trees grown without pesticides hardly developed apple scab. This made the fruit highly marketable. However, it is still impossible to control apple scab with only the microbial inoculant. A method is needed to control apple scab without using pesticides up to the young fruit stage, and a method to control *Alternaria* blotch in a year with excess rain in July. For insect pests such as the infestation of moth larvae, the control method varies depending on how much the grower can allow the infestation to develop. With chemical cultivation, the grower must try to reduce the cost of pesticides by controlling the pesticide application rate according to the condition of the trees and the extent of infestation. If damage caused by insect pests was allowed to a certain level by reducing the rate and amount of pesticides, the benefits from the resulting increases in beneficial organisms, including natural predators of pest insects, would compensate this loss by insect pests from reduction of pesticide costs. Apples are now imported by Japan from the United States and New Zealand. It then becomes increasingly important to encourage consumers to change their way of thinking on fruit safety and quality and to select apples based on cultivation methods rather than on fruit size and appearance. Although there are many difficulties for nature farming practices, concerns on food safety and quality and environmental protection are prompting scientists, farmers and policymakers to change their thinking on current market standards, reconsider limitations of modern chemical agriculture, and recognize the benefits of nature farming practices to food safety and quality, human health and environmental preservation.

REFERENCES

Anonymous. (1986). *Special Bulletin: History, Varieties, and Pest Controls*. Association of Consumers, Yokohama, pp. 52-61.

Anonymous. (1994). *Application Manual for Kyuse Effective Microbial Inoculant*. Atami-city (Japan): International Nature Farming Research Center, 24 p.

Fukuda, H. (1994). Apple. In *Horticulture in Japan,* Japanese Society for Horticultural Science. Tokyo: Japanese Society for Horticultural Science, pp. 23-27.

Higa, T. and J.F. Parr. (1994). *Beneficial and Effective Microorganisms for a Sustainable Agriculture and Environment*. International Nature Farming Research Center, Atami, Japan, 16 p.

Nakata, H. (1994). Waves of the importation of overseas apples. *News Asahi*, 17 June 1994.

Matsumoto Hiland, J.A. (1994). *Controls of Apple Pests*. Matsumoto-city (Japan): Japanese Agricultural Cooperative.

Statistics Bureau, *Management and Coordination Agency, 1995*. Japan Statistical Yearbook. 4th Ed. Tokyo: Japan Statistical Association, 872 p.

Effects of Organic Farming Practices on Photosynthesis, Transpiration and Water Relations, and Their Contributions to Fruit Yield and the Incidence of Leaf-Scorch in Pear Trees

Hui-lian Xu
Xiaoju Wang
Masao Fujita

SUMMARY. This study involved a comparison of two pear (*Pyrus communis* L. cv. Nijuseki) orchards in the South Nagano area that were managed according to either chemical farming (CF) or organic farming (OF) practices. Leaf-scorch was not observed in the pear orchard under OF but was severe under CF. Fruit yield was 62% higher for the pear orchard farmed organically. Photosynthetic capacity (P_C) was 33% higher under OF. Leaf symplastic water fraction (z_{sym}) and osmotic concentration (C_{FT}) were higher for pear trees under OF than for those under CF. Both stomatal (a) and cuticular (b) transpiration were faster for excised pear leaves under CF and, as a consequence, the leaf water retention ability (WRA) of leaves under CF was lower than for those under OF. Results suggest that the higher yield of pear trees with organic

Hui-lian Xu is Senior Crop Scientist, Xiaoju Wang, Soil Scientist, and Masao Fujita is Soil Animal Scientist, International Nature Farming Research Center, 5632 Hata, Nagano 390-1401, Japan (E-mail: huilian@janis.or.jp).

[Haworth co-indexing entry note]: "Effects of Organic Farming Practices on Photosynthesis, Transpiration and Water Relations, and Their Contributions to Fruit Yield and the Incidence of Leaf-Scorch in Pear Trees." Xu, Hui-lian, Xiaoju Wang, and Masao Fujita. Co-published simultaneously in *Journal of Crop Production* (Food Products Press, an imprint of The Haworth Press, Inc.) Vol. 3, No. 1 (#5), 2000, pp. 127-138; and: *Nature Farming and Microbial Applications* (ed: Hui-lian Xu, James F. Parr, and Hiroshi Umemura) Food Products Press, an imprint of The Haworth Press, Inc., 2000, pp. 127-138. Single or multiple copies of this article are available for a fee from The Haworth Document Delivery Service [1-800-342-9678, 9:00 a.m. - 5:00 p.m. (EST). E-mail address: getinfo@haworthpressinc.com].

farming management, compared with chemical farming, is attributed to high P_C and that the absence of leaf-scorch is attributed to a high water retention ability that is associated with a high symplastic cell water fraction and high tissue osmotic concentration.

KEYWORDS. Effective microorganisms, leaf-scorch, organic farming, pear, photosynthesis, transpiration, water relations

INTRODUCTION

There are many difficulties in orchard fruit production without pesticides and chemical fertilizers. This is especially true in Japan, where the appearance and color of fruit are very important, and the fruit will not be marketable if there is even a small blemish or scar on the surface or if the fruit is less than standard size (Fukuda, 1994). The consumer demand for these fruit quality parameters forces growers to use pesticides and fertilizers to ensure a large harvest of marketable fruits of large size and attractive appearance. However, chemicals can not solve all of these problems. For example, in the pear production area of Japan, a severe problem in orchard management is leaf-scorch and/or anomalous defoliation before harvest, which affects fruit yield and quality in both the current and future years by reducing photosynthate accumulation (Nagano Prefecture, 1987). This problem is attributed to leaf tissue dehydration caused by the unbalance between excessive transpiration and water uptake on hot-dry days, even under ideal soil conditions. To date, there has been no effective chemical practice to prevent or alleviate this problem. Therefore, scientists are prompted to elucidate the mechanism of leaf-scorch and to devise suitable cultural management practices that would prevent this problem. Some farmers have suggested that nature farming or organic farming practices could minimize or eliminate this problem. This paper reports the results of a comparison between two pear orchards, one managed by organic farming practices and the other by chemical farming. The effects of photosynthetic capacity, cell water compartmenting and leaf water retention ability on the incidence of leaf-scorch are assessed, and possible mechanisms for the leaf-scorch avoidance are discussed.

MATERIALS AND METHODS

Orchard Conditions

Two orchards adjacent to the Tenryu River in the South Nagano area were selected for the study. The surface soil was an Andosol and the lower layer was gravel. Pear trees (*Pyrus communis* L. cv. Nijuseki) were planted in 1984 and both orchards were managed conventionally (i.e., chemically). The problem of leaf-scorch or leaf-burn had occurred several years before the research study began. Starting in 1995, one orchard continued to receive chemical fertilizers and pesticides while the other orchard received organic fertilizers and limited pesticide applications, if necessary.

Fertilization and Pest Control

The organically managed orchard received a microbial inoculant called "Effective Microorganisms" (EM-1) that was applied to a mixture of organic materials and fermented to yield a product called EM-Bokashi (International Nature Farming Research Center, Atami, Japan) with an N-P-K of 5.5-5.5-1.0 and applied each September at a rate of 4000 kg ha^{-1}. Pig manure was applied each December at a rate of 20000 kg ha^{-1}. Magnesium sulfate was applied in March and June at a rate of 300 kg ha^{-1}. The chemically-managed orchard received compound chemical fertilizers applied at equivalent rates of N-P-K and at the same time that biofertilizers were applied to the organically-managed orchard. Magnesium sulfate was applied at the same rate and time to both the chemical and organic orchards. Pesticide chemical sprays were applied nine times a year to the organic orchard and 18 times to the chemical orchard.

Water Retention Curve for Excised Leaves

The leaf water retention curve or transpiration decline curve as well as photosynthesis and cell water compartmenting were examined and analyzed two weeks before harvest in 1997. The third leaf from the base of a fruit cluster was excised and the cut end was placed in water with the leaf blade in air at saturated vapor pressure in a transparent acrylic bucket. The bucket was covered and kept in a cool room at 15°C. The leaves were rehydrated to a fully-turgid state. The leaf samples were then placed under light in air saturated with vapor in a closed acrylic container for 30 min to allow the stomata to open. Then the leaves were placed on a net fixed on a light paper box under 500 μmol m^{-2} s^{-1} of photosynthetic photon flux (PPF) at 60% of relative humidity. The leaf sample was weighed at 2-15 min intervals using an electronic balance. Changes in fresh weight were recorded and water loss was determined on a relative basis (Quisenberry et al., 1982; Xu et al., 1995). A curve was obtained by plotting leaf relative water content against time (Figure 1). This curve exhibits two different phases, the initial steep-sloped part and the gentle-sloped part. The slope of the initial phase shows the rapid rate of water loss by both stomatal transpiration and cuticular transpiration. The slope of the second phase or gentle-sloped part shows the rate of water loss by only cuticular transpiration, assuming that the stomata were completely closed. Usually a tangent line method (Quisenberry et al., 1982; Xu et al., 1994) is used to analyze the water retention curve of excised leaves. However, an arbitrary factor is involved in drawing the tangent lines by hand. Therefore, in the present research we used the mathematical model as follows to fit the water retention curve:

$$\zeta = [\zeta_{sat} - \zeta_{SC}(1 - \beta t)]\, e^{-\alpha t} + \zeta_{SC}(1 - \beta t) \qquad [1],$$

where ζ is relative water content of the excised leaf; ζ_{sat} is the leaf relative water content at saturation status before drying started; ζ_{SC} is the leaf relative water content at the time when stomata are closed; t is the time from beginning of the transpiration course at a PPF of 500 μmol m^{-2} s^{-1}; β is the constant showing the slope of the second gentle-loped part of the curve; α is the constant showing the slope of the first steep-sloped part of the curve. Water retention ability (WRA) was expressed as

$$\mathrm{WRA} = 1/\alpha + 1/\beta \qquad [2].$$

FIGURE 1. Leaf water retention curved modeled by $\zeta = [\zeta_{sat} - \zeta_{SC}(1 - \beta t)]\, e^{-\alpha t} + \zeta_{SC}(1 - \beta t)$. ζ is leaf relative water content; ζ_{sat} is the leaf water content at saturated status; ζ_{ASC} is the leaf relative water content at the condition of average stomatal closure; β is the constant related to cuticular conductance; t is the time; and α is the constant related to stomatal conductance.

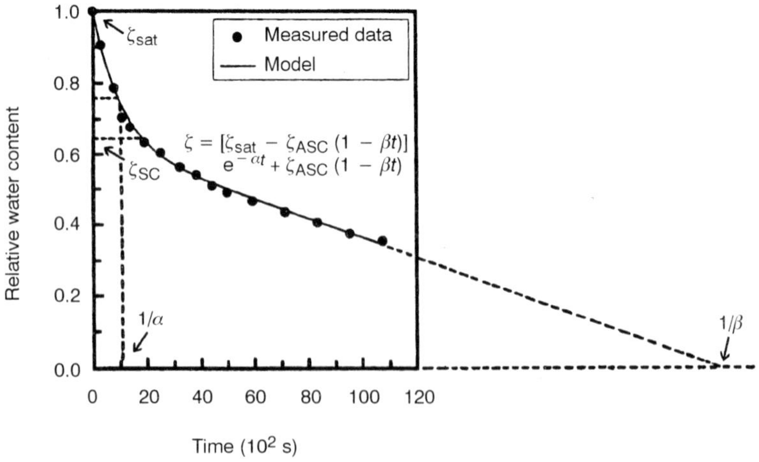

Analysis of Cell Water Compartmenting

Symplastic water fraction (ζ_{sym}), and osmotic potential at a fully-turgid state (π_{FT}) and at incipient plasmolysis (π_{IP}) were obtained from the pressure-volume curve (P-V curve). This curve was obtained according to Roberts and Knoerr (1977) with some modifications. After intact transpiration was measured, the leaf was excised under water. The excised leaf was rehydrated to saturation with the cut end in water in a dark, cool room (15°) to prevent possible biomass loss from respiration. After rehydration, a turgid leaf was weighed and mounted in a pressure chamber (Model 3000, Soilmoisture Equipment Corp., Santa Barbara, California, USA) with the leaf blades inserted in a thin polyethylene bag to prevent transpirational water loss. The exposed cut end of the petiole was about 1 cm beyond the rubber stopper and covered with a pre-weighed Eppendorf vial filled with tissue paper to absorb expressed sap. Sap expressions were made successively by increasing pressure at about 0.3 MPa per increment. This was done until the water potential reached a level of −3.0 to −3.5 MPa. The expressed sap in each increment was immediately weighed. About 30 min was needed for each increment at low pressures and about 1 to 2 h for each increment at high pressures. After a series of measurements, the leaf was weighed, then dried in an oven at 80°C for 48 h. The value of the collected sap at each increment was calibrated in accordance with the total expressed sap since about 10% of the total water content was lost by evaporation during the measurement period. The P-V curve was obtained by plotting the reciprocal of the pressure against the expressed water volume on a relative basis at each increment. The osmotic potential was obtained from the point on the curve that corresponds to the measured water potential. The turgor potential was calculated as the difference between the osmotic potential and water potential. The osmotic potential at a fully turgid state, ζ at which leaf turgor reaches zero (incipient plasmolysis or zero turgor point) and ζ in the symplastic fraction, were obtained as shown in Figure 2. The solute concentrations at the fully turgid status and at incipient plasmolysis were calculated using the Van't Hoff relation (Jones, 1992):

$$\pi = -RTC_S \qquad [3]$$

where C_S is concentration of solute; R is the gas constant; and T is absolute temperature with a value for RT of 2437 J mol^{-1}. Therefore, the total cell sap solute concentration at 1 MPa is $-(-10^6/2437) \approx 410$ osmol m^{-3}.

Analysis of Photosynthesis

Net photosynthetic rate (P_N) of the third leaf from the base of a fruit cluster was measured under different PPF (i) using a gas exchange system

FIGURE 2. Pressure-volume curve. Here, π_{FT} is the osmotic potential at a fully-turgid state and π_{IP} is the osmotic potential at the incipient plasmolysis (zero turgor point).

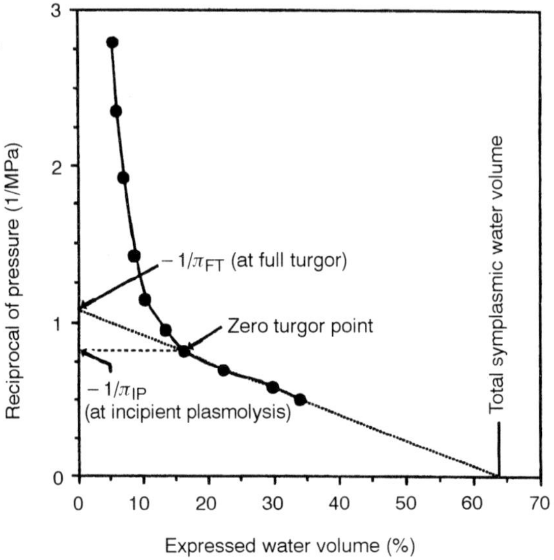

(Li-6400, Li-cor Inc. Lincoln, Nebraska, USA). Photosynthetic capacity (P_C), respiration rate (R_D) and quantum yield ($Y_Q = KP_C$) were obtained from the light-response curve:

$$P_N = P_C(1 - e^{-Ki}) - R_D \qquad [4],$$

where K is constant and i is the photosynthetic photon flux.

Measurements of Soil Mineral Nutrients

Soil samples were taken from both orchards at a soil layer of 0-20 cm. The samples were diluted with distilled water at a 1:4 ratio and pH of the soil solution extract was read with a pH meter (F-7AD, Horiba, Ltd., Tokyo, Japan) and electrical conductivity (EC) was read with a EC meter (Model CM-20E, Toa Electronics, Ltd., Tokyo, Japan). The concentrations of mineral nitrate-nitrogen and ammonium nitrogen as well as the concentration of soil available phosphorus and plant phosphorus were measured using colorimetric method (Ishitsuka, 1985). Total nitrogen and total organic carbon were measured by a carbon-nitrogen recorder (Vanaco MT-700, Yanakotosa Ltd., Tokyo, Japan). Minerals of K, Mg, Ca, Na, and Mn were measured with an atomic absorption spectrophotometer (180-30, Hitachi Ltd., Tokyo, Japan) according to Donohue and Aho (1992).

RESULTS AND DISCUSSION

Leaf Scorch

Leaf scorch or leaf burn was not observed in the pear orchard under organic farming management but was severe in the orchard subjected to chemical farming practices. The leaf scorch showed the symptom with the whole leaf dead and dried. This is a special local problem in the middle mountain area of Nagano Prefecture and in areas with a high evapotranspiration demand in the summer. There have been very few documents explaining the mechanism of this leaf-scorch problem. Some suggestions on causative effects were found in a booklet entitled "Fruit Production Guidance" published by Nagano Prefecture (1987). This problem is attributed to the insensitivity of stomatal regulation in response to suddenly increased transpiration. On a hot and dry summer day, especially after a rain, the leaf transpiration rate increases in response to the evapotranspiration demand shown by high light intensity, high temperature and low air humidity. In such a case, the stomata usually close to adjust the balance between water influx to the leaf and transpiration water loss from the leaf. If the physiological activity associated with stomatal function decreases, the stomata cannot adjust in response to increased evapotranspiration demand so that transpiration water loss is greater than water influx to the leaf and the leaf dies because of sudden water loss and desiccation. No leaf scorch was found in the pear trees managed by organic farming. This suggested that physiological activity associated with stomatal function was higher in leaves of pear trees with organic farming management than leaves of pear trees subjected to chemical farming practices. However, the mechanisms have not been clearly elucidated. Consequently, we examined the leaf water relations and stomata related variables to support the observation of leaf-scorch avoidance. The results are described as follows.

Water Retention Ability

Water supply was completely stopped in the case of excised leaves. When the excised leaves were placed under light, transpiration continued at the same speed as at the beginning so that the leaves lose water and the stomata close at different speeds in response to the leaf water deficits shown by the leaf relative water content. At the beginning, much water was lost by transpiration and the leaf relative water content sharply decreased following a pattern of exponential function, shown by the first phase of the curve. When the stomata had completely closed, water loss occurred only by cuticular transpiration. The rate of cuticular transpiration water loss is indicated by the slope of the second phase of the curve. This water retention curve was fitted with an exponential equation (Equation 1). In this equation, α is proportional to the speed of stomatal transpiration water loss and β to the cuticular transpiration water loss speed. Both α and β were larger for leaves of pear trees under chemical farming than for those under organic farming. As a consequence, water retention ability, shown by the sum of reciprocals of α and β, was higher for leaves of pear trees under organic farming than for those under chemical farming (Table 1). Moreover, the leaf relative water content at the time point when stomata were completely closed (z_{SC}) was lower in leaves of organically-farmed pear trees than in leaves of chemically-farmed pear trees. The result of α suggested that the physiological activity associated with stomatal function was higher in pear tree leaves under organic farming than in those under chemical farming. The result of β suggested that the leaves of pear trees under organic farming practices had thicker cuticles or more wax on the leaf surface, compared with the leaves of chemically-farmed trees, which prevented water loss from the cuticular membrane. Compared with the pear leaves from chemical farming practices, the stomata of pear leaves from organic farming could adjust the opening aperture in response to decreasing leaf water content so that more tissue water was retained as water was lost by transpiration. This was consistent with the incidence of leaf scorch. Because the leaves of pear trees in the chemical farming orchard lost their physiological activity associated with stomatal adjustment, stomata could not close when water uptake could not balance the transpiration water loss during hot, dry days, and especially when a period of intense sunshine followed heavy summer rainfall. Consequently, the leaves of chemically-farmed pear trees

TABLE 1. Relative leaf relative water content at stomatal closure (ζ_{SC}), speed constants of stomatal (α) and cuticular (β) transpiration and leaf water retention ability (WRA) as well as the symplastic water fraction (ζ_{sym}), osmotic potentials at fully turgid status (π_{FT}) and at incipient plasmolysis (π_{IP}), and osmotic concentrations (C) at FT and at IP for pear trees under chemical and organic farming.

Farming	ζ_{SC}	α (10^{-4} s^{-1})	β	WRA (10^4 s)	ζ_{sym}	π_{FT} (MPa)	π_{IP} (osmol m^{-3})	C_{FT}	C_{IP}
Chemical	0.58 a	0.64 a	11.7 a	1.64 b	80 b	−2.000 a	−2.647 a	820 b	1085 b
Organic	0.52 b	0.35 b	6.8 b	3.00 a	88 a	−2.093 b	−2.857 b	858 a	1171 a

Column values sharing the same letters are not significantly different at P = 0.05 according to Duncan's multiple range test.

lost excessive amounts of water, desiccated and died, as evidenced by leaf scorch. This is the possible hypothesis of the mechanism for leaf scorch. Our data on water retention ability support this hypothesis. Moreover, the value of z_{SC} suggested that the stomatal function of leaves of pear trees managed organically could tolerate leaf water deficit to a greater extent and recover more readily, compared with leaves of chemically-farmed pear trees. This explains the absence of leaf-scorch on pear trees in the orchard under organic farming.

Cell Water Compartmenting

Osmotic potentials at a fully-turgid state (π_{FT}) and at incipient plasmolysis (π_{IP}) as well as the leaf relative water contents of symplastic (ζ_{sym}) and apoplastic ($\zeta_{apo} = 100 - \zeta_{sym}$) fractions were analyzed from the pressure volume curve. Values of π_{FT} and π_{IP} were lower and consequently the cell sap solute concentration (C_{FT}) were higher in leaves of pear trees under organic farming than in those under chemical farming (Table 2). The cell sap solutes include sugars, ions and other soluble substances. Under well-watered conditions, cell sap concentration may be proportional to the physiological activity of a leaf. The ζ_{sym} was higher and ζ_{apo} was lower in leaves of pear trees under organic farming than in those under chemical farming. This suggested that leaves of pear trees under organic farming hold more active water in symplast, compared with leaves of pear leaves under chemical farming, where most metabolic reactions takes place. The above-mentioned results of leaf water relations and cell water compartmenting might, at least in part, account for the leaf-scorch avoidance of pear trees under organic farming practices.

Measurement of Photosynthesis

Photosynthetic capacity was significantly higher in leaves of pear trees under organic farming than for those under chemical farming. Compared with the chemical orchard, the organically-managed pear tree leaves had a

TABLE 2. Single leaf are, photosynthetic capacity (P_C), respiration (R_D) and quantum yield (Y_Q) as well as fruit yield and size of pear trees under chemical and organic farming practices.

Farming	Leaf area (cm^2)	P_C ($\mu mol \cdot m^{-2}s^{-1}$)	R_D ($\mu mol \cdot m^{-2}s^{-1}$)	Y_Q ($mol \cdot mol^{-1}$)	Yield ($Mg \cdot ha^{-1}$)	Fruit Size (g)
Chemical	89.9 b	16.8 b	3.09 a	0.072 a	21 b	235.2 b
Organic	106.7 a	22.4 a	2.96 a	0.069 a	34 a	343.8 a

Column values sharing the same letters are not significantly different at P = 0.05 according to Duncan's multiple range test.

higher photosynthetic rate per unit of leaf area (33% higher) and a larger leaf area per leaf (19% higher). Consequently, the total photosynthetic output was much higher in leaves of pear trees under organic farming than for those under chemical farming. This rather striking difference might be associated with the physiological activities and factors mentioned earlier. No large differences in dark respiration and quantum yield were found between leaves of pear trees under different farming practices.

Soil Nutrition

For several years, organic materials and pig manure had been applied to the organic farming orchard. Therefore, the soil organic matter content was higher in the organic farming orchard than in the chemical farming orchard. This can be shown by the total carbon and total nitrogen content of the soils which were significantly different between these two farming systems (Table 3). The C:N ratio was lower in soil of the organic farming orchard. Available phosphorus and other cations such as K, Mg, Ca, Na, and Mn were also significantly higher in the soil of the organic farming orchard than in soil of the chemical farming orchard. Although chemical fertilizers were supplied to soil of the chemical farming orchard with the same contents of N, P, K and Mg as in the organic materials that were supplied to soil of the organic farming orchard, nutrients in the chemical orchard might have been lost by leaching or runoff from heavy rains which occurred every year in this area. The organic materials applied to the organic orchard contain nutrients that are in an organic form and become available to plants over time, thus the organic orchard soil would have higher residual nutrient levels. Moreover, elements other than N, P, K, and Mg were not intentionally applied to the chemical farming orchard while most of the micronutrients (Cu, B, Mn, Mo, Cl) were present in organic materials and manure. Compared with chemical fertilizers, organic materials can also improve soil physical properties (Vogtmann, 1984). Unfortunately, soil physical properties were not analyzed. The higher nutrient content of soil in the organic farming orchard could also help to

TABLE 3. Mineral nutrient concentration in the soils chemical and organic farming pear orchards.

	$g\ kg^{-1}$			$mg\ kg^{-1}$									
Farming	T-N	T-C	C:N	NH_4-N	NO_3-N	Av.-P	K	Mg	Ca	Na	Mn	pH	EC
Chemical	5.1 b	78 b	15.4 a	22.8 a	31.5 b	300 b	380 b	390 b	160 b	62 b	17 b	5.8 a	0.06 b
Organic	6.5 a	88 a	13.6 b	18.9 b	77.0 a	760 a	900 a	460 a	220 a	71 a	69 a	5.2 b	0.20 a

Column values sharing the same letters are not significantly different at P = 0.05 according to Duncan's multiple range test.

enhance the higher physiological activities that prevented leaf scorch as mentioned earlier.

Fruit Yield

Compared with the chemically-farmed orchard, fruit yield was 62% higher in the organically farm orchard (Table 1). This higher fruit yield for latter orchard was attributed mainly to the larger fruit size (46% larger) and also in part to higher fruit number (11% higher). Fruit yield was very consistent with the leaf photosynthesis.

CONCLUSIONS

The higher photosynthetic capacity and higher physiological activities associated with cell water relations and water retention ability of pear trees managed by organic farming practices, as well as the higher levels of plant nutrients supplied in organic materials, were important factors that contributed to leaf scorch avoidance and higher fruit yield from organic farming practices compared with chemical farming. Further study is needed to (a) equalize the macronutrient and micronutrient levels in both farming systems, (b) determine changes in soil physical properties that might contribute to the differences in yield and disease incidence of organic vs. chemical farming of pear trees, (c) determine the extent to which chemical pesticide sprays can be reduced or avoided entirely from pear orchards managed by organic practices, (d) determine the extent to which Effective Microorganisms (EM) can substitute for chemical sprays, and (e) conduct a comparative study on the cost: benefit relationship, i.e., an economic analysis of pear production under organic vs. chemical management practices. Three years of data would be advisable on which to draw firm conclusions.

REFERENCES

Donohue, J.S. and D.W. Aho. (1992). Determination of P, K, Ca, Mg, Mn, Fe, Al, B, Cu, and Zn in plant tissue by inductively coupled plasma (ICP) emission spectroscopy. In. *Plant Analysis Reference Procedures for the Southern Region of the United States*, ed. C.O. Plank. Atlanta: Univ. Georgia, Atlanta, pp. 37-40.

Fukuda, H. (1994). Apple. In *Horticulture in Japan*, Japanese Society for Horticultural Science. Tokyo: Japanese Society for Horticultural Science, pp. 23-27.

Ishitsuka, J. (1985). Measurements of minerals and nitrogen compounds. In *Experiment Methods of Crop Physiology*, eds. Y. Kitajo and J. Ishitsuka. Tokyo: Association of. Agricultural Technology, pp. 281-316.

Jones, H.G. (1992). *Plant and Microclimate–A Quantitative Approach to Environmental Plant Physiology.* London: Cambridge University Press, pp. 1-323.

Nagano Prefecture. (1987). *Fruit Production Guidance*. Nagano (Japan): Agriculture Cooperation, pp. 577-578.

Quisenberry, J.E., B. Roark and D.A. Johnson. (1982). Use of transpiration decline curve to identify drought-tolerant cotton germplasm. *Crop Science* 22: 918-922.

Roberts, S. and K. R. Knoerr. (1977). Components of water potential estimated from xylem pressure measurements in five tree species. *Oecologia* (Berl) 28:191-202.

Vogtmann, H. (1984). Organic farming practices and research in Europe. In *Organic Farming: Current Technology and Its Role in a Sustainable Agriculture*, eds. D.F. Bezdicek, J.F. Power, D.R. Keeny and M.J. Wright MJ. Madison: American Society of Agronomy, pp. 19-36.

Xu, H.L., L. Gauthier and A. Gosselin. (1995). Stomatal and cuticular transpiration of greenhouse tomato plants in response to high solution electrical conductivity and low soil water content . *Journal of American Society for Horticultural Science* 120: 417-422.

Xu, H.L., L. Gauthier, P.A. Dubé and A. Gosselin. (1997). Effects of fertigation management on water relations of tomato plants grown in peat, rockwool and NFT. *Journal of Japanese Society for Horticultural Science* 66: 359-370.

PART II: MICROBIAL APPLICATIONS

Soil-Root Interface Water Potential in Sweet Corn as Affected by Organic Fertilizer and a Microbial Inoculant

Hui-lian Xu

SUMMARY. Effects of organic fertilization and application of a microbial inoculant (with EM as commercial name) on soil-root interface water potential (Ψ_{S-r}) of sweet corn (*Zea mays* L. cv. Honey-Bantam) were examined. The microbial inoculant includes about 80 species of microbes, such as *Lactobacillus*, *Rhodopseudomonas*, *Streptomyces*, and *Aspergillus*. The contributions to Ψ_{S-r} from root amount and root activity were analyzed using the Ohm's law. Plants were potted with an

Hui-lian Xu is Senior Crop Scientist, International Nature Farming Research Center, 5632 Hata, Nagano 390-1401, Japan (E-mail: huilian@janis.or.jp).

The author thanks S. Kato, K. Yamada, M. Fujita, K. Katase and Dr. T. Higa for their technical assistance and advice.

[Haworth co-indexing entry note]: "Soil-Root Interface Water Potential in Sweet Corn as Affected by Organic Fertilizer and a Microbial Inoculant." Xu, Hui-lian. Co-published simultaneously in *Journal of Crop Production* (Food Products Press, an imprint of The Haworth Press, Inc.) Vol. 3, No. 1 (#5), 2000, pp. 139-156; and: *Nature Farming and Microbial Applications* (ed: Hui-lian Xu, James F. Parr, and Hiroshi Umemura) Food Products Press, an imprint of The Haworth Press, Inc., 2000, pp. 139-156. Single or multiple copies of this article are available for a fee from The Haworth Document Delivery Service [1-800-342-9678, 9:00 a.m. - 5:00 p.m. (EST). E-mail address: getinfo@haworthpressinc.com].

Andosol soil fertilized using anaerobically- or aerobically-fermented organic materials with or without EM application, and with chemical fertilizers as control. One month after sowing, as soil matric water potential decreased, Ψ_{s-r} was higher in plants with organic fertilizer than those with chemical fertilizer and was also higher in plants treated with EM than in those without EM. The higher Ψ_{s-r} was attributed to the larger root volume and higher root activity shown by root respiration rate. Consequently, photosynthetic rates under soil water deficits were also maintained higher in these plants. This suggested that maintenance of a high Ψ_{s-r} favored plants to resist water deficits. The methodology is a practical means of analyzing the soil-plant water status under undisturbed conditions. *[Article copies available for a fee from The Haworth Document Delivery Service: 1-800-342-9678. E-mail address: getinfo@haworthpressinc.com <Website: http://www.HaworthPress.com>]*

KEYWORDS. Effective microorganisms, EM, microbial inoculant, organic fertilizer, soil-root interface water potential, drought resistance, *Zea mays*

ABBREVIATIONS AND SYMBOLS. A_L, leaf area per plant; DM_R, root dry mass per plant; E, transpiration rate per plant; E_A, average leaf transpiration rate; g, leaf conductance; L_R, total root length per plant; P_C, photosynthetic capacity; r, leaf resistance; r_p, average radius of soil particles; r_R, average radius of the root; R_{s-r}, soil-root interface hydraulic resistance; R/T, root to total ratio on dry mass basis; S_R, root surface area; S_{s-r}, soil-root interface area; V_R, root volume; Γ_{s-r}, soil-root interface hydraulic conductivity; r_{s-r}, soil-root interface hydraulic resistivity; Ψ_{s-r}, soil-root interface water potential; ι_{s-r}, soil-root interface distance

INTRODUCTION

Organic farming practices have been proposed as alternatives to chemical agriculture to protect the environment, reduce production costs and improve food quality. Organic farming and nature farming have been practiced by farmers in Japan for many years. Recently, a technology of applications of a microbial inoculant (with Effective Microorganisms or EM as the commercial name, International Nature Farming Research Center, Atami, Japan) is introduced to the organic or nature farming (Higa, 1994, 1996). The microbial inoculant consists of a group of beneficial microbes containing about 80 species such as *Lactobacillus*, *Rhodopseudomonas*, *Streptomyces*, and *Aspergillus*. Research has shown that organic fertilizers improve soil physical properties, whereby the soil retains more water and the crops growing there resist a stronger water stress compared with the cases of chemical fertilizer (Letey, 1977). However, there are very few scientific reports to support the mechanisms for effects of EM although farmers have found that applications of this microbial inoculant with organic fertilizers improve crop yield and quality (Higa, 1994, 1996). Some reports show that EM applications to crops are more effective in drought regions (Li and Ni, 1996) than in humid areas. This suggests that effects of EM might be associated with increased water stress resistance caused by root water uptake ability.

Moreover, in research on plant response to drought conditions, measurements of plant water status such as leaf water potential (Ψ_{leaf}) are suggested to be more appropriate as a water stress indicator than measurements of soil water conditions such as soil water potential (Ψ_{soil}) and water content (Hsiao, 1973; Jackson, 1974). However, the fluctuation of environmental factors on a hot-dry day can cause a short-term change in plant water status such as Ψ_{leaf}. On the other hand, some plants with very inadequate water supply can have the same Ψ_{leaf} as in well-watered plants if the stomata close efficiently in response to changes in environmental factors. This leads to a disadvantage in using Ψ_{leaf} as an indicator of water stress. Fortunately, Jones (1983a) proposed a method to estimate the Y_{soil} at the root surface or, the so-called, soil-root interface water potential (Ψ_{s-r}) using the familiar Ohm's law. The Ψ_{s-r} is associated with both soil physical properties and plant water root activity and water consumption. It can be considered as an appropriate indicator for the status of soil water that is available or ready to enter the plant. Therefore, in the present research, Ψ_{s-r} of sweet corn plants and the contribution of root amount and root activity to Ψ_{s-r} were examined and the effects of organic fertilizations and microbial inoculation were investigated.

MATERIALS AND METHODS

Plant Materials

Sweet corn (*Zea mays* L. cv. Honey-Bantam) plants were grown in Wagner's pots (16 cm in diameter and 20 cm high) in June 1996. The pots, each with one plant remaining after thinning, were placed in a Latin Square design in a glasshouse.

Soil

A fine textured Andosol soil was collected from a field where soybean was previously cultivated. The total soil nitrogen, available phosphorus, and potassium were 3.4, 0.025 and 0.44 g kg^{-1}, respectively, with a C:N ratio of 13. The field capacity (or capillary capacity) was 80% on a gravimetric basis, i.e., there was 80 g water in 100 g dry soil when it was saturated with water. Each pot was filled with 3 kg fresh soil with a water content of 38% of the field capacity.

Fertilizers

Ammonium sulfate, superphosphate and potassium sulfate were used for the chemical fertilizer treatments and the total quantities of N, P and K applied to the chemical treatments were equivalent to the total contents of N, P, K contained in the organic fertilizers applied to the organic treatments. Organic materials such as oilseed sludge, rice bran and fish-processing by-product were fermented anaerobically or aerobically with or without EM. The total nitrogen, available phosphorus, and potassium concentrations were 58, 30 and 2 g kg^{-1}, respectively, for both anaerobic and aerobic organic fertilizers.

Microbial Inoculant

The microbial inoculant used in this study, with a commercial name as "Effective Microorganisms" or EM (International Nature Farming Research Center, Atami, Japan), contains a group of beneficial microorganisms that can co-exist under the same culture. About 80 species are reportedly present in the product liquid of this microbial inoculant. The main species comprising EM are summarized as follows: (1) Lactic acid bacteria: *Lactobacillus plantarum, Lactobacillus casei, Streptococcus lactis*; (2) Photosynthetic bacteria: *Rhodopseudomonas palustris, Rhodobacter sphaeroides*; (3) Yeasts: *Saccharomyces cerevisiae, Candida utilis*; (4) Ray fungi: *Streptomyces albus, Streptomyces griseus*; (5) Fungi: *Aspergillus oryzae, Mucor hiemalis*; (6) Others: Some microorganisms that are naturally occurring and combined into the inoculant in the manufacturing process and can survive in the inoculant liquid at pH 3.5 or below (Pathogenic microorganisms do not usually survive solution at pH below 3.5). The density of most of the above-mentioned microbes are in the range of 1×10^6 to 1×10^8 per ml.

Treatments

Six fertilization treatments were designed as follows: (1) EM-Organic 1–anaerobically fermented organic materials 80 g, in which EM was added to the materials before fermentation; (2) Organic 1–anaerobically fermented organic materials 80 g; (3) EM-Organic 2–aerobically fermented organic materials 80 g, in which EM was added before fermentation; (4) Organic 2–aerobically fermented organic materials 80 g; (5) EM-Fertilizer–chemical fertilizers (ammonium sulfate 5.3 g, long-period coated urea called LP coat-70 2.8 g, superphosphate 13 g and potassium sulfate 4.95 g), with EM 80 ml, applied into soil at the same time before sowing; the total amounts of nitrogen, phosphorus and potassium were the same as in the above-mentioned organic materials; and (6) Fertilizer–the same fertilizers as in (5) without EM application.

Soil-Root Interface Water Potential

Figure 1 shows an electrical circuit analogue of the pathway of water flow from soil through the plant to the atmosphere. The pathway is also called soil-plant-air continuum (Nobel and Jordan, 1983). With the exception of the tissue storage capacity (C_S) and resistance (R_S), parameters in the analogue will be used to derive an equation to estimate Ψ_{s-r}. In this pathway, Ψ_{soil} plays a dominant role in controlling plant water status and Ψ_{s-r} is proposed as a better indicator for the status of the water available or ready to enter the plant (Jones 1983a and b) because it is related not only to soil hydraulic properties, which are determined by soil texture and physical properties, but also to the plant root amount and water uptake ability. Usually, it is difficult to

FIGURE 1. An electric circuit analogue for water flow from the soil through the plant to the air. Ψ_{soil}, soil water potential; Ψ_{s-r}, soil-root interface water potential; R_{s-r}, soil-root interface hydraulic resistance; R_p, the total hydraulic resistance within the plants; Ψ_x, leaf xylem water potential; r, leaf resistance; E, total plant transpiration; R_s, water storage resistance; C_s, water storage capacitance.

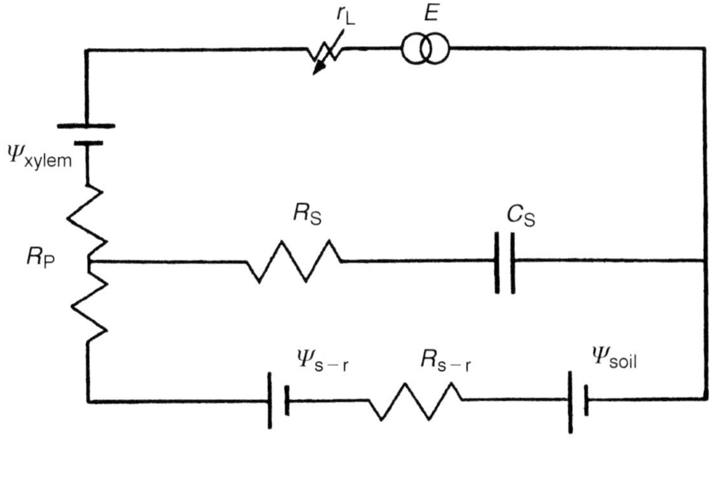

measure Ψ_{s-r} routinely. However, it can be estimated using the concept of an electrical circuit analogue (Figure 1). In the present study we estimated Ψ_{s-r} as proposed by Jones (1983a). According to the first equation of Ohm's Law, in a series-connected circuit,

$$V = V_1 + V_2 + \ldots + V_i + V_n = I_1R_1 + I_2R_2 + \ldots + I_iR_i + \ldots + I_nR_n, \quad (1)$$

where V, I, and R are the electrical potential or voltage (Volt), current (Ampere) and resistance (Ohm) in an electric circuit. Accordingly, Ψ_{soil} and Ψ_{s-r} can be expressed as follows:

$$\Psi_{soil} = \Psi_{s-r} + ER_{s-r}, \quad (2)$$

and

$$\Psi_{s-r} = \Psi_X + ER_P, \quad (3)$$

where E is the plant transpiration rate (kg s^{-1}), corresponding to I_i in (1); R_{s-r} and R_P are the hydraulic resistance between the soil and root or called soil-root interface hydraulic resistance (MPa s kg^{-1}) and the total plant hydraulic resistance (MPa s kg^{-1}), respectively, corresponding to R_i in (1); Ψ_X is leaf xylem water potential (MPa), corresponding to V_i in (1). Because R_{s-r} cannot be measured directly, it is not possible to estimate Ψ_{s-r} using Equation (2), although Ψ_{soil} and E can be determined. Therefore, Equation (3) should be considered. It is known that transpiration rate per unit of leaf area, E_A (kg m^{-2} s^{-1}), is proportional to water vapor concentration difference, $(e_s - e_a)$ (kg m^{-3}), between the evaporating surface and the ambient air. According to the Ohm's law, it can be expressed as follows:

$$E_A = (e_s - e_a)/r \quad (4)$$

$$E_A = (e_s - e_a)g \quad (5)$$

where r (s m^{-1}) is the leaf resistance and g (m s^{-1}) is leaf conductance shown as the reciprocal of r. Substituting (5) into (3) gives the following equation:

$$\Psi_{s-r} = \Psi_X + (e_s - e_a)gAR_P \quad (6)$$

where A is the total leaf area (m^2) of the plant. Because R_P and A can not be easily measured in undisturbed conditions, comparison with the well watered control plant is used to eliminate these variables in the equation. First, for convenience, let $b = (e_s - e_a)AR_P$, and thus

$$\Psi_{s-r} = \Psi_X + gb. \quad (7)$$

Ψ_{s-r} for a well watered plant is 0.8 kPa in the present work and negligible as 0 in comparison to the values (10-320 kPa) for plants with water deficit treatments. Hence, equation (7) will be

$$\Psi_{X[W]} = -bg_{[W]} \tag{8}$$

where the subscript [W] refers to a well-watered control plant. If the differences in $(e_s - e_a)$, A and R_P between moderately drought stressed [D]) and well-watered plants change little during a short term treatment, b is assumed the same for the former and latter (Jones, 1983a). Consequently, the following equation is derived from (7) and (8):

$$\Psi_{s-r[D]} = \Psi_{X[D]} - \Psi_{X[W]}g_{[D]}/g_{[W]}. \tag{9}$$

Equation (9) can be used to estimate $\Psi_{s-r[D]}$ using only g and Ψ_X without disturbing or damaging the plant and soil, except a half leaf blade is cut for Ψ_X measurement. However, leaf area would differ with a severe water deficit or a long period of stress. In this case leaf area should be measured and a factor of $(A_{[D]}/A_{[W]})$ should be included and (9) will be

$$\Psi_{s-r[D]} = \Psi_{X[D]} - \Psi_{X[W]}g_{[D]}A_{[D]}/g_{[W]}A_{[W]}. \tag{10}$$

Moreover, $(e_s - e_a)$ may also vary with water stress because stomatal closure can increase leaf temperature and consequently increase e_s. Therefore, values of $\Psi_{s-r[D]}$ should be corrected using the data of e_a and boundary layer conductance (Jones, 1983b), when the leaf temperature difference is large. Variance of another variable that has been neglected in the present work is R_P because many documents have shown that the effect of short-term water deficit on R_P are usually small (Simmelsgaard, 1976; Jones, 1978; Passioura, 1980).

Soil-Root Interface Conductivity

Maintenance of Ψ_{s-r} is, in a sense, attributed to the root volume on one side and root water uptake activity and/or soil physical properties on the other side. In order to elucidate the mechanism, I separately analyzed the contributions of root amount, shown here by total root surface area or soil-root interface area (S_{s-r}) and of the root activity and/or soil physical properties, shown here by the soil-root interface conductivity (Γ_{s-r}, g m MPa^{-1} m^{-2} s^{-1}). The determination of Γ_{s-r} was made using data including those from disturbed measurement, which was different from the determination of the above-mentioned Ψ_{s-r}. Soil-root interface hydraulic resistance, R_{s-r} (MPa m^2 s kg^{-1}), in relations to S_{s-r} (m^2), soil-root interface resistivity, ρ_{s-r} (MPa m^2 m kg^{-1}), and soil-root interface distance, ι_{s-r} (m), were analyzed according the second equation of Ohm's laws:

$$R = \rho\iota/S \tag{11}$$

where ρ, ι, and S are the resistivity that is determined by the quality of the conductor, conductor length, and conductor crosscut section surface area, respectively. Therefore, in the case of soil-root interface,

$$R_{s-r} = \rho\iota_{s-r}/S_{s-r} \tag{12}$$

where S_{s-r} (m^2) is the total contacting surface area between the soil and root and equal to root surface area (S_R). Therefore, (12) will be

$$R_{s-r} = \rho_{s-r}\iota_{s-r}/S_R. \tag{13}$$

In electrical physics, ρ is the resistivity of conductor determined by the material property of that conductor. Therefore, the analogue variable ρ_{s-r} is the resistivity of the soil-root surface determined by root properties on one side and soil properties on the other side. In convenience of proportion to Ψ_{s-r}, the reciprocal of ρ_{s-r}, named as the soil-root interface conductivity, Γ_{s-r} (g m MPa^{-1} m^{-2} s^{-1}), was used to show the contribution to Ψ_{s-r} from the soil-root interface properties (the root activity and soil physical properties). It is expressed as follows:

$$\Gamma_{s-r} = 1/\rho_{s-r} = \iota_{s-r}/R_{s-r}S_R. \tag{14}$$

It must be mentioned here that Γ_{s-r} is different in definition from both the reciprocal of R_{s-r} (hydraulic conductance) and the soil hydraulic conductivity obtained from the desorption curve (Hillel, 1980). Then Ψ_{s-r} can be written as

$$-\Psi_{s-r} = ER_{s-r} = E\iota_{s-r}/\Gamma_{s-r}S_R \tag{15}$$

and Γ_{s-r} can be calculated as

$$\Gamma_{s-r} = -E\iota_{s-r}/\Psi_{s-r}S_R. \tag{16}$$

Plant total transpiration (E) was calculated as

$$E = E_A A \tag{17}$$

Here, E_A is the average transpiration rate of the leaves at different positions on the plant. The length of the soil-root interface pathway is assumed as the average distance from the soil particles to the root surface and calculated as

$$\iota_{s-r} = r_p + r_R - 4r_p/3\pi - [\int f(r_R)dr_R]/r_p \; (r_R \geq r_p) \tag{18}$$

where r_p and r_R are the average radii of the soil particles and root crosscut section surfaces; and $f(r_R)$ are the circle area function for the root crosscut section surface with r_R as the radius. Equation (18) was derived from Figure 2 on the assumption that spheres of soil particles and cylinders of roots are tangent with each other. Actually, i_{s-r} is equal to the average height of the cubic space between the soil particle and the overlapped part of the root cylinder. The magnitude of this cubic space is graphically shown by the broken line shaded parts in Figure 2. The S_R was estimated from the root volume (V_R) and root length (L_R) as follows:

$$S_R = 2\pi L_R \sqrt{V_R/\pi L_R} \tag{19}$$

Root volume was estimated with the water replacement method. Washed fresh roots were placed in a volumetric cylinder with a recorded volume of water in. The increase in volume of the cylinder content was recorded as the root volume. Root length was measured with the cross-line method (Morita et al., 1992).

Photosynthesis, Transpiration and Leaf Conductance

Five weeks after sowing, photosynthetic rates (P_N) in the leaf just above the ear were measured using a gas exchange system (LI-6400, LI-COR, Inc., Lincoln, Nebraska, USA) at different photosynthetic photon flux densities (0, 100, 250, 500, 800, 1200, 1600, 2000 μmol m^{-2} s^{-1}) under different soil water conditions. The maximum photosynthetic potential or photosynthetic capacity (P_C) was obtained from the light-response curve as

$$P_N = P_C\,(1 - e^{-Ki}) - R_D \tag{20}$$

FIGURE 2. Crosscut section (left) and vertical section (right) of the root in connection with a soil particle. The broken line sheltered parts show the areas of the space between the root and the soil particle.

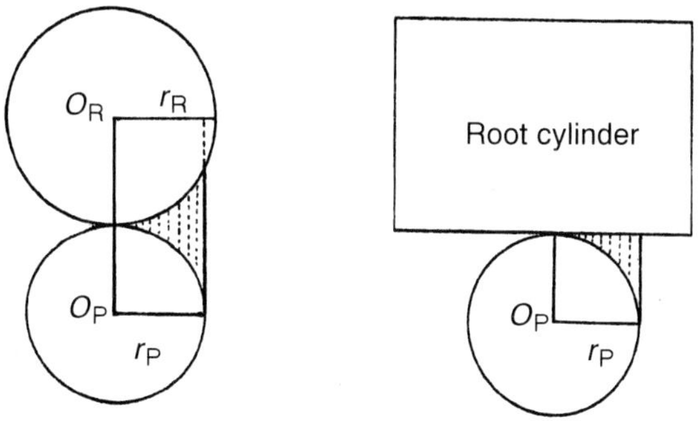

where P_N is the net photosynthetic rate; K, the half-time constant; i, the photosynthetic photon flux; and R_D, the dark respiration rate. Leaf conductance and transpiration rate used for calculation of Ψ_{s-r} and Γ_{s-r} were measured with this gas exchange system under natural light condition (about 1600 μmol m^{-2} s^{-1}) when Ψ_{s-r} was measured.

Root Respiration Rate

Tip parts of the root with a length of 5 cm were cut and collected as root samples (each with 1 g of fresh mass) for respiration measurement. The fresh root sample was inserted in the assimilation chamber and measured under a relative humidity of 85% at 25°C using the same gas exchange system as for photosynthetic measurement. After measurement, the root sample was dried and the respiration rate was expressed on a dry mass basis. The root respiration was used to indicate the root physiological activity.

Leaf Xylem Water Potential

Because the leaf blade of corn is much broader and longer than those of other cereal crops such as wheat and rice, a particular procedure was adopted for the measurement. The tip half part of the leaf blade after photosynthetic measurement was cut for measurement of leaf xylem water potential. The leaf blade was aligned on the adhesive surface of a 60 cm wide Scotch tape and folded along the main vein. The leaf blade was then enclosed in the Scotch tape to prevent evaporation water loss and pressure gas penetration into the tissue. The prepared leaf blade sample was set in a pressure chamber (Model 3000, Soilmoisture Ltd., Santa Barbara, California, USA). The tip of the leaf blade with the tape was cut and the cut end was left 1 cm over the rubber stop. Other processes were the same as in usual cases as mentioned by Turner (1988).

Soil Matric Water Potential

Soil matric water potential was measured using tensiometers (SPAD PF-33 with sensors of SPAD 2124 and 2127, Fujiwara Corp., Tokyo) with a capacity range up to 350 kPa. Three tensiometer sensors were placed with the sensor heads in different places within the soil volume of the sampled pots. The average value of the tensiometer readings was recorded as the soil matric water potential. Pots were weighed at the time when transpiration and leaf conductance were measured and the soil water content was calculated. The correlation ($y = 7011.9\, e^{-0.0767x}$; $r^2 = 0.83$; n = 28) between soil matric water potential (y) and soil water content (x) was obtained and confirmed in the laboratory. It is noteworthy that the soil matric water potential measured *in situ* here does not represent Ψ_{soil} in the analogue of Figure 1 because Ψ_{soil} includes osmotic potential and pressure potential in addition to the matric potential. In the present study, we used the *in situ* measured soil matric water potential just to show the decreasing extent of soil water. If soil water potential is involved in the calculation and analysis, the one measured with a psychrometer should be used.

RESULTS

Soil-Root Interface Water Potential

As soil matric water potential decreased, soil-root interface water potential (Ψ_{s-r}) maintained higher in organic-fertilized sweet corn plants than in chemical-fertilized ones (Figure 3). For example, at the soil matric water potential of 70 kPa, Ψ_{s-r} decreased to -0.61 MPa in chemical-fertilized plants while it maintained at -0.45 and -0.43 MPa in plants fertilized with anaerobic and aerobic organic materials, respectively (Table 1). There was no difference in Ψ_{s-r} found between anaerobic and aerobic organic fertilizer. Under both organic and chemical fertilization conditions, Ψ_{s-r} was 0.08 MPa higher in plants with EM application than in those without EM. This result suggests that microbial inoculation diminished decreases in Ψ_{s-r} under soil water deficit conditions although the differences between plants with EM and without EM application were not as large as those between chemical and organic fertilizations. There were only simply additive effects without synergistic interactions between the treatments of organic fertilization and EM application.

Contributions of Root Amount and Root Activity to Ψ_{s-r}

The contributions to Ψ_{s-r} from root amount and root physiological activity were analyzed using the second equation ($R = \rho l/S$) of Ohm's law. The related variables are presented in Table 1. First, it was found that root/top ratio of biomass was higher for organic-fertilized plants than chemical-fertilized plants and also higher for plants with EM application than for those without EM (Table 1). However, the absolute values of root dry mass, total root length, total root volume, and total root surface area were higher in chemical-fertilized plants than organic-fertilized plants because the chemical-fertilized plants grew better as a whole. The differences in above-mentioned variables between plants with and without EM application were clear in both organic and fertilizer treatments. The calculated value of soil-root

FIGURE 3. Maintenance of soil-root interface water potential (Ψ_{s-r}) and relative photosynthetic capacity (P_C) in sweet corn plants grown with different fertilizations as soil matric water potential decreased.

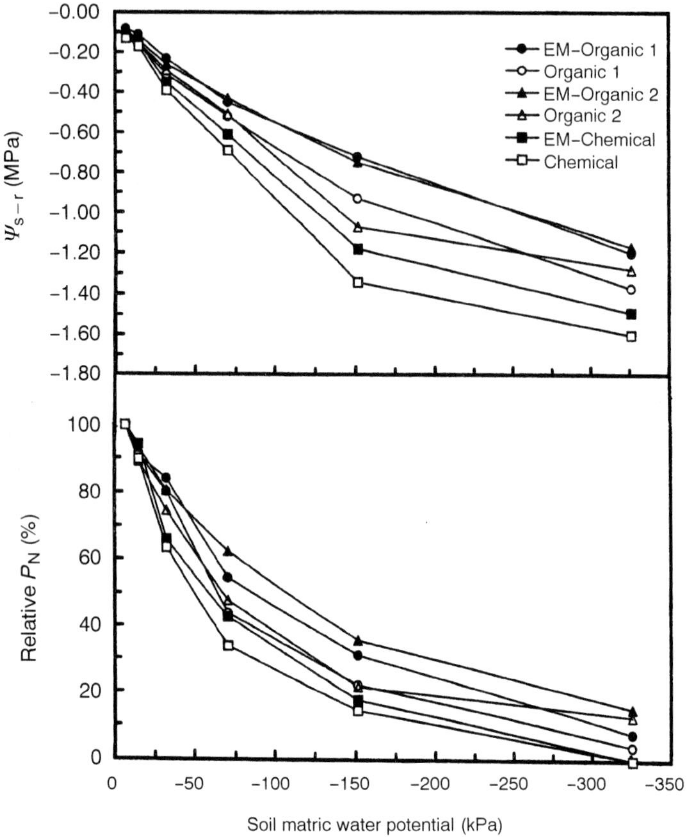

interface resistance (R_{s-r}) at the soil matric water potential at 70 kPa was higher in chemical-fertilized than organic-fertilized plants and also higher for non-inoculation than for microbial inoculation treatments. The differences between treatments in root radius and the length of the soil-root interface pathway are very small. The little differences in ι_{s-r} was from the differences in root amount. The soil-root interface hydraulic conductivity, shown as the reciprocal of the resistivity, was higher in organic-fertilized than chemical-fertilized plants and also higher in plants of microbial inoculation treatment in plants without. This result suggests that Γ_{s-r} contributes to Ψ_{s-r}.

Root Respiration Activity

The respiration rate in the tip part of the root measured at 25° was higher in organic-fertilized plants than chemical-fertilized plants and higher for plants with EM application than those without EM (Table 2). This result is consistent with that of Γ_{s-r} analyzed using Ohm's law. This suggests that the relatively high Ψ_{s-r} in organic-fertilized plants or in EM-applied plants is attributed, at least in part, to the relatively high root physiological activity.

TABLE 1. Soil-root interface water potential (Ψ_{s-r}) and related parameters at the soil matric potential of 70 kPa under organic and chemical fertilizations with (MI+) or without (MI−) MI applications.

Variable		Organic 1		Organic 2		Chemical	
		MI+	MI−	MI+	MI−	MI+	MI−
Ψ_{s-r}	(MPa)	−0.45 a	−0.52 b	−0.43 a	−0.51 b	−0.61 c	−0.69 d
E_A	(10^{-5} kg m^{-2} s^{-1})	3.31 c	3.03 d	3.59 b	3.18 c	2.64 a	2.34 b
A_L	(m^2)	0.108 e	0.106 e	0.113 c	0.107 d	0.146 a	0.129 c
DM_R	(10^{-3} kg)	1.83 bc	1.57 d	1.99 b	1.68 c	2.30 a	1.91 b
R/T	(%)	27 a	24 b	28 a	24 b	20 c	18 d
L_R	(m)	440 b	393 c	457 b	397 c	501 a	438 b
V_R	(10^{-5} m^3)	1.91 a	1.72 b	2.29 a	1.88 b	2.65 c	2.16 d
S_R	(m^2)	0.325 a	0.291 b	0.362 a	0.316 b	0.408 c	0.344 d
R_{s-r}	(10^5 MPa s kg^{-1})	1.26 a	1.61 b	1.06 a	1.49 b	1.63 c	2.24 d
r_R	(10^{-4} m)	1.18 cc	1.18 cc	1.26 b	1.23 b	1.30 a	1.23 b
l_{s-r}	(10^{-5} m)	4.59 a	4.59 a	4.55 a	4.56 a	4.53 a	4.56 a
ρ_{s-r}	(10^8 MPa s m^2 m^{-1} kg^{-1})	8.94 d	10.28 c	8.44 e	10.06 c	14.69 b	16.90 a
Γ_{s-r}	(10^{-8} kg m m^{-2} MPa^{-1} s^{-1})	11.19 b	9.23 c	11.84 a	9.94 c	6.81 d	5.92 e

See "Abbreviations and symbols" for definitions of the variables. The data followed by the same letter are not significantly different with each other according to Waller-Duncan comparison.

TABLE 2. The photosynthetic capacity under well-watered conditions ($P_{C[W]}$) and at soil matric water potential of 70 kPa ($P_{C[70]}$) and dark respiration rate of the root at 25°C (R_D) in leaves of sweet corn plants with different fertilization treatments.

Treatment	$P_{C[W]}$	$P_{N[70]}$	(% of $P_{C[W]}$)	$R_{D[25]}$
	(μmol m^{-2} s^{-1})			(μmol kg^{-2} s^{-1})
MI-Organic 1	25.8 bc	14.1 b	(54.5)	5.2 c
Organic 1	24.8 c	10.9 d	(43.9)	5.8 b
MI-Organic 2	26.6 b	16.6 a	(62.4)	6.0 b
Organic 2	23.8 c	11.3 cd	(47.7)	6.8 a
MI-Chemicals	29.0 a	12.4 c	(4.8)	4.8 dc
Chemicals	26.8 b	9.0 e	(33.7)	5.4 c

Photosynthetic Maintenance Under Soil Water Deficit Conditions

Since Ψ_{s-r} maintained higher in organic-fertilized and EM-applied plants as soil matric water potential decreased, photosynthetic capacity (P_C) was consequently higher in these plants with higher Ψ_{s-r} on both absolute and relative bases (Figure 3, Table 2). The proportional association between Ψ_{s-r} and P_C was only apparent when soil matric water potential decreased, i.e., this relationship does not hold under well-watered conditions.

DISCUSSION

In the pathway of water flow from soil through the plant to the atmosphere or soil-plant-air continuum (SPAC), Ψ_{soil} plays a dominant role in controlling plant water status (Nobel and Jordan, 1983). However, in some specific cases, whether or not plants can absorb sufficient water from the soil is not only dependent on water amount in the soil but also dependent on the water uptake ability and the ability of the soil to transfer water from the soil to the root surface. The soil-root interface water potential (Ψ_{s-r}) is not only related to soil properties but also associated with plant root activity and plant water consumption. That is why Ψ_{s-r} is considered as a better indicator of soil water available to the plants (Jones 1983a). As mentioned by Jones (1983a), Ψ_{s-r} is actually the average water potential or effective soil water potential at the root surface. It shows the status of water that is available or ready to enter the plant. In magnitude, Ψ_{s-r} is close to the plant water potential at predawn but far lower than Ψ_{soil}, especially that measured with a tensiometer. This suggests that there is a large resistance at the soil-root interface. The results of the present study indicated that the calculation of Ψ_{s-r} can be a useful additional method to estimate plant water stress although there are limitations from the fluctuations of related variables. In every treatment, measurements are made on comparable leaves of water-stressed and well-watered control plants that errors from the Ψ_X and γ could be minimized. There is no denying the fact that leaf temperature and leaf area (in case the plant is vegetatively growing) might change in response to a severe and/or a long-term water deficit. In this case, the estimated Ψ_{s-r} should be corrected using data of leaf temperature, boundary conductance and leaf area as described by Jones (1978, 1983a, b). The results showed a good trend among the treatments. The Ψ_{s-r} was higher for both anaerobic and aerobic organic fertilizers than chemical fertilizers. In both organic- and chemical-fertilized plots, EM application increased Ψ_{s-r} almost equally. The relatively high Ψ_{s-r} in organic-fertilized plants or in plants applied with EM was attributed to both the relatively large root surface area and relatively high soil-root interface hydraulic conductivity (Γ_{S-r}). By inferences as shown in Equation (14) from Ohm's law, Ψ_{S-r} is determined by four factors, plant transpiration, the length of the soil-root interface conductor, the root surface area and the soil-root interface hydraulic conductivity (its reciprocal is called resistivity). The length of the soil-root interface is actually the distance from the soil to root surface and associated with the contact between soil and root. It is affected by the soil texture and root morphology. In the present study, an Andosol was used with a soil particle diameter of 0.18 mm for all treatments. Therefore, the distance between soil and root varied just slightly with treatments because of the small difference in root diameter. From the result of root surface area it is concluded that the relatively high Ψ_{S-r} in organic-fertilized plants or in EM-applied plants is attributed, on the one hand, to the promoted root amount. Another main factor that contributes to Ψ_{S-r} is Γ_{S-r}, which is determined by the properties of the soil-root interface pathway. One of the terminals of the conducting pathway of the soil-root interface is the soil and the other is the root of the plant. Therefore, Γ_{S-r} is determined by both soil physical properties and root physiological activities. Compared with plants under chemical fertilization and without EM application as control, Γ_{S-r} is larger for organic-fertilized plants than chemical-fertilized plants and also larger for plants with EM applications than for those without EM. We measured root respiration rate, which is supposed to be an indicator of the physiological activity in the root (Huck, 1982). It was found that root respiration rate at 25°C was relatively high in organic-fertilized or EM-applied plants. Respiration rate in the root indicates the physiological activity for ion and water uptakes (Huck, 1982; Lauchli, 1982). This result is consistent with the result of Γ_{S-r} with analysis of Ohm's law. Another factor associated with Γ_{S-r} might be the physical property, which determines the ease of water to flow onto the root surface. In the present study, the author did not measure the physical properties of the soil. So, it is not known whether just one season of fertilization with a small quantity of organic materials or application of EM can change soil physical properties. From our results, however, it can be concluded that Ψ_{S-r} is attributed to the developed root system and promoted root physiological activity. In methodology, the analysis with the second equation of Ohm's law showed a good results for the experiment. However, the disadvantage was that many data used for calculations were obtained from disturbed measurements. Here, it was different from the estimation of Ψ_{S-r}.

Because of higher Ψ_{S-r} under soil water deficit conditions, plants fertilized with organic materials and applied with EM maintained higher P_C than plants fertilized with chemicals and those without EM application. It is logical that, compared to that of a small root system, a plant with a large root system and high root activity shows a higher water stress resistance ability.

Although the absolute values of root biomass were larger in chemical-fertilized than organic-fertilized plants, the root/top ratio was lower in the former than in the latter. The results in the present study also support observations that EM is more effective to crops under conditions of water deficit and under other stresses (Li and Ni, 1996).

It has been known that long-term fertilization with organic materials improves soil physical, chemical and biological properties (Hillel, 1980). Growth and activity of the root system are promoted by the improved soil properties and as a consequence the plants become more resistant to soil water deficits by their strong root system (Jones, 1983b). Up to now, there have been considerable research on organic farming (USDA, 1980; Lockeretz and Kohl, 1981; Harwood, 1984; Vogtmann, 1984). However, the exact mechanisms of the effects of EM are not precisely or conclusively known. For example, it is not known whether EM applications affect soil physical properties and how they are affected if so. It is also not clear whether the effects are due to microbes themselves or from the substances produced by microbes during the product manufacture or after it is applied. There has been extensive research on the individual microbes that comprise the EM used in the present study. Some phytohormones and their derivatives can be produced by soil microbes including some species existing in EM used in this study (Arshad and Frankenberger Jr, 1992). Barea et al. (1976) have found that among 50 bacterial isolates obtained from the rhizosphere of various plants, 86, 58, and 90% produce auxins, gibberellins, and kinetin-like substances, respectively. Another report has shown that 55% of bacteria and 86% of fungi isolated from the rhizosphere of *Pinus silvestris* are capable of producing gibberellins and their derivatives (Kampert et al. 1975). There have been many reports showing that *Actinomyces* and *Streptomyces* like those included in the EM used in the present study, produce auxins and similar substances (Purushothaman et al., 1974; Mahmoud et al., 1984), gibberellins (Arshad and Frankenberger Jr, 1992), and cytokinins (Bermudez de Castro et al., 1977; Henson and Wheeler, 1977). Some fungi, like those included in EM of this study (*Aspergillus niger*) produce gibberellins (El-Bahrawy, 1983). Therefore, the promotion of root development and activity by EM applications might be due to the production of plant growth regulators by inoculated microbes. However, there are not available data to support this hypothesis. Further studies are needed to examine the mechanistic basis for the effects of EM on plant growth, characteristics associated with root water uptake, and soil physical properties. Another fact is that EM has been used in agriculture, especially in organic farming systems, in advance of fundamental researches. Therefore, crop scientists and agronomists are encouraged to elucidate the problems that they may encounter using the EM technology.

REFERENCES

Arshad, M. and W. T. Frankenberger Jr. (1992). Microbial production of plant growth regulators. In *Soil Microbial Ecology*, ed. F. B. Metting Jr. New York: Marcel Dekker, Inc, pp. 307-348.

Barea, J. M., E. Navarro and E. Montoya. (1976). Production of plant growth regulators by rhizosphere phosphate-solubilizing bacteria. *Journal of Applied Bacteriology* 40: 129-134.

Bermudez de Castro, F., A. Canizo, A. Costa, C. Miguel and C. Rodriguez-Barrueco. (1977). Cytokinins and nodulation of the non-legumes *Alnus glutinosa* and *Myrica gale*. In *Recent Developments in Nitrogen Fixation*, eds. W. Newton, J. R. Postgate and C. Rodriguez. London: Academic Press, pp. 539-550.

El-Bahrawy, S. A. (1983). Associative effect of mixed cultures of *Azotobacter* and different rhizosphere fungi determined by gas chromatography-mass spectrometry. *New Phytologists* 94: 401-407.

Harwood, R. R. (1984). Organic farming research at the Rodale Research Center. In *Organic Farming: Current Technology and Its Role in a Sustainable Agriculture*. eds. D. F. Bezdicek, J. F. Power, D. R. Keeney, M. J. Wright. Madison: American Society of Agronomy, pp. 1-18.

Henson, I. E. and C. T. Wheeler. (1977). Hormones in plants bearing nitrogen-fixing root nodules: Cytokinins in roots and root nodules of some non-leguminous plants. *Journal of Plant Physiology* 84: 179-782.

Higa, T. (1994). *The Completest Data of EM Encyclopedia*. Tokyo: Sogo-Unicom, pp. 1-385 (in Japanese).

Higa, T. (1996). Effective microorganisms: their role in Kyusei Nature Farming and sustainable agriculture. In *Proceedings of the Third International Conference on Kyuusei Nature Farming*, eds. J. F. Parr, S. B. Hornick and M. E. Simpson. Washington DC: U.S. Department of Agriculture.

Hillel, D. (1980) *Fundamentals of Soil Physics*. New York: Academic Press, pp. 195-222.

Hsiao, T. C. (1973). Plant responses to water stress. *Annual Review Plant Physiology* 24: 519-570.

Huck, M. G. (1982). Water flux in the soil-root continuum. In *Roots, Nutrient and Water Flux, and Plant Growth*, eds. S. A. Baber, D. R. Bouldin, D. M. Kral and S. L. Hawkins. Madison: *American Society of Agronomy*, pp. 47-64.

Jackson, D. K. (1974). The course and magnitude of water stress in *Lolium perenne* and *Dactylis glomerata*. *Journal Applied Ecology* 15: 613-626.

Jones, H. G. (1978). Modeling diurnal trends of leaf water potential in transpiring wheat. *Journal Applied Ecology* 15: 613-626.

Jones, H. G. (1983a). Estimation of an effective soil water potential at the root surface of transpiring plants. *Plant, Cell and Environment* 6: 671-674.

Jones, H. G. (1983b). *Plant and Microclimate (A Quantitative Approach to Environmental Plant Physiology*. London: Cambridge Univ Press, pp. 1-323.

Kampert, M., E. Strzelczyk and A. Pokojska. (1975). Production of gibberellin-like substances by bacteria and fungi isolated from the roots of pine seedlings (*Pinus sylvesreis* L.). *Acta Microbiologica* 7: 157-166.

Lauchli, A. (1982) Mechanisms of nutrient fluxes at membranes of the root surface and their regulation in the whole plant. In *Roots, Nutrient and Water Flux, and Plant Growth,* eds. S. A. Baber, D. R. Bouldin, D. M. Kral and S. L. Hawkins. Madison: *American Society of Agronomy,* pp. 1-26.

Letey, Jr. J. (1977). Physical properties of soils. In *Soils for Management of Organic Wastes and Waste Waters*, ed. American Society of Agronomy. Madison: American Society of Agronomy, pp. 101-114.

Li, W.J. and Y. Z. Ni. (1996). *Researches and Applications of Microbial Technology.* Beijing: China Press of Agric Sci Tech, pp. 42-102.

Lockeretz, W. and D. H. Kohl. (1981). Organic farming in the corn belt. *Science* 211: 540-547.

Mahmoud, S. A. Z., E. M. Ramadan, F. M. Thabet and T. Khater. (1984). Production of plant growth promoting substances by rhizosphere microorganisms. *Journal of Microbiology* 139: 227-232.

Morita, S., S. Thongpae, J. Abe, T. Nakamoto and Yamazaki. (1992). Root branching in maize. I. "Branching index" and methods for measuring root length. *Japanese Journal of Crop Science* 61: 101-106.

Nobel, P. S., P. and W. Jordan. (1983). Transpiration stream of desert species: resistances and capacitances for a C_3, a C_4, and a CAM plant. *Journal of Experimental Botany* 34: 1379-1391.

Passioura, J. B. (1980). The transport of water from soil to shoot in wheat seedlings. *Journal of Experimental Botany* 31: 333-345.

Purushothaman, D., T. Marimuthu, C. V. Venkataramanan and R. Kesavan. (1974). Role of actinomycetes in the biosynthesis of indole acetic acid in soil. *Current Science* 43: 413-414.

Simmelsgaard, S. E. (1976). Adaptation to water stress in wheat. *Physiologia Plantarum* 37: 167-174.

Turner, N.C. (1988). Measurement of plant water status by the pressure chamber technique. *Irrigation Science* 9: 289-308.

U.S. Department of Agriculture. (1980) *Report and Recommendations on Organic Farming. A Special Report Prepared for the Secretary of Agriculture.* Washington DC: U.S. Government Printing Office.

Vogtmann, H. (1984). Organic farming practices and research in Europe. In *Organic Farming: Current Technology and Its Role in a Sustainable Agriculture*, eds. D. F. Bezdicek, J. F. Power, D. R. Keeny and M. J. Wright. Madison: American Society of Agronomy, pp. 19-36.

Biological Control of Common Bunt (*Tilletia tritici*)

Anders Borgen
Mehrnaz Davanlou

SUMMARY. Common bunt (*Tilletia tritici* syn. *T. caries*) is a significant seed-borne plant disease in organic agriculture. General measures in ecological crop protection like crop rotation and manuring have in practice failed to control this disease, and direct seed treatment may be necessary to ensure yield and food quality. The present study indicates that biological control can be successfully used without negative effects on seed germination and vigor. A combination of biocontrol agents and milk powder improves the efficacy of biocontrol. Application of compost to the soil increases the frequency of this disease. The potential role of biocontrol within the principles of ecological disease management is discussed. *[Article copies available for a fee from The Haworth Document Delivery Service: 1-800-342-9678. E-mail address: getinfo@haworthpressinc.com <Website: http://www.HaworthPress.com>]*

Anders Borgen is PhD Student, Department of Agricultural Sciences, Royal Veterinary and Agricultural University, Agrovej10, 2630 Taastrup, Denmark (E-mail: anders@ecoweb.dk).

Mehrnaz Davanlou is Green Guide, Kongens Enghave Bydel, Wagnersvej 19, 2450 Copenhagen SV, Denmark (E-mail: md.kgse@ipost.kk.dk).

Address correspondence to: Anders Borgen at the above address.

The authors thank sponsors and the companies who delivered the products knowing that the purpose was not to promote or depreciate any specific product or method, but only to screen the agents not previously tested with this disease. They specially thank Erik Nielsen (Sonnerup Eco-Advisors) and Berndt Gerhardson (SLU) also for information about EM and *Pseudomonas* and Birgit Jensen and Inge Knudsen (Department of Plant Biology, KVL) for help with application of *Gliocladium* and other supports. The authors also thank the staff of the Department of Agricultural Sciences, KVL for practical assistance in the experimental process.

[Haworth co-indexing entry note]: "Biological Control of Common Bunt (*Tilletia tritici*)." Borgen, Anders, and Mehrnaz Davanlou. Co-published simultaneously in *Journal of Crop Production* (Food Products Press, an imprint of The Haworth Press, Inc.) Vol. 3, No. 1 (#5), 2000, pp. 157-171; and: *Nature Farming and Microbial Applications* (ed: Hui-lian Xu, James F. Parr, and Hiroshi Umemura) Food Products Press, an imprint of The Haworth Press, Inc., 2000, pp. 157-171. Single or multiple copies of this article are available for a fee from The Haworth Document Delivery Service [1-800-342-9678, 9:00 a.m. - 5:00 p.m. (EST). E-mail address: getinfo@haworthpressinc.com].

KEYWORDS. Common bunt, compost, biocontrol, ecological agriculture, effective microorganisms, EM. seed treatment, stinking smut, *Tilletia caries*, *Tilletia tritici*

INTRODUCTION

Common bunt (*Tilletia tritici* syn. *T. caries*) is potentially one of the most devastating plant diseases. The fungus grows systemically in infected wheat plants (*Triticum aestivum*) and develops ovaries filled with fungal spores (bunt balls). During threshing, the bunt balls break and the spores attach to the healthy seeds during seed handling. When spore-contaminated seeds are sown, the spores germinate synchronously with the seeds and infect the germinating plants.

Simple nitrogen components like trimethylamine volatilize from the fungal spores (Ettel and Halbsguth, 1963), giving bunt-infested grain a smell like rotten fish. This is also the reason for the disease to be called stinking smut. Only a small number of infected heads in the field will reduce the quality of the wheat because of the stench of the bunt spores. Control of common bunt is therefore crucial for the production of quality wheat. In a Swedish experiment, 50% of the people could smell the presence of only 1000 spores per gram of seeds (Johnsson, 1991). This level can occur with a field frequency of less than 0.1% infected heads (Borgen et al., 1992). Ingestion of contaminated grain may also be hazardous to human health as it is for some animals (Westermann et al., 1988).

Common bunt is an ancient plant disease. Spores have been found on 4000 year old seeds from the ancient Mesopotamia (Johnsson, 1990), and it is likely that the disease has been a problem for wheat production ever since its domestication. Since then, the disease has been one of the most intensively treated in plant protection (Woolman and Humphrey, 1924; Buttress and Dennis, 1959; Sharvelle, 1979). Since cheap and effective seed treatments with organic mercury started in the 1920s, research in this disease has been limited. Mercury is now banned in most industrialized countries for environmental reasons, and modern synthetic pesticides have taken its place in the control of bunt.

The multiplication of bunt frequency from year to year, will, in untreated seeds, depends on the wheat variety and meteorological conditions especially during germination, but is often about 100-fold under Danish weather conditions in susceptible varieties (Borgen et al., 1992). In order to prevent multiplication of the bunt infections from year to year, the sum of all involved control measures must therefore have an efficiency of more than 99%. In conventional agriculture, this control level is exclusively reached by seed treatments, and at least in Denmark almost all seed lots of winter wheat are seed treated with pesticides mainly because of this disease (Nielsen et al., 1998). In organic agriculture, seed treatments with synthetic pesticides are excluded. Common bunt is therefore potentially a very serious disease for wheat production in organic cropping systems (Piorr, 1991; Borgen et al., 1992).

Research on alternative control of common bunt is going on in Europe, mainly focussing on different seed treatments like uses of plant extracts, cereal flour, milk powder and other organic compounds, hot water and hot air treatments and antagonistic bacteria and fungi (Spiess and Dutschke, 1991; Becker and Weltzien, 1993; Heyden, 1993, 1997; Borgen et al., 1995; Knudsen et al., 1995; Bergman, 1996; Borgen and Kristensen, 1996; Gerhadson, 1997). In the tradition of research in organic agriculture, the use of compost and EM (Effective Microorganisms) have played a central role in plant protection, since these products simultaneously attempt to improve soil fertility and create a beneficial microbial soil flora, and thereby prevent development of some plant diseases (Tränkner, 1992, 1993; Sangakkara et al., 1999). Furthermore, one of the negative effects of seed treatments is often that seed vigor in terms of germination speed is reduced. Low doses of EM have been shown to improve germination speed (Sangakkara and Attanayake, 1993). Fast and even germination is essential for yield potential also in conventional agriculture, but in organic agriculture the influence is likely to be higher since it also influences the competition with weeds and some pests and pathogens which are not easily be regulated without pesticides.

Our current study investigated different biological methods for their potential to control common bunt in organic agriculture.

MATERIALS AND METHODS

Trials were conducted over three years to screen different biological material and products for their effect on bunt infection and for side effects on germination vigor.

In the first year (1994-1995) two field trials were conducted. Spore-contaminated seeds of winter wheat (*Triticum aestivum*, cultivar 'Kosack') were treated either with liquid manure in a dose of 30 ml/kg, or with 130 g milk powder per kg or with both. Seed treatments were done in a spinning-wheel seed-dresser (Hege no. 11). In the second experiment in year 1994-5, seeds were treated with 21.7 g/kg of the commercial product Mycostop containing *Streptomyces griseoviridis* stain K61 which contains 10^8 CFU/gram (Lahdenperä et al. 1991). Water (26 ml/kg) was used as carrier.

Field trials were conducted at Højbakkegård, an experimental farm of the Royal Veterinary and Agricultural University and located 18 km east of Copenhagen, Denmark (55°40′ N, 12°18′ E, 28 m above mean sea level).

The soil type is a moreanic sandy loam, and common bunt had not been recorded in the experimental area beforehand. Untreated seeds were sown in 8 replications, and seeds for each treatment were sown in four replications in 4 m × 5 m plots. Heads in the first and second row in each plot were diagnosed for bunt infection. An average of 2392 heads were diagnosed for each treatment.

In the second year (1996-7) a field trial was also conducted in which seeds of winter wheat (cultivar 'Pepital') were contaminated with spores of *T. tritici* at a dose of 5 g per kg resulting in a spore contamination of 1,975,000 spores per gram seed tested by the ISTA Haemocytometer Method (Kietreiber, 1984). Where nothing else is noted, seed samples of 100 g were treated after contamination in a spinning-wheel seed-dresser (Hege no. 11) with milk powder and the biological agents listed below.

A bacterial suspension containing 10^9 spores per ml of *Pseudomonas chororaphidis* strain MA 342, was applied in 300 ml water per kg seed for two hours in a closed plastic bag and afterwards dried with cold air (Gerhadson, 1997). We compared this treatment with the application of lower doses of 150 ml/kg and 40 ml/kg of the same suspension applied in the seed dresser, the last dose both with and without 20 g milk powder per kg seed.

EM (Effective Microorganisms) is a commercial product containing 80 species of beneficial microorganisms, the major part being yeast and lactic acid bacteria. EM was applied in a dose of 40 and 150 ml/kg seed, and the lower dose was applied with and without 20 g milk powder per kg seed.

To see the effect of metabolites and growth media in the suspensions the doses of 150 ml/kg of both water, EM and the *Pseudomonas* suspension were also applied after the suspensions had been autoclaved for 20 minutes at 120°C at 0.2 MPa.

'Symbioplex' is a commercial product which according to the list of ingredients contains 5 × 10^8 spores per gram of each of the lactic acid bacteria *Lactobacillus acidophilus, Bifidobacterium bifidus* and *Streptococcus thermophillus.* This product was applied with and without 20 g milk powder per kg seed in a dose of 5 g + 20 ml water per kg seed (total 7.5 × 10^9 spores/kg).

One strain of *Trichoderma harzianum* in the commercial product Supresivit, containing 10^9 bacterial spores per g, was tested at a dose of 2 g/kg. The bacterial powder was added with 30 ml water/kg.

The isolate of *Gliocladium roseum* IK726 was applied in a lower dose of 2.5 × 10^9 CFU per kg seed, and a higher dose at 4.5 × 10^9 CFU per kg. Three ml of sterile water was in both cases used as carrier. The lower dose was applied with and without milk powder.

Two samples of compost were used. One type was composted garden waste which had been stored for more than one year; the other compost type was from a fresh compost heap of cattle manure and straw. The temperature in the cattle compost heap was about 60°C at the time of collection, which was three days before seed treatment. Compost extracts were made by mixing the compost with water to give a thick paste (50% D.M.) which was filtered through a 0.5-mm sieve. The concentrations and composition of active microorganisms in the compost extracts were not measured.

Rumen juice was taken from a cow through a rumen fistula and was applied at a dose of 40 and 150 ml/kg to the seeds. The lower dose was applied with and without milk powder.

After treatment, seeds were dried in the open air at room temperature, and where more than 40 ml/kg liquid had been added, the treatment was divided into a series of treatments with each application of no more than 20 ml water. The seeds were then dried with the help of a cold air stream. After treatment, the seeds were stored at 5°C in paper bags. Samples removed for a field test which took place 2-6 days after seed treatment. Germination tests were conducted 1-3 months thereafter.

After sowing, 100 g of compost was put into each row of an untreated control, equivalent to 6 t/ha. The compost was placed by hand directly into the rows in close contact with the seeds in the rows.

Germination tests were conducted in plastic plates containing 1.5 kg sand with water (65 ml water/kg quartz sand). One hundred seeds were sown in each of three replicates at a depth of 1.5 cm and at a temperature of 10°C. Emerged plants were counted over 3 consecutive days.

Each treatment was sown in rows of 1.25 m with 10 replicates at Højbakkegård. In each treatment, an average of 1869 heads were diagnosed for bunt infection based on visible macro-symptoms after heading.

In 1998-1999 a container experiment was conducted using combinations of seed and soil treatments with EM. One ml EM was applied on 18 August 1998 to 30 cm × 40 cm boxes (= 83 liters/ha) containing 18 liters pre-fertilized peat soil (Pinstrup whole mixture no. 2). Application was repeated on 26 August. EM was diluted 1:500 with water before applications; another treatment received the same volume of water but no EM. On 21 September, seeds of the spring wheat cultivar 'Cadenza' contaminated with 951,000 spores per gram of seeds, were treated with EM, and two days later 16 seeds were sown in each of 20 boxes for each treatment. Seed treatment included concentrated fresh EM at a dose of 20 ml/kg and a water-diluted solution of 1:200 at the same dose. The boxes were placed in open air from 18 August when the first soil application was done until the end of November, when the boxes were brought into a greenhouse. Plants were then grown to the heading stage and were then diagnosed for bunt infection. The same germination test was conducted as in the field trial.

RESULTS AND DISCUSSION

Results from the field experiments are presented in Tables 1 and 2. In addition to the presented results in Table 2 the recommended dose of *Pseudomonas* MA 342 of 300 ml/kg gave a reduction in bunt infection of 96.2% and the autoclaved treatment 150 ml/kg a reduction of 58.1%. The autoclaved treatment of EM 150 ml/kg resulted in a reduction of 76.5%. All these effects were statistically significant ($p < 0.001$) when tested against untreated controls by a Generalized Linear Model (GENMOD-procedure in SAS ver. 6.12.), as was a decrease in germination vigour for the seed treatments when tested by the Generalized Linear Mixed Model (GLIMMIX procedure in the software SAS ver. 6.12).

Results from the container experiments are presented in Table 3. Seed treatments resulted in a minor, but non-significant decrease in germination speed.

The principles of organic agriculture are driven by an attempt to promote beneficial life forms, rather than directly to kill damaging ones. Killing organisms always creates a biological vacuum, which may be the basis for migration of other possibly damaging organisms. The killing of pathogens also often includes killing of beneficial organisms which may help in protecting the plants from pathogens (Neergaard, 1977). Hence, the fundamental principles of ecological disease control is to optimize growth conditions for the plants rather than to kill the pathogens. This principle includes adequate crop rotation, resistant varieties, mixed cropping, balanced plant nutrition and the aerobic composting of manure (IFOAM, 1998). However, these principles only have a limited potential in the control of common bunt.

Being a seed-borne disease, crop rotation only has a very limited effect

TABLE 1. Results from two field trials with treatments of biologically based products in control of common bunt (*Tilletia tritici*). Treatments statistically tested against untreated control. Bunt frequency in the control was 53.2% for the first five treatments, and 35.2% for the last treatment.

Treatment	Dose	Reduction of bunted heads (%)
Control	0	0
Milk powder	130 g/kg	95.8 ***
Milk powder + Liquid manure	130 g/kg + 30 ml/kg	94.5 ***
H_2O	30 ml/kg	13.1 *
Diluted liquid manure	15 + 15 ml/kg	13.4 *
Liquid manure	30 ml/kg	÷2.3 (n.s.)
Streptomyces griseoviridis + H_2O	2.2×10^9 CFU/kg + 30 ml/kg	45.7 ***

*, ** and *** mean significance at P = 0.05, P = 0.001 and P = 0.001, respectively

TABLE 2. Control of common bunt with biological agents in combination with a low dose of milk powder. The doses (low and high) are not based on previous trials or recommendations for this disease, and the effectiveness and potentials of the products are therefore not comparable. Treatments statistically tested against untreated control. Bunt frequency in the control was 27.4%. Regression lines for logarithmically-transformed data for the germination curve is tested against untreated control by a Generalized Linear Mixed Model.

		Low dose (1)	higher dose (2)	Low dose(*) +2% milkpowder
Control + H_2O	Reduction of bunted heads	0%	9.6% n.s.	~42% (3)
	Mean Germination Time, days	9.08	8.81 n.s.	
Uncontaminated seeds	Reduction of bunted heads	97.4% ***		
	Mean Germination Time, days	8.85 n.s.		
Compost in field	Reduction of bunted heads	−61.1% ***		
Pseudomonas MA342	Reduction of bunted heads	63.7% ***	87.6% ***	97.2% ***
	Mean Germination Time, days	9.19 n.s.	9.68 **	9.16 n.s.
Thricoderma harzianum	Reduction of bunted heads	0.4% n.s.		
	Mean Germination Time, days	8.70 n.s.		
Gliocladium roseum	Reduction of bunted heads	14.7% n.s.	45.9% **	86.6% ***
	Mean Germination Time, days	8.50*	8.68 n.s.	8.77 n.s.
Lactic acid bacteria	Reduction of bunted heads	3.0% n.s.		59.3% ***
	Mean Germination Time, days	9.05 n.s.		9.02 n.s.
EM	Reduction of bunted heads	0.9% n.s.	87.6% ****	72.8% ***
	Mean Germination Time, days	9.11 n.s.	9.83 ***	9.35 n.s.
Garden compost	Reduction of bunted heads	8.6% n.s.	10.3% n.s.	64.7% ***
	Mean Germination Time, days	8.73 n.s.	8.93 n.s.	9.03 n.s.
Thermo compost	Reduction of bunted heads	0.1% n.s.	40.8% ***	45.3% ***
	Mean Germination Time, days	9.04 n.s.	9.09 n.s.	9.13 n.s.
Rumen juice	Reduction of bunted heads	3.9% n.s.	76.3% ***	47.0% ***
	Mean Germination Time, days	9.01 n.s.	9.13 n.s.	9.16 n.s.

(1) low dose is for *Gliocladium roseum* = 2.5 × 10^9 CFU/kg, for Symbioplex = 7.5 × 10^9 spores/kg, for *Thricoderma harzianum* 2 × 10^9 spores/kg and for *Pseudomonas* MA342 40 × 10^9 spores/kg. For the rest of the treatments the microbial content was not estimated and they were dosed by 40 ml/kg seeds except for the control, where nothing was applied.
(2) High dose is for *Gliocladium roseum* = 4.5 × 10^9, CFU/kg, for *Pseudomonas* MA342 150 × 10^9 spores/kg. For the rest of the treatments 150 ml/kg seeds.
(3) Result for milk powder and water are extrapolated from a logistic dose-response curve. 1% milk powder application reduced the bunt infection by 12.2% (n.s.) and 3% applied reduced attack by 74.1% (***) (Borgen and Kristensen, unpublished).
*, ** and *** mean significance at P = 0.05, P = 0.001 and P = 0.001, respectively

against common bunt. Spores can survive in the soil for an extended period of time and infect wheat later in a rotation (Johnsson, 1990; Borgen and Kristensen, 1997). Crop rotation is therefore important to prevent uninfected crops from being infected, but it has no influence in preventing the multiplication of the disease from year to year in an already infected seed lot. Being a systemic disease, plant nutrition has a very limited effect on the disease once

TABLE 3. Semi field trial with control of common bunt with EM as seed treatment and as soil amendment. Bunt infection in the untreated control was 67.7%. Seed treatment caused minor statistical insignificant reductions in germination speed. Treatments statistically tested against untreated control.

	Reduction of bunted heads (%)
Control	0
1 × soil treatment (83 l/ha)	5.0 n.s.
2 × soil treatment (2 × 83 l/ha)	− 10.2 n.s. (P = 0.08)
2 soil treatment +	
Seed treatment (2% diluted EM 1:200)	8.6 n.s.
Seed treatment (2% diluted EM 1:200)	− 1.4 n.s.
2 × soil treatment +	
Seed treatment (2% concentrated EM)	42.4 ***
Seed treatment (2% concentrated EM)	49.8 ***

*, ** and *** mean significance at P = 0.05, P = 0.001 and P = 0.001, respectively

the plant is infected. Breeding for resistant varieties has a potential, but for the time being, the variety, 'Stava', is the only variety available in Denmark known to be fully resistant, and this variety is not optimal for Danish meteorological conditions. In many regions of the world the situation is the same. Wheat grown for baking purpose also sets strong limitations for the choice of varieties and the possibilities for crop mixing.

The use of biological control of pathogens in organic agriculture implies a dilemma. On the one hand it is a pesticide-free control measure that promotes beneficial life-forms rather than actively kills pathogens. On the other hand there may be problems connected to the use of some biological control measures parallel to problems related with pesticides. In organic agriculture, e.g., plants juices can be used in plant protection, but single chemical compounds isolated from plants or copies hereof are unwanted because they are not used in the concentration and in a chemical and organic environment where they naturally occur. In the same way the use of a single or a very limited number of microorganisms may disturb the existing balance in the soil flora, and the use of non-indigenous species not already present in the local soil is certainly questionable in organic agriculture. If a biological agent is cheap, easy and effective against a specific problem, the use of this agent has the same problem as synthetic pesticides; it removes the problem, but does not resolve the cause of the problem. If problems are treated rather than prevented, they are likely to show up in another form. If, for example, a problem with a disease related to mono-cropping is removed by a treatment, other problems related to the mono-cropping system will remain. In the case of common bunt, the mono-cropping system has been still more intensive with narrower crop rotations and less focus on resistance in the period where pesticide seed treatment has been used. Now, where the sustainability of the pesticides are in question, we are 70 years behind in development of sustainable control methods against this disease. With a few exceptions, the use of specific microorganisms in biocontrol of pathogens is therefore excluded in the standards of the Danish Organic Farmers Association (LØJ, 1999) and many other organic standards, although the international minimum standards which apply in Denmark are open for the possibility (IFOAM, 1998; EU, 1991). The claim for a specific biological control agent to be included in the organic standards is being discussed among organic farmers in Denmark and in IFOAM. Beside the existing risk assessment from the authorities the discussion centres around the following principles

- the problem to be treated is significant for the production,
- the use of biocontrol is necessary in control, meaning that the problem can not be prevented with known methods or be substituted by less critical methods,
- the microorganisms are naturally-occurring in the local soil already in the actual form. Any use of genetically manipulated organisms are excluded,
- the microorganisms are non-toxic for higher organisms including the farmer, and
- the use will not irreversibly change the balance of the soil microflora.

Even seed treatment is not an optimal way to control plant diseases according to the principles of organic agriculture, it may be necessary until more resistant varieties have been bred and marketed. Our current study shows that seed treatments with biological agents may have a potential against this pathogen in organic agriculture.

The frequency of infections by common bunt are affected by some of the treatments presented in this study, others not. The most promising results are obtained by a combination of milk powder and some of the specific microorganisms, of which *Pseudomonas* strain MA342 gave almost full control without significant reduction of germination vigour. *Pseudomonas* MA 342 alone had a significant effect on bunt even at the low dose of 40×10^9 spores/kg, and the effect was improved considerably in combination with milk powder and came out with a result better than the recommended high dose of 300 ml/kg (300×10^9 spores/kg) and without significant negative effect on germination vigor.

Pseudomonas MA 342 has been shown in previous studies to control bunt fully at the full dose (300 ml/kg) and with 92.3-98.0% in half dose (Gerhadson, 1997). Considering the small amount of bunt infection in the uncontaminated control plots in 1997 which indicates a minor background infection from soil or machinery, the result confirms previous studies. For commercial purposes the formulation used in this study is now exchanged in favor of an oily suspension using only about 6 ml/kg. This product is sold in many European countries as "Cedomon." However, this product is not approved for winter cereals today (1999), since problems with germination vigor are observed with this formulation. This study indicates that a combination of *Pseudomonas* MA 342 with milk powder may lead to a solution of the problems with germination vigor in the product. *Pseudomonas* MA 342 was isolated from cereal roots in a Swedish soil and has been found in many other European soils as well (Gerhadson, personal communication).

Gliocladium roseum isolate IK726 reduced the bunt frequency in the higher dose of 4.5×10^9 CFU per kg, but did not significantly do so at the lower dose. In neither doses were negative side effects on germination observed. This indicates that the dose could be increased further and thereby improve the product efficacy. This experiment can therefore not exclude the potential of this isolate of *Gliocladium roseum* as a control agent against common bunt, and further experiments are planned to evaluate this. The current strian of *Gliocladium roseum* IK726 was isolated from barley roots in a Danish field soil (Knudsen et al., 1995).

Streptomyces griseoviridis strain K61 in the product Mycostop reduced the bunt frequency at the dose of 2.2×10^9 CFU/kg tested, but this was inadequate to give full control. There might therefore be a potential for this product, but further studies with higher doses or combinations with other treatments are needed to determine the full potential.

Trichoderma harzianum in the product Supresivit did not show any effect on bunt frequency or germination vigour. The dose of 2×10^9 spores/kg may be too low for the purpose. Further studies with higher doses or combinations with other treatments are needed to determine whether this product has a potential as seed treatment in control of common bunt.

EM1 reduced the frequency of common bunt, but only in concentrations where germination vigour was also reduced. The effect of the product can therefore not be improved by increasing the dose. The product has a low pH, about 3.5 and contains various other metabolites from the microbial activity. The treatment with the high dose of fresh EM only gave an insignificant improvement in effect as compared with autoclaved EM. Lactic acid bacteria are the major group of organisms in EM. The results with Symbioplex containing only lactic acid bacteria shows that this group of bacteria has no effect on bunt infection at least in the concentration of 7.5×10^9 spores/kg used here. This indicates that the major effect of EM as seed treatment in this study is not biological, but probably chemical.

The container experiment with EM basically confirms the results from the field experiment that common bunt cannot be fully controlled by EM used as a seed treatment, even it has a reducing effect when used in concentrated form. Seed treatment with diluted EM seemed to have no influence on bunt frequency or on germination speed. In this study it increased mean germination time, but this effect was not statistically significant. In previous studies on the effect of diluted EM on germination, EM has been applied to the water used in the germination test. In this study it has been used as an application to the seed, and fresh water is applied in the germination test. The dose applied in this study is therefore not comparable with previous studies on the effect of EM on germination of seeds which may explain the contradictory results reported earlier (Sinqueira et al., 1993; Sangakkara and Attanayake, 1993). The use of EM as a soil inoculant seems to have no influence on bunt infection of the seedlings. The microorganisms in the EM product are formulated in Japan, and information on the occurrence of the same strains of organisms in other regions is limited, which makes the use questionable in organic agriculture elsewhere. In Denmark the Demeter Association has decided to ban the use in biodynamic agriculture, while the Organic Farmers Association are still discussing the issue.

Milk powder can reduce infection of common bunt, but full control is often related to problems with field emergence (Winter et al., 1997; Borgen et al., 1995; Becker and Weltzien, 1993; Tränkner, 1993). Tränkner (1993) concluded that even the effectiveness of the milk powder in control of the disease could be improved by mixing with compost extracts as shown by Becker and Weltzien (1993), the effect is limited and could be better obtained by an increase of the amount of milk powder. This conclusion is generally confirmed by this study. Compost and rumen juice used as seed treatment had a very limited effect on bunt infection. Microorganisms growing in the fresh compost at 60°C must be thermofillic. Soil temperature under Danish conditions at the start of October is about 5-10°C, and the bacteria in the fresh compost may therefore not be well-adapted to this environment. The same can be expected for the rumen juice, where the microorganisms are mainly obligate and facultatively anaerobic and selected for growth at 39°C in the cow rumen (Van Soest, 1982). The well-composted garden waste had no effect when used alone, but had a tendency to improve the effect of the milk powder. A higher concentration of organisms in this treatment may have improved the effect.

Liquid manure has been used for centuries in Europe as a seed treatment against common bunt (Buttress and Dennis, 1959). Also modern research has included treatments with liquid manure (Borgen et al., 1995; Heyden, 1993), but in these trials seeds were washed in liquid manure. The effect of the treatments were therefore a combination of the washing and the biological/chemical effect of the liquid manure. In the current experiment the seeds were not been dipped in the liquid, but only added as a surface amendment.

This treatment seems to have no effect on bunt infection which indicates that the effect shown in previous experiments is mainly a washing effect.

The application of compost to the soil significantly increased the infection of common bunt. This confirms previous field trials by Rabien (1928), even if it could not be confirmed in greenhouse experiments by Voss (1938). Rabien explained the effect of compost by the fact that oxygen in the seed-surface environment is a liming factor for the infection of bunt, and compost increases the air volume in the soil compared with normal mineral field soil. Even application of compost to the soil has many beneficial effects on plant production, including the increasing effect of air volume, it may therefore have a negative side effect on bunt infection in cases where seeds or soil are contaminated with bunt spores. In the current experiment the compost was deliberately put into the row close to the seeds, while in normal farming practice the physical distance between seeds and compost is likely to be greater. On the other hand, 6 t compost per hectare as used in this experiment is relatively low compared with the manuring practice for cereals on many organic farms. Whether the normal use of compost will have practical implications for bunt infection can not be decided by this experiment.

Among the well characterized products in this screening test, *Pseudomonas* MA 342 is the only product previously tested against common bunt. The high effect of this product compared with the others must be seen in relation to the 10-20 fold higher number of organisms applied to the seeds in the treatments. The doses chosen to test the effect of the other treatments are therefore not qualified, but must be characterized as preliminary screenings. "No effect" for a product does not necessarily mean "no potential effect," but may indicate that the dose tested was too low for the control of this pathogen. Evaluation of the products at this stage must therefore be viewed in combination with the side-effects on seed germination and vigor.

CONCLUSION

In organic agriculture infection by plant pathogens should if possible be prevented by general means to improve growth conditions for the plants. In the case of common bunt (*Tilletia tritici*) these general means are at the present stage inadequate to control the disease, and other control measures are therefore necessary. The current study shows that the selected general substances with high concentration of microorganisms like liquid manure, compost extracts, rumen juice and EM (Effective Microorganisms) have a limited potential in the control of common bunt. The application of compost to the soil even increases the infection of common bunt. Some selected microorganisms used for biological control of other seed pathogens seem to have a potential also in the control of common bunt. *Pseudomonas* MA 342 can give full control of common bunt in combination with milk powder with no negative effect on germination vigor of the seeds, while further studies on *Trichoderma harzianum, Streptomyces griseoviridis* and *Gliocladium roseum* are needed to evaluate the potential of these fungi. Combinations of active biological agents with milk powder indicates that this combination can improve the potential of the biological treatments, i.e., in cases where physical limitations hinders the application of the dose needed to obtain adequate control of the pathogen.

REFERENCES

Becker, J. and H.C. Weltzien. (1993). Bekämpfung des Weizensteinbrandes (*Tilletia caries* (D.C.) Tul. & C. Tul.) mit organischen Nährstoffen. *Zeitschrift für Pflanzenkrankheiten und Pflanzenschutz*. 100(1):49-57.

Bergman, S. (1996). Värmebehandling mot utsädesburne svampsjukdomar. *Forskningsnytt om øekologisk landbruk i Norden* 2:6-7.

Borgen, A., L. Kristensen, and P. Kølster. (1995). Direkte bekæmpelse af hvedens stinkbrand uden brug af pesticider. *Control of common bunt (Tilletia caries) without use of pesticides. 12th Danish Plant Protection Conference. SP-report* 5:149-158.

Borgen, A. and L. Kristensen. (1996). Stinkbrand må reguleres. *Forskningsnytt om øekologisk landbruk i Norden* 2:9-11.

Borgen, A. and L. Kristensen. (1997). Markforsøg med flerårrig overlevelse af stinkbrand (*Tilletia tritici*) i jord. *Perennial survival of common bunt (Tilletia tritici) in soil. 14th Danish Plant Protection Conference.* SP-rapport 8:113-120.

Borgen, A., C. Markussen and L. Kristensen. (1992). *Frøbårne svampesygdomme på byg og hvede i økologisk landbrug-udbredelse og betydning i 1990.* Msc. thesis at Dept. of Plant Biology and Agricultural sciences. Copenhagen 137 p.

Buttress, F.A. and R.W. Dennis. (1959). The early history of cereal seed treatment in England. *Agricultural History* 21:93-103.

Denish Organic Association (LØJ). (1999). *Avlsregler for økologisk Jordbrug.* Landsforeningen Økologisk Jordbrug, Denmark.

Ettel, G.E. and W. Halbsguth. (1963). Über die Wirkung von Trimethylamin, Calciumnitrat und licht bei der Keimung der Brandsporen von *Tilletia tritici* (Bjerk.) Winter. *Beiträge zur Biologi der Pflanzen.* 39:451-488.

European Union. (1991). *Regulation nr. 2092/91 on organic agricultural methods and reference hereto on agricultural products and foodstuff.*

Gerhadson, B. (1997). Biologisk bekämpning af svamsjukdomar-aktuella möjligheter och långsiktiga perspektiv. *Proceedings from the 14th Danish Plant Protection Conference, SP-rapport* 8:17-26.

Heyden, B. (1993). Zum Pilzproblem. *Mitteilungen aud der Arbeit J. und C. Graf Keyserlingk-Institut.* 9:5-32.

Heyden, B. (1997). Unsere Versuche in der Vegationsperiode 96/97. *Mitteilungen aud der Arbeit J. und C.Graf Keyserlingk-Institut.* 12:12-25.

IFOAM. (1998). *Basic Standards for Organic Production and Processing.* ISBN 3-934055-00-1. p 62.

Johnsson, L. (1990). Brandkorn i Bibeln, stinksot i vetet och *Tilletia* i litteraturen-en kortfattad historik från svensk horisont. *Växtskyddsnotiser* 54:76-80.

Johnsson, L. (1990). Survival of common bunt (*Tilletia caries* (DC)Tul.) in soil and manure. *Zeitschrift für Pflanzenkrankheiten und Pflanzenschutz*. 97:502-507.

Johnsson, L. (1991). Dvärgstinksot (*Tilletia controversa*) og vanligt stinksot (*Tilletia caries*) i svenskt vete. *Växtskyddsrapporter. Jordbruk* 6:1-19.

Kietreiber, M. (1984). ISTA handbook of seed health testing. *Working sheet no. 53*. Bundesanstalt für pflanzenbau, Wien, Austria.

Knudsen, I.M.B., J. Hockenhull and D.F. Jensen. (1995). Biocontrol of seedborne diseases of barley and wheat caused by *Fusarium culmorum* and *Bipolaris sorokiniana*: Effects of selected fungal antagonists on growth and yield components. *Plant Pathology* 44:467-77.

Lahdenperä, M.L., E. Simon and J. Uoti. (1991). Mycostop a novel biofungicide based on Streptomyces bacteria. In *Developments in Agricultural and Managed-Forest Ecology, Vol. 23. Biotic Interactions and Soil-Borne Diseases*; *First Conference of the European Foundation for Plant Pathology*, ed. A.B.R. Beemster. Wageningen, Netherlands, February 26-March 2, 1990. 428 pp. Elsevier Science pp. 258-263.

Neergaard, P. (1977). *Seed Pathology*, vol I and II. The MacMillian Press Ltd. London Baingtoke 1187 pp.

Nielsen, B.J., A. Borgen, G.C. Nielsen and C. Scheel. (1998). Strategies for controlling seed-borne diseases in cereals and possibilities for reducing fungicide seed treatments. *The Brighton Conference-Pest and Diseases*, pp. 893-900.

Piorr, H.P. (1991). Bedeutung und Kontrolle saatgutübertragbarer Schaderreger an Winterweizen im Organischen Landbau. *Dissertation, Bonn University*. pp. 166.

Rabien, H.. (1928). Ueber Keimungs- und Infektionsbedingungen von *Tilletia tritici*. *Arbeiten aus der Biologischen Reichanstalt für Land- und Forstwirtschaft* 5:297-353.

Sangakkara, U.R. and A.M.U. Attanayake. (1993). Effect of EM on germination and seedling growth of rice. In: *Proceedings of the Third International Conference on Nature Farming*, eds. J.F. Parr, S.B. Hornick and M.E. Simpson. Washington DC: US Department of Agriculture, pp. 223-227.

Sangakkara, U.R., T. Higa and P. Weerasekera. (1999). Effective microorganisms: A modern technology for organic systems. In: *Organic Agriculture. The Credible Solution for the XXIst Century*. Eds. D. Foguelman and W. Lockeretz. pp. 205-211.

Sharvelle, E.G. (1979). *Plant Disease Control*. The AVI Publishing Inc., Westport, Connecticut. ISBN 0-87055-335-6, 331 p.

Sinqueira, M.F.B, C.P. Sudré, L.H. Almeida, A.P.R. Pegorer and F. Akiba. (1993). Influence of effective microorganisms on seed germination and plantlet vigor of selected crops. In: *Proceedings of the Third International Conference on Nature Farming*, eds. J.F. Parr, S.B. Hornick and M.E. Simpson. Washington, DC: US Department of Agriculture, pp. 222-45.

Spiess, H. and J. Dutschke. (1991). Bekämpfung des Weizensteinbrandes (*Tilletia caries*) im biologisch-dynamischen Landbau unter experimentellen und praktischen Bedingungen. *Gesunde Pflanzen*, 43:264-270.

Tränkner, A. (1992). Biologische Weizensteinbrandbekämpfung-mehrjährige prakstsche

Erfahrungen mit der Milchpulverbehandlung. *Mitteilungen aus der Biologischen Bundesanstalt für Land- und Forstwirtschaft Berlin-Dahlem*. 321:417

Tränkner, A. (1993). *Kompost und Pflanzengesundheit: Möglichkeiten und Auswirkungen der biologischen Beeinflussung pflanzlicher Oberflächen zur Krankheitsbekämpfung*. PhD. Bonn Univ. Verlag Dr. Kova, Hamburg.

Van Soest, P.J. (1982). *Nutritional Ecology of the Ruminant*. O & B Books Oregon, USA, 374 p.

Voss, J. (1938). Zur Methodik der Prüfung von Weizensorten auf ihre Wiederstandsfähigkeit gegen Steinbrand (*Tilletia tritici*). *Pflanzenbau* 14:113-153.

Westermann, H.-D., H. Barnikol, E. Fiedler, H. Rang, and A. Thalmann. (1988). Gesundtheitsliche Risiken bei Verfüttering von Brandweizen (Weizensteinbrand und Zwergbrand). *Landwirtschaftliche Forschung* 41:169-176.

Winter, W., C. Rogger, I. Bänziger, H. Krebs, A. Rüegger, P. Frei, D. Gindrat and L. Tamm. (1997). Weizenstinkbrand: Bekämpfung mit Magermilchpulver. *Agrarforschung* 4:153-156.

Woolman, H.M. and H.B. Humphrey. (1924). Studies in the physiology and control of bunt, or stinking smut, of wheat. *United States Department of Agriculture, Department Bulletin* 1239:1-29.

Effects of Organic Fertilizers and a Microbial Inoculant on Leaf Photosynthesis and Fruit Yield and Quality of Tomato Plants

Hui-lian Xu
Ran Wang
Md. Amin U. Mridha

SUMMARY. An experiment was conducted to examine the effects of applications of an organic fertilizer (bokashi), and chicken manure as well as inoculation of a microbial inoculant (commercial name, EM) to bokashi and chicken manure on photosynthesis and fruit yield and quality of tomato plants. EM inoculation to both bokashi and chicken manure increased photosynthesis, fruit yield of tomato plants. Concentrations of sugars and organic acids were higher in fruit of plants fertilized with bokashi than in fruit of other treatments. Vitamin C concentration was higher in fruit from chicken manure and bokashi plots than in those from chemical fertilizer plots. EM inoculation increased vitamin C concentration in fruit from all fertilization treatments. It is concluded that both fruit quality and yield could be signifi-

---

Hui-lian Xu is Senior Crop Scientist, International Nature Farming Research Center, 5632 Hata, Nagano 390-1401, Japan. Ran Wang is Professor, Laiyang Agricultural University, Laiyang, Shandong, China. Md. Amin U. Mridha is Professor, Department of Botany, Chittagong University, Chittagong, Bangladesh.

Address correspondence to: Hui-lian Xu at the above address (E-mail: huilian@janis.or.jp).

[Haworth co-indexing entry note]: "Effects of Organic Fertilizers and a Microbial Inoculant on Leaf Photosynthesis and Fruit Yield and Quality of Tomato Plants." Xu, Hui-lian, Ran Wang, and Md. Amin U. Mridha. Co-published simultaneously in *Journal of Crop Production* (Food Products Press, an imprint of The Haworth Press, Inc.) Vol. 3, No. 1 (#5), 2000, pp. 173-182; and: *Nature Farming and Microbial Applications* (ed: Hui-lian Xu, James F. Parr, and Hiroshi Umemura) Food Products Press, an imprint of The Haworth Press, Inc., 2000, pp. 173-182. Single or multiple copies of this article are available for a fee from The Haworth Document Delivery Service [1-800-342-9678, 9:00 a.m. - 5:00 p.m. (EST). E-mail address: getinfo@haworthpressinc.com].

cantly increased by EM inoculation to the organic fertilizers and application directly to the soil. *[Article copies available for a fee from The Haworth Document Delivery Service: 1-800-342-9678. E-mail address: getinfo@haworthpressinc.com <Website: http://www.HaworthPress.com>]*

KEYWORDS. Effective microorganisms, EM, nature farming, organic farming, sugar, organic acid, tomato

INTRODUCTION

Excessive use of chemical fertilizers has caused many problems in environmental pollution and soil degradation. With these concerns, many farmers in Japan have adopted nature farming practices. The concept and principles of nature farming were proposed by Mokichi Okada more than 60 years ago (Okada, 1993). Because chemical fertilizers and untreated animal products are not allowed in nature farming systems for crop and vegetable production, it is not easy to achieve yields equal to or higher than those obtained with chemical fertilizers. First, the growers must seek an alternative nutrient source. An organic fertilizer often used by farmers is called bokashi, which is a fermented mixture of oilseed cake, rice bran and fish-processing by-product (Yamada et al., 1996). A microbial culture called Effective Microorganisms or EM is often inoculated into bokashi before fermentation (Higa, 1994). This kind of organic fertilizer with EM inoculated is called EM bokashi. EM bokashi has been to be a useful nutrient source, but the other aspects of EM bokashi need to be elucidated. Therefore, a research project was initiated to examine the performance of EM bokashi in vegetable production. The first experiment was conducted to examine the effects of EM inoculation to bokashi and chicken manure on fruit yield and quality of tomato plants.

MATERIALS AND METHODS

Materials and Treatments

Tomato (*L. esculentum* L. cv. Momotaro T 96) seedlings with 5 leaves were transplanted into plastic pots each with a surface area of 0.02 m^2 and a height of 0.25 m. The pots were arranged randomly in a glasshouse. Six fertilization treatments each with 33 pots were as follows: (1) chicken manure; (2) chicken manure with EM (effective microorganisms, EM1) inoculated before fermentation; (3) anaerobic bokashi (anaerobically fermented organic materials such as rice bran, rapeseed mill cake and fish processing by-product); (4) anaerobic bokashi with EM inoculated before fermentation; (5) chemical fertilizer (ammonium sulfate 5.3 g, superphosphate 13 g and potassium sulfate 5 g per pot; and (6) the same amount of chemical fertilizer as in treatment (5) with 80 ml EM applied together. The amounts of N-P-K were adjusted to the same levels for all treatments.

Photosynthetic Measurement

Photosynthesis was measured using Li-6400 Portable Photosynthesis System (LI-COR Inc. Lincoln, Nebraska, USA) at 50 and 90 days after tomato plants were transplanted. The 5th leaf from the top was used for measurements for each sampled plant. The maximum gross photosynthetic capacity (P_C), the quantum yield ($Y_Q = KP_C$) and dark respiration rate (R_D) were analyzed from a light response curve modeled using an exponential equation, $P_N = P_C (1 - e^{-KI}) - R_D$, where K is a constant and I is the photosynthetic photon flux (Xu et al., 1995).

Preparation of Plant, Soil and Fruit Samples

The whole plant was sampled with leaves and stem separated on the 50 and 90 days after tomato plants were transplanted. The samples were dried in an oven at 105°C for 2 h and under 85°C over 24 h. Dry mass of a whole plant was recorded and the dry material was ground with a vibrating sample mill. A prepared sample of 5 g was used for measurements of mineral salts and other nutrients. The tomato fruits were picked once a week when tomato fruits began to ripen 2.5 months after being transplanted. The fruit yield, the crack rate and single fruit mass were calculated. A slice representing the whole fruit was used for fruit quality analysis. The duration involved in fruit development from pollination to maturity were divided into 4 stages as (1) 15 days after pollination with small size and green color; (2) fruit begin to turn color from green to white; (3) fruit in orange color; and (4) fruit in red color. The soil samples were taken at the same time as the plant samples.

Mineral Analyses of Soil and Plant Samples

Concentrations of K, Mg and Ca in plant and fruits were determined with an atomic absorption spectrophotometer (180-30, HITACHI, Japan); concentrations of total N and C were measured by MT-700 CN CORDER (Yanaco, Japan); and concentrations of nitrate-N and phosphorus were determined by colorimetry.

Analysis of Fruit Quality

Fresh fruit tissue was homogenized with distilled water in a ratio of 1:4. The homogenate was centrifuged at 8000 × g for 15 min at 4°C and the supernatant passed through 0.45-μm filter. Sugars were measured by HPLC (Jasco) with RI-930 Detector and a column of Shodex SC1011) at a column temperature of 80°C and a flow rate of 1 ml min^{-1}. Organic acids were measured by HPLC (Jasco) with UV-970 Detector and column of Shodex RSPark KC-811 at a column temperature of 40°C and a flow rate of 0.75 ml min^{-1}. Vitamin C was determined by a reflectometer (RQflex, Merck).

RESULTS AND DISCUSSION

Plant Growth and Fruit Yield

Organic Fertilization. At the early growth stage, plant growth or fruit yield was lower in the bokashi plots but turned higher at later growth stages, compared with the chemical-fertilized plants (Figure 1, Table 1). This might be due to the low nutrient availability at the beginning, which limited the plant growth. In the present study, the organic fertilizer is an anaerobically fermented mixture of organic materials. Nutrients, especially nitrogen, are not mineralized immediately after fermentation. The mineralization of the nutrients takes a period of time even when applied to the soil. That is why the plants fertilized with organic materials grew more slowly than those fertilized with chemical fertilizers at earlier stages. Therefore, the growers should take some measures to make the nutrients in organic materials available before plants begin their rapid growth. Nutrients in chemical fertilizers are immediately available when applied to the soil but the sustainability is low. As shown in Table 2, 50 days after the seedlings were transplanted, nitrate and available

FIGURE 1. Fruit yield at different stages of tomato plants with different fertilization treatments.

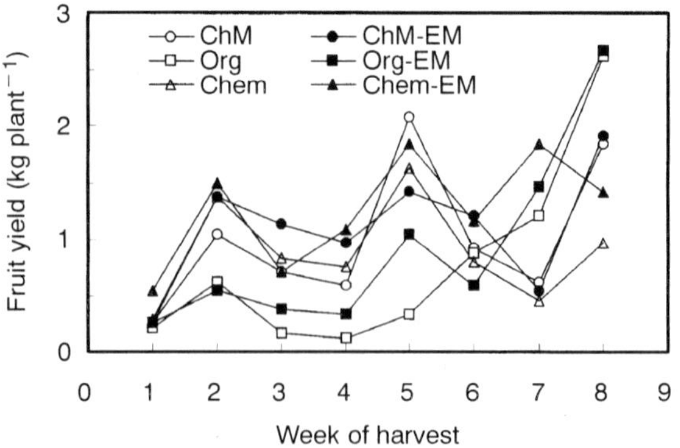

TABLE 1. Fruit yield and number, abnormal and green fruit and fruit size as well as photosynthetic capacity (P_C), respiration (R_D) and quantum yield (Y_Q) at later growth stage of tomato plants under different fertilizations.

Treatment	Fruit characteristics					Photosynthetic parameter		
	Yield (g plt^{-1})	Numb. (plt^{-1})	Abnormal (%)	Green (%)	Size (g)	P_C (μmol m^{-2} s^{-1})	R_D (μmol m^{-2} s^{-1})	Y_Q (mmol mol^{-1})
ChM	823 ±32	8.0 ±0.4	9.8 ±3.4	8.1 ±1.7	94.5 ±2.5	17.6 ±1.2	1.02 ±0.02	19.5 ±2.7
ChM + EM	935 ±63	8.8 ±0.6	9.5 ±2.0	5.9 ±1.3	100.4 ±3.0	19.3 ±1.7	1.31 ±0.02	24.3 ±2.9
Org	622 ±36	6.2 ±0.4	17.3 ±2.8	20.0 ±3.9	81.1 ±3.9	20.4 ±2.1	1.05 ±0.01	31.2 ±3.4
Org + EM	723 ±59	7.3 ±0.7	17.2 ±3.9	18.5 ±2.7	81.6 ±2.5	23.1 ±1.6	1.84 ±0.02	34.4 ±1.8
Che	818 ±21	7.1 ±0.2	12.4 ±3.5	11.4 ±2.9	102.4 ±2.5	18.4 ±1.3	1.12 ±0.03	30.7 ±2.5
Che + EM	1012 ±30	10.1 ±0.3	8.7 ±1.8	9.4 ±1.8	91.2 ±2.8	20.2 ±0.9	1.65 ±0.02	34.5 ±2.9

Data showing means ± SE (n = 9). ChM = chicken manure; Org = bokashi; Che = chemical fertilizer.

phosphorus concentrations were higher with bokashi and chicken manure treatments than for the chemical fertilizer treatment. The nutrients in the chemical treatment might be lost by leaching from the soil-root zone in irrigation water at the early growth stages. On the contrary, organic materials could sustain the nutrients for a longer time than chemical fertilizers. Moreover, organic materials also contain micronutrients in addition to the macronutrients that are available in chemical fertilizers. Some macronutrients such as calcium and magnesium were included in organic fertilizers but not chemical fertilizers and consequently were more in soils fertilized with bokashi and chicken manure than soils treated with chemical fertilizer (Table 2). Therefore, at the later growth stages, plants fertilized with organic materials grew better than those fertilized with chemical fertilizers. The chicken manure used in the present study was aerobically treated before application and no growth limitation was observed at the early stages. However, the nutrients could not be sustained at the later growth stage of plants with either chicken manure or chemical fertilizer treatment, compared with bokashi.

Effects of EM Inoculation. EM inoculation increased plant growth and fruit yield in all treatments. EM was inoculated to the organic materials or chicken manure before anaerobic fermentation. The microorganisms were reproduced and changed the properties of the organic materials. Some microorganisms produce plant growth regulators (Arshad and Frankenberger Jr.,

TABLE 2. The concentration (mg kg^{-1}) of mineral salts and C:N in soil in 50 days and 90 days after tomato seedlings were transplanted.

Treat.	T-C	T-N	C:N	NH_4^+	NO_3^-	Av.-P	K_2O	CaO	MgO
	50 days after transplanting								
ChM	58.5 ±1.36	4.5 ±0.01	12.9	8.2 ±0.41	174.2 ±30.36	517.9 ±7.67	939 ±111.5	4367 ±419.2	960 ±67.2
ChM + EM	58.0 ±1.36	4.6 ±0.07	12.6	8.2 ±1.22	175.7 ±12.22	573.5 ±83.73	880 ±35.3	4879 ±262	1020 ±37.2
Org	60.2 ±0.82	5.0 ±0.07	12.0	11.4 ±3.04	367.0 ±154.1	318.0 ±35.12	630 ±194.0	4030 ±41.9	1161 ±13.8
Org + EM	60.3 ±0.88	5.0 ±0.18	12.0	10.2 ±2.75	279.0 ±163.8	356.7 ±43.33	358 ±97.4	3720 ±149.1	1070 ±87.9
Che	54.9 ±0.96	4.2 ±0.03	13.1	32.6 ±10.75	152.3 ±17.60	209.8 ±28.48	327 ±10.8	5110 ±282.0	825 ±34.5
Che + EM	55.7 ±0.50	4.3 ±0.03	13.1	35.53 ±9.77	115.6 ±34.48	271.1 ±8.82	288 ±22.5	5112 ±319.0	761 ±64.5
	90 days after transplanting								
ChM	57.1 ±0.93	4.7 ±0.04	12.0	10.1 ±0.88	21.2 ±2.91	726.8 ±75.35	601 ±91.5	5481 ±44.4	1034 ±22.9
ChM + EM	58.1 ±0.59	4.7 ±0.03	12.3	10.5 ±0.70	20.89 ±1.62	674.8 ±63.60	511 ±74.6	5233 ±155.4	1010 ±29.6
Org	58.8 ±1.10	4.86 ±0.0	12.3	16.8 ±1.49	75.5 ±4.3	370.8 ±24.93	252 ±33.59	3356 ±201.2	953 ±33.5
Org + EM	58.6 ±0.53	4.8 ±0.06	12.2	23.6 ±0.82	51.8 ±2.12	365.8 ±16.67	141 ±1.63	3351 ±193.2	883 ±51.5
Che	54.0 ±0.70	4.2 ±0.03	13.1	33.0 ±4.46	30.5 ±3.51	287.8 ±15.28	204 ±38.6	4907 ±110.1	690 ±22.1
Che + EM	53.9 ±0.36	4.1 ±0.04	13.1	44.3 ±2.77	16.7 ±1.35	312.8 ±8.82	163 ±23.7	4642 ±130.1	634 ±10.74

Data showing means ± SE (n = 9). ChM = chicken manure; Org = bokashi; Che = chemical fertilizer.

1992). In this study, available phosphorus concentration on 50 days after transplanting were higher in EM treated soils than untreated soils. This might be associated with the activities of the EM microbes. However, on 90 days after planting, the nitrogen and available phosphorus concentrations were lower in EM-treated soils. This might be associated with more absorption of the nutrients by the plants that showed faster growth and higher fruit yield in EM-treated plots than untreated plots. Even if the EM liquid was directly applied to soil at the same time with chemical fertilizers, it also showed growth promotion and yield increasing effects.

Photosynthetic Activity

Effect of Fertilization. Photosynthetic capacity (P_C) and dark respiration were maintained higher at the later growth stage in plants of the bokashi treatment than in those of chemical and chicken manure treatments. This result was visible from the plant appearance at the later growth stage. Plants of bokashi treatment maintained more active young leaves and developed more young fruit than plants in other two treatments. This was due to more nutrients sustained in the soil of bokashi treatment than in soils of other treatments (Table 2). Quantum yield was higher in plants of bokashi and fertilizer treatments than plants of chicken manure treatment. The reason for this result is not clear.

Effect of EM Inoculation. EM inoculated to bokashi and chicken manure and directly applied to soil together with chemical fertilizer increased photosynthetic activity, dark respiration and quantum yield. This result was consistent with plant growth, plants appearance and fruit yield at the later growth stages.

Sugars, Organic Acids and Vitamin C

Effects of Fertilizations. The sugars in tomato fruit are mainly glucose, fructose and sucrose. The concentration of sugars in fruit varied with the fertilizers. As shown in Table 3, the fruit in plots fertilized with chicken manure contained the highest concentration of sugars and those in chemical fertilizer plots had the lowest concentrations of sugars. Compared with the chemical fertilization treatment, the organic acid concentration of fruit in bokashi-fertilized plot was high, followed by the treatment with chicken manure. The ratio of sugars to organic acids was higher in fruit with the chicken manure treatment, resulting in a sweeter taste of fruit. The ratio of sugars to organic acids in the bokashi treatment was similar to that in the chemical fertilizer treatment, but the fruit grown with bokashi-fertilizer was more tasteful since both the sugars and organic acids were higher. As shown in Table 3, vitamin C (ascorbic acid) concentration was lower for the chemical fertilizer treatment than for the two organic fertilizer treatments. The results suggested clearly that organic fertilization improved fruit quality shown by sugar, organic acid and vitamin C concentrations. The bokashi and chicken manure contain not only nutrients but also organic matter required for plants. Organic matter plays important roles in plant growth and development by releasing nutrients, improving soil physical and chemical properties and promoting root activity. The nutrients released from the organic materials are in balance with various elements. Some physiologically active substances released from organic matter can increase root activity (Yamada, 1997). Tomato plants with a high root activity can penetrate more deeply in the soil

TABLE 3. Sugars and organic acids in the ripe tomato fruit from different fertilization treatments

Treatment	Sugars ($g\ kg^{-1}$)				Organic acids ($g\ kg^{-1}$)				Sugars/acids
	Sucrose	Glucose	Fructose	Total	Citric	Malic	Total	Ascorbic	
ChM	1.01 ±0.62	33.4 ±1.8	29.8 ±1.0	64.2	6.32 ±0.42	1.80 ±0.42	8.12	0.16 ±0.019	7.91
ChM + EM	0.68 ±0.45	33.0 ±0.4	30.7 ±0.5	64.4	6.41 ±0.09	1.99 ±0.28	8.40	0.19 ±0.010	7.66
Org	1.43 ±0.21	30.7 ±4.3	27.0 ±1.5	59.1	6.98 ±1.22	1.85 ±0.85	8.83	0.12 ±0.0004	6.70
Org + EM	1.70 ±0.55	29.5 ±1.8	29.1 ±1.1	60.3	6.96 ±1.35	1.69 ±0.03	8.65	0.14 ±0.007	6.97
Che	0.24 ±0.11	25.1 ±3.7	25.2 ±2.1	50.5	6.57 ±1.06	1.48 ±0.25	8.05	0.11 ±0.004	6.28
Che + EM	0.64 ±0.17	26.6 ±4.2	26.9 ±1.9	54.1	6.69 ±0.78	1.24 ±0.18	7.93	0.12 ±0.009	6.78

Data showing means ± SE (n = 9). ChM = chicken manure; Org = bokashi; Che = chemical fertilizer.

with improved physical and chemical properties. This enable tomato plant to absorb water in deep soil layers an irrigation is reduced. Reduced irrigation or low water management can increase sugar concentration of fruit (Yamada, 1997).

Effects of EM Inoculation

EM inoculated to the organic materials or applied directly to the soil with chemical fertilizers did not have a significant effect on fruit sugar and organic acid concentrations per unit of dry mass. However, the EM treatment increased fruit yield. Moreover, increasing the fresh yield might also dilute the active substances such as sugars and organic acids. If the fruit sugar and organic acid concentrations are calculated on per plant, the effect of EM in increasing sugar concentration becomes apparent. As shown in Table 3, EM did increase the fruit vitamin C concentration. In all the treatments of organic and chemical fertilizer, EM inoculation increased vitamin C concentration in tomato fruit, although the mechanism for this effect of EM is not clear.

Dynamic Changes in Sugars and Organic Acids in Developing Fruit. The changes in concentrations of sugars during fruit development in different treatments showed the same trend, as in Figure 2. The concentrations of sugars increased steadily from stage 1 (green fruit) to stage 4 (ripe fruit), but the increasing extent was lower before stage 2 than thereafter. This suggested that different kinds of fertilizers affected the fruit sugar concentration mainly

FIGURE 2. Dynamic changes in sugar and organic acid concentrations of tomato plants fertilized with bokashi and chemical fertilizers.

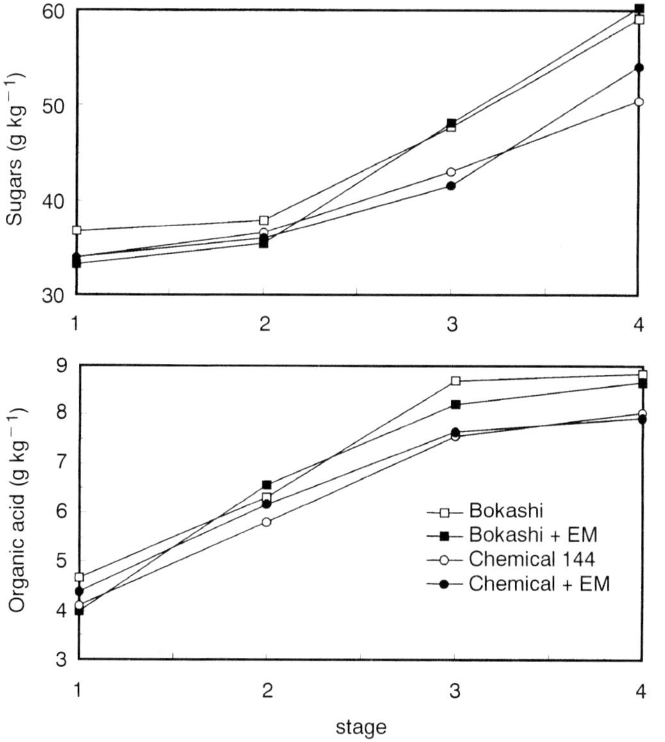

at the later development stage. Though the organic acid in all treatments steadily increased from beginning to the end, there was very little increment after Stage 3. This suggested that the effect of different fertilizers on organic acids were mainly at early stages of fruit development.

The integrated results showed that if the nutrients in organic materials were available at the early growth stages, both chicken manure and bokashi could be used as substitute for chemical fertilizer with a comparable yield and higher quality. Both quality and yield increasing effects could be expected from EM inoculation to the organic fertilizers and application to the soil directly.

CONCLUSIONS

Plant growth and fruit yield were low in the bokashi-applied plots at earlier growth stages but were higher at later growth stage because of the low nutrient availability at the beginning and high nutrient sustainability at the later stage. Concentrations of sugars were highest in fruit of plants fertilized with chicken manure and lowest in fruit of plants with chemical fertilizers. Organic acid concentration was higher in fruit of bokashi-fertilized plants than in fruit of other plots. Vitamin C (ascorbic acid) concentration was higher in fruit of plants fertilized with chicken manure and bokashi than in those fertilized with chemical fertilizer. EM inoculation increased fruit yield and vitamin C concentration. If the nutrients in organic materials were available, both chicken manure and bokashi could be used as substitutes for chemical fertilizer with a comparable yield and higher quality. Increased yields and improved quality could be expected from EM inoculation either to the organic materials or to the soil directly.

REFERENCES

Arshad, M. and W. T. Frankenberger Jr. (1992). Microbial production of plant growth regulators. In *Soil Microbial Ecology*, ed. FB Metting Jr. New York: Marcel Dekker, Inc, pp. 307-348.

Higa, T. (1994). *The Completest Data of MI Encyclopedia* (in Japanese). Tokyo: Sogo-Unicom, 385 p.

Okada M. (1993). *The Basis of Paradise-Kyusei Nature Farming*. Atami (Japan): Seikai Kyusei Kyo Press, pp. 331-393.

Yamada, K. (1997). Fertilizations for high sugar concentration of tomato fruit. In *Agricultural Technology-Vegetables II: Tomato*, ed. Y. Kondo. Tokyo: Rural Culture Association, p. 373.

Yamada, K., S., Kato, M. Fujita, H.L. Xu, K. Katase and H. Umemura. (1996). An organic fertilizer inoculated with EM used in nature farming practices. Ann. Asia-Pacific Nature Agriculture Network, Oct. 8-12, 1996, Bangkok, Thailand.

Xu, H. L., L. Gauthier and A. Gosselin (1995). Effects of fertigation management on growth and photosynthesis of tomato plants grown in peat, rockwool and MFT. *Scientia Horticulturae* 63: 11-20.

Effects of a Microbial Inoculant and Organic Fertilizers on the Growth, Photosynthesis and Yield of Sweet Corn

Hui-lian Xu

SUMMARY. A microbial inoculant known as Effective Microorganisms or EM is a mixed culture of naturally-occurring, beneficial microorganisms (predominantly lactic acid bacteria, yeast, actinomycetes, photosynthetic bacteria and certain fungi) that has been used with considerable success to improve soil quality and the growth and yield of crops, particularly in nature farming and organic farming systems. Despite this success, the exact mechanisms of how this EM elicits such beneficial effects is largely unknown. Consequently, a study was conducted to determine the effects of EM and organic fertilizer on the growth, photosynthesis, and yield of sweet corn (*Zea mays* L.) under glasshouse conditions, compared with chemical fertilizer. An organic fertilizer consisting of a mixture of oilseed mill sludge, rice husk and bran, and fish processing waste, was inoculated and fermented with EM as the microbial inoculant. The organic fertilizer and chemical fertilizer were then applied to respective pots to compare the growth, yield and physiological response of sweet corn plants. EM applied with the organic fertilizer was shown to promote root growth and activity, and to enhance photosynthetic efficiency and capacity, which resulted in in-

---

Hui-lian Xu is Senior Crop Scientist, International Nature Farming Research Center, 5632 Hata, Nagano 390-1401, Japan (E-mail: huilian@janis.or.jp).

The author thanks Dr. H. Umemura, K. Katase, M. Fujita, K. Yamada and S. Kato for their technical assistance and advice in the research.

[Haworth co-indexing entry note]: "Effects of a Microbial Inoculant and Organic Fertilizers on the Growth, Photosynthesis and Yield of Sweet Corn." Xu, Hui-lian. Co-published simultaneously in *Journal of Crop Production* (Food Products Press, an imprint of The Haworth Press, Inc.) Vol. 3, No. 1 (#5), 2000, pp. 183-214; and: *Nature Farming and Microbial Applications* (ed: Hui-lian Xu, James F. Parr, and Hiroshi Umemura) Food Products Press, an imprint of The Haworth Press, Inc., 2000, pp. 183-214. Single or multiple copies of this article are available for a fee from The Haworth Document Delivery Service [1-800-342-9678, 9:00 a.m. - 5:00 p.m. (EST). E-mail address: getinfo@haworthpressinc.com].

creased grain yield. This was attributed largely to a higher level of nutrient availability facilitated by EM application over time. Interestingly, during the early stage of the experiment, the growth and dry matter yield of plants that received organic fertilizer were actually lower than those treated with chemical fertilizer that provided higher initial levels of macronutrients. However, during the intermediate and late growth stages, EM increased the nutrient availability of the organic fertilizer to a higher level, than the chemical fertilizer. Consequently, even though there was an early lower growth rate for plants that received EM-fermented organic fertilizer compared with chemical fertilizer, the final biomass and grain yield from organic fertilizer was equal to or higher than from chemical fertilizer. *[Article copies available for a fee from The Haworth Document Delivery Service: 1-800-342-9678. E-mail address: getinfo@haworthpressinc.com <Website: http://www.HaworthPress.com>]*

KEYWORDS. Effective Microorganisms, EM, growth, nature farming, organic farming, organic fertilizer, photosynthesis, root, sweet corn

INTRODUCTION

Many problems of environmental pollution have been caused by excessive applications of chemical fertilizers and pesticides in conventional or traditional farming systems. This is now seen as a threat to the earth and existence of human life on the earth. This has caused increased interest in nature farming and organic farming systems. The concept and principles of nature farming was proposed in Japan by Mokichi Okada more than 60 years ago (Okada, 1987). One of the strict principles of nature farming according to Okada was the prohibition of using both chemical fertilizer and untreated animal manure fertilizers (Okada, 1987). Therefore, nature farming practitioners and researchers have long sought various kinds of organic materials as nutrient sources to substitute for chemical fertilizers. Farmers in Japan have extensively used a fermented organic fertilizer consisting of such organic materials as oil mill sludge, rice husk and bran, and fish processing byproduct. Such organic fertilizer is often referred to as bokashi. Compared with the untreated animal manure fertilizers that are used in organic farming, this kind of organic fertilizer is devoid of such problems as pathogens, heavy metals, antibiotics and animal growth-promoting hormones. Recently, the application of a microbial inoculant known as EM (Effective Microorganisms) has been used successfully in nature farming systems. Research has shown that EM has been effective in promoting crop growth and yield and protecting the environment. However, special details on how EM actually achieves these benefits are not known. Generally, the adoption and use of EM in nature farming systems has preceded the essential fundamental research on the exact mechanisms or modes-of-action of how EM affects the soil-plant ecosystem. Moreover, with organic fertilizers, it is difficult to achieve crop yields equal to or higher than those obtained with chemical fertilizers. A partial explanation is that soil physical and chemical properties have been changed by intensive chemical practices and the soil and the land environment have become dependent on chemicals. Therefore, in the present study, the effects of organic fertilizers and EM applications on plant growth, photosynthetic capacity, grain yield and soil-plant nutrient availability were studied compared with chemical fertilizer applications.

MATERIALS AND METHODS

Plant Materials

Experiment I. Sweet corn (*Zea mays* L. cv. Honey Bantam) with a growth period of about 80 days was used in this study. Plants were grown in Wagner pots (0.02 m^2 of surface area) in June 1996. The pots, each with one plant remaining after thinning, were placed in a Latin Square design in a glasshouse. A fine-textured Andosol was used having a total nitrogen content, available phosphorus, and available potassium levels of 3.4, 0.025 and 0.44 g kg^{-1}, respectively, and a C:N ratio of 13. The field capacity of the soil was 80% on a gravimetric basis. Each pot was filled with 3 kg fresh soil at a water content of 38% of field capacity.

Experiment II. Seeds of the same sweet corn as Experiment II were sown in August 1996. The management of the plants was the same as in Experiment I.

Fertilizers

For both experiments, ammonium sulfate, superphosphate and potassium sulfate were used as the chemical fertilizer treatment and the quantities of N, P and K were equivalent to the total macronutrient content in the organic fertilizer treatments described later. Organic materials such as oilseed mill sludge, rice husk and bran, and fish-processing by-product were fermented anaerobically or aerobically with or without EM (Effective Microorganisms) as the microbial inoculant. The total nitrogen, available phosphorus, and potassium concentrations were 58, 30 and 2 g kg^{-1}, respectively, for both anaerobic and aerobic organic fertilizers. For Experiment I, only anaerobic organic fertilizer was used with the combined chemical fertilizer as a control. Both anaerobic and aerobic organic fertilizers were used in Experiment II.

Microbe Inoculant

The microbial inoculant used in this study, known as EM (Effective Microorganisms), is a mixed culture of beneficial microorganisms containing about 80 different species. The main species included in EM are as follows. (1) Lactic acid bacteria: *Lactobacillus plantarum, Lactobacillus casei, Streptococcus lactis*; (2) Photosynthetic bacteria: *Rhodopseudomonas palustris, Rhodobacter sphaeroides*; (3) Yeasts: *Saccharomyces cerevisiae, Candida utilis*; (4) Actinomycetes: *Streptomyces albus, Streptomyces griseus*; (5) Fungi: *Aspergillus oryzae, Mucor hiemalis*; and (6) Others: some microorganisms that are naturally-occurring and introduced into EM in the manufacturing process which can survive in EM liquid at pH 3.5 and below. The concentration of most of these microorganisms in the liquid phase range from 1×10^6 to 1×10^8 per ml. The same microbial inoculant, EM, was used for both Experiments I and II with the same quantity.

Treatments. Four treatments were made for Experiment I without Treatment (3) and (4) of the following six treatments, and six treatments for Experiment II as follows: (1) EM + Organic 1–anaerobically fermented organic materials 80 g, in which EM was added to the materials before fermented; (2) Organic 1– anaerobically fermented organic materials 80 g; (3) EM + Organic 2–aerobically fermented organic materials 80 g, in which EM was added before fermented; (4) Organic 2–aerobically fermented organic materials 80 g; (5) EM + Fertilizer–chemical fertilizers (ammonium sulfate 5.3 g, a long-period coated urea called LP coat-70 2.8 g, superphosphate 13 g and potassium sulfate 4.95 g), with EM, 80 ml, applied into the soil at the same time before sowing; the total amounts of nitrogen, phosphorus and potassium were the same as in the above mentioned organic materials; and (6) Fertilizer–the same fertilizers as in (5). In experiment I, the amount of chemical fertilizers applied was half of that in Experiment II.

Relative Growth Rate

Samples of the total plant were taken 10, 20, 40, 60, and 80 days after sowing for Experiment I, and 8, 25, 50, and 120 days after sowing for Experiment II. Relative Growth Rate (RGR) was calculated according to Nakaseko (1985) as follows:

$$\mathrm{RGR} = (\ln M_2 - \ln M_1)/(t_2 - t_1) \quad (1)$$

where M_2 and M_1 are dry mass at time t_2 and t_1.

Net Assimilation Rate

The net assimilation rate (NAR) was calculated as follows:

$$\text{NAR} = \{(M_2 - M_1)/(t_2 - t_1)\} \times (\ln A_2 - \ln A_1)/(A_2 - A_1) \qquad (2)$$

where A_2 and A_1 are leaf areas at time t_2 and t_1.

Root Length

Roots in the whole pot soil volume were carefully washed out, cut into 2 to 3 cm long pieces, and distributed onto a wire sieve with colored lines. The root lengths were measured using the line interception method according to Morita et al. (1992) and calculated as follows:

$$L_R = N \times G \qquad (3)$$

where L_R is the root length; N the interception number; and G the interception constant depending on the square size. In the present study the square was 5 mm and the interception constant was 0.393.

Branching Index

The branching index (BI) of the root system was calculated according to Morita et al. (1992) as follows:

$$\text{BI} = (L_T - L_P)/L_P \qquad (4)$$

where L_T and L_P are the total root length and total primary root length.

Root Respiration Rate

Tip parts of the root with a length of 5 cm were cut and collected as root samples (each with 1 g of fresh mass) for respiration measurement. The fresh root sample was inserted in the assimilation chamber and measured under a relative humidity of 85% at 15, 20, 25, 30 and 35°C using the same gas exchange system (LI-6400) as for photosynthetic measurements. After measurement, the root sample was dried and the respiration rate was expressed on a dry mass basis. The temperature response curve for the root respiration rate (R_D) was modeled and analyzed using the following equation:

$$R_D = R_0\, e^{\alpha T} \qquad (5)$$

where R_0 is the theoretical value of the respiration rate at 0°C; α the constant showing temperature response sensitivity; T the tissue temperature of the root. The increment of R_D by increasing temperature from 20 to 30°C was defined as Q_{10}, an indicator of the response of respiration temperature.

Photosynthetic Measurements

The fully expanded youngest leaf was used at early growth stages and the leaf just above the ear was used at later growth stages for photosynthetic measurements. The net photosynthetic rate (P_N) were measured using an open gas exchange system (LI-6400, Li-Cor, Inc., Lincoln, Nebraska, USA). Leaf temperature was fixed at 23°C and the relative humidity was maintained at 65% in the assimilation chamber during the photosynthetic measurement. Photosynthetic light response curve was obtained by measuring the photosynthetic rate at different photosynthetic photon fluxes (PPF) under a LED light source.

Photosynthetic Light-Response Curve

The response curve of photosynthetic rate to the increasing PPF was modeled by an exponential equation as follows:

$$P_N = P_C(1 - e^{-Ki}) - R_D \quad (6)$$

where P_C is photosynthetic capacity; R_D is respiration rate; K is the half time constant and the maximum quantum yield (Y_Q) is shown as $Y_Q = KP_C$; and i the PPF.

Photosynthetic Hysteresis

Hysteresis in general physics is described by Poulovassilis (1962) as follows. For many physical properties, the curve for decreasing x does not coincide with that for increasing x, i.e., the relationship between y and x is not unique and irreversible. In the present experiment, P_N was measured as PPF increased from 0 to 2000 μmol m^{-2} s^{-1} and then PPF was decreased from 2000 to 0. P_N during the periods of increasing and decreasing PPF were plotted against the PPF. P_N always showed higher values during the decreasing cycle than during the increasing cycle. The difference between the two curves was defined as photosynthetic hysteresis. The extent of hysteresis (H_P) could be calculated as

$$H_P = 1 - \int f(i_L) / \int f(i_H) \quad (7)$$

where the i_L and i_H are the PPF which was started from low to high, and from high to low, respectively (Xu et al., 1994).

Senescence Analysis in Photosynthetic Aspects

Photosynthetic rate was measured in leaves at different positions at the late growth stage before harvesting. Photosynthesis in the senescent lower leaves was compared with the leaf where the photosynthesis was highest as the control.

Measurements of Mineral Nitrogen and Available Phosphorus in the Soil

After the plant samples were taken, the soil samples were collected. The concentrations of mineral nitrate-nitrogen and ammonium nitrogen as well as the concentration of available phosphorus in the soil and plant were measured using the colorimetric method (Ishitsuka, 1985).

RESULTS

Dry Mass and Leaf Area at Different Stages

The upper part of Figure 1 shows the absolute levels of plant dry mass for four different treatments in Experiment I. The lower part of this figure shows relative dry mass levels with the non-EM chemical treatment as 100%. Figure 2 shows the result of dry mass for the six treatments of Experiment II with the same style as in Figure 1. Chemical fertilizers were applied with an N-P-K amount half of that in organic fertilizer for Experiment I, but with the same amount as in organic fertilizers for Experiment II. In both experiments, plants in EM plots showed slightly higher dry mass under all fertilizations at most growth stages. At the early growth stages, dry mass of organic-fertilized plants was significantly lower than that of chemical-fertilized plants. However, as the growth stage progressed, dry mass of organic-fertilized plants reached a similar level to that of chemical-fertilized plants. Figures 3 and 4 show the results of leaf area in Experiment I and Experiment II, respectively. Except for the anaerobic organic fertilizer plot in Experiment II, leaf area was increased by EM addition treatments. In Experiment II, leaf area of organic-fertilized plants did not reach the same level as that of chemical-fertilized plants, although the dry mass reached a level close to that of the chemical-fertilized plants at the harvesting stage. This suggested that less leaf area of organic-fertilized plants produced a similar dry mass compared to the chemical-fertilized plants.

FIGURE 1. Total dry mass of sweet corn plants fertilized with organic or chemical fertilizers with or without microbial inoculant application at different growth stages (Experiment I).

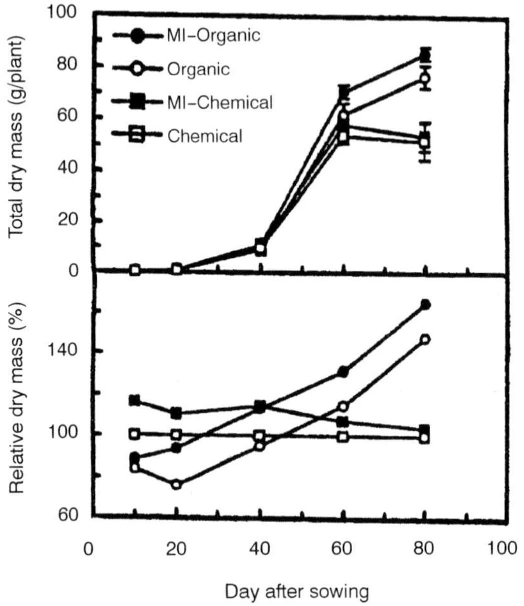

Relative Growth Rate and Net Assimilation Rate

Figures 5 and 6 show the relative growth rate (RGR) for Experiment I and Experiment II, respectively. The upper parts of the figures show the absolute values of RGR. Because there were large fluctuations in the values among growth stages, it is difficult to tell the differences from absolute values in these figures. The lower parts show the percentage values relative to the non-EM-chemical treatment. For both experiments, at the early stage, RGR in the organic-fertilized plants was lower, but then increased very fast and became much higher than that of chemical-fertilized plants at middle and later stages. Compared with that for chemical-fertilized plants with larger dry mass at the early stages, greater RGR for organic-fertilized plants from the middle stages was because the RGR was calculated from the dry mass on an exponential basis. EM application showed a positive effect on RGR at all growth stages under both organic and chemical fertilizations. Plants fertilized with aerobically-fermented organic materials grew better than plants with anaerobically-fermented organic materials. The results of net assimilation

FIGURE 2. Total dry mass of sweet corn plants fertilized with organic or chemical fertilizers with or without microbial inoculant application at different growth stages (Experiment II).

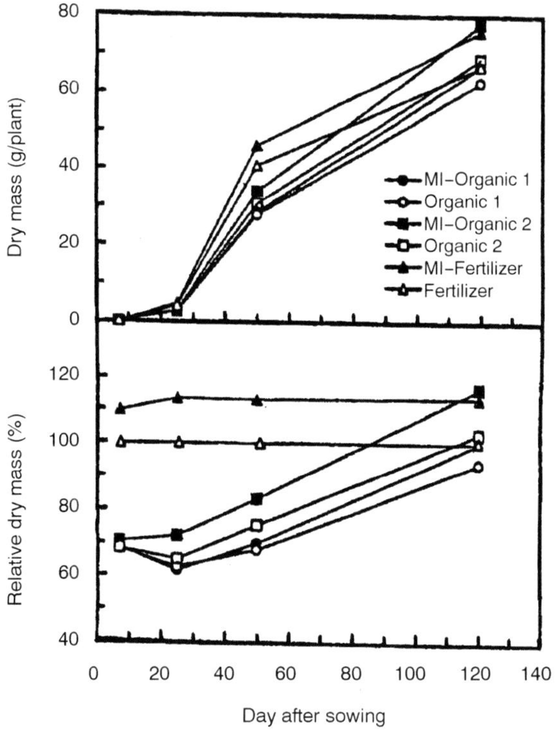

rate (NAR) were presented in Figures 7 and 8 for Experiment I and Experiment II, respectively. NAR showed similar tends to those of RGR.

The Final Dry Mass and Grain Yield

Table 1 shows the final dry mass and grain yield for Experiment I and Experiment II. Although dry mass at early stages was low in organic-fertilized sweet corn plants, the final dry mass was not low. Moreover, organic-fertilized plants show a slightly higher grain yield and higher harvest index. EM treatments in most cases increased the grain yield and harvest index. In Experiment I, because the fertilizers were applied at half the rate of N-P-K as that in organic fertilizers, and the soil was very poor, plants in chemical-fertilized plots could not produce normal ears. Therefore, the conclusion was limited from the data of Experiment I.

FIGURE 3. Leaf area of sweet corn grown with organic or chemical fertilizers with or without microbial inoculant application at different stages (Experiment I).

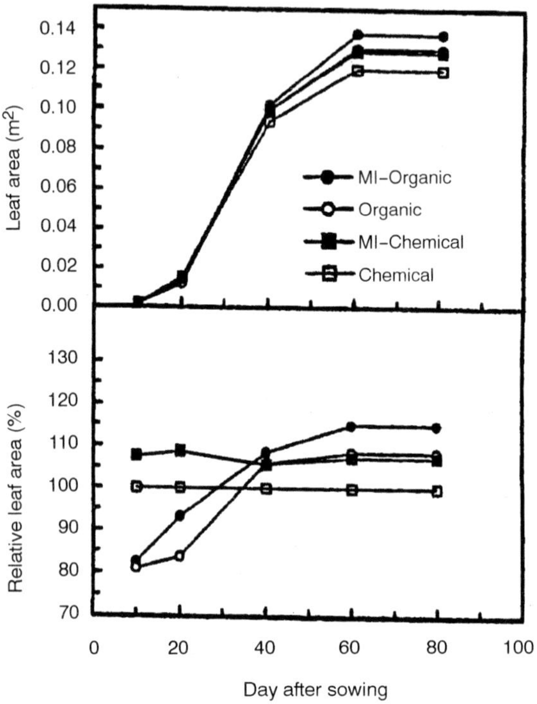

Root Number and Root Length at Different Stages

Compared with plants under chemical fertilizations, organic-fertilized sweet corn plants developed more and longer seminal roots and nodal roots in Experiment I (Figure 9). EM applications showed positive effects on both root number and root length. In Experiment II, more seminal roots were found in organic-fertilized plants than chemical-fertilized ones, but no significant differences in seminal root length were found between organic and chemical-fertilized plants (Figure 10). As in Experiment I, EM applications increased both the number and length of seminal roots. Nodal root number was higher for chemical-fertilized plants but the nodal length was not higher compared with organic-fertilized plants. EM applications also increased nodal root number and length slightly in all fertilization treatments in Experiment II.

FIGURE 4. Plant total leaf area of sweet corn fertilized with organic or chemical fertilizers with or without microbial inoculant application at different growth stages (Experiment II).

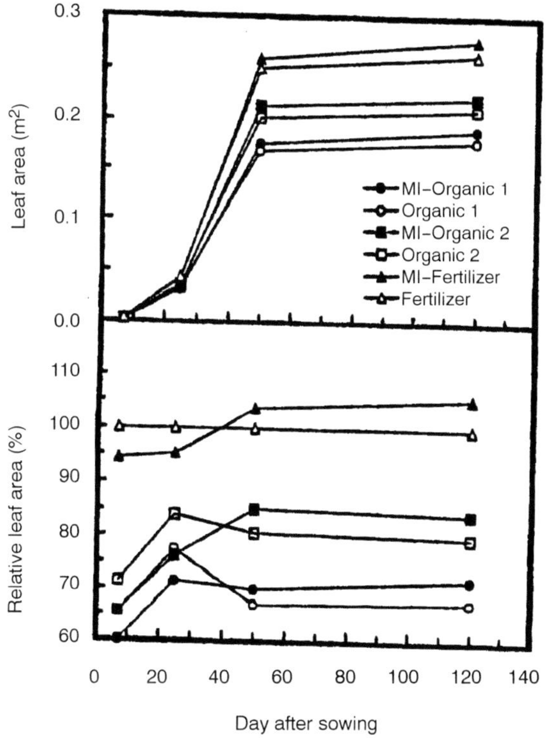

Dry Mass Ratio of Root to Total at Different Stages

The differences in root dry mass can be clearly understood from the relative values and the root/total ratios at different stages (Figure 11 and Figure 12). The difference in root/total dry mass ratio was significant at early stages but declined at later stages between organic-fertilized and chemical-fertilized plants or between EM applied plants and plants without EM applications. The absolute root dry mass was higher in chemical-fertilized than organic-fertilized plants because the total dry mass at most stages was higher for the former than the latter.

FIGURE 5. Relative growth rate (RGR) sweet corn plants grown with organic or chemical fertilizers with or without microbial inoculant application at different growth stages (Experiment I).

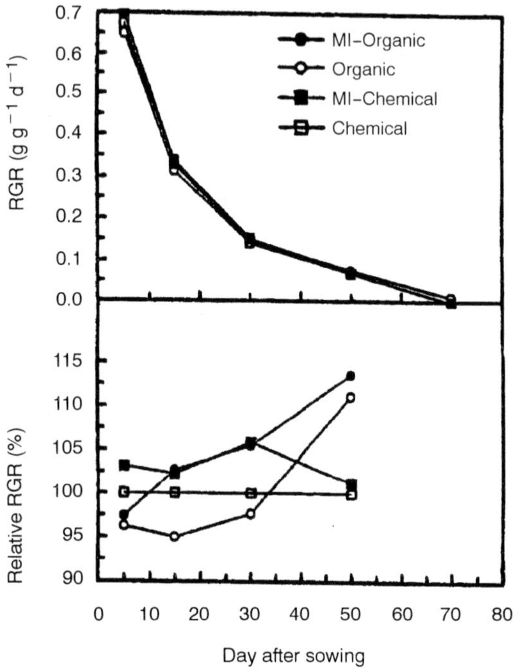

The Final Root Dry Mass, Root Length and Branching Index

The total root length and branch index were examined in Experiment II and presented in Table 2. The total root length and branch index were greater for organic-fertilized plants than for chemical-fertilized plants and also greater for EM-treated plants than for plants without EM applications (Table 2).

Root Respiration

Figure 13 shows the response of root respiration rate to increase in temperature for organic and chemical-fertilized plants. Respiration rate in organic-fertilized or EM-treated plants was more sensitive than for plants of other treatments (shown by the *a* value in Table 3). The respiration rate in the root tips measured at 25°C was higher in organic-fertilized plants than chemical-

FIGURE 6. Relative growth rate (RGR) of sweet corn plants grown with organic or chemical fertilizers with or without microbial inoculant application at different growth stages (Experiment II).

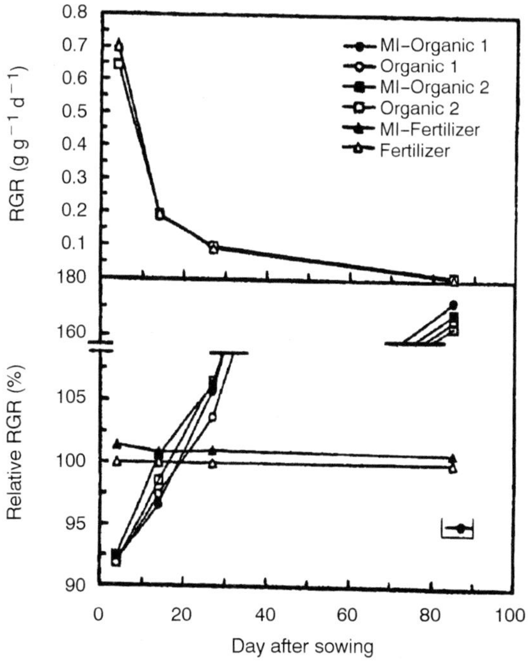

fertilized plants and higher for plants with EM application than those without EM application.

Photosynthetic Capacity, Quantum Yield and Respiration at Different Stages

Photosynthesis was only examined in Experiment II. The photosynthetic capacity (P_C) and the maximum quantum yield (Y_Q) were analyzed with an exponential model as shown in Figure 14. At early growth stages, the photosynthetic capacity (P_C) was lower in plants under organic fertilization than those under chemical fertilization. However, P_C became higher in the former than in the latter from the middle growth stages (Figure 15 and Figure 16). EM applications increased P_C at most growth stages. Leaf dark respiration rate and the quantum yield showed no differences between the fertilizers (Figure 17 and Figure 18). EM application increased Y_Q slightly at some stages but showed no effect on dark respiration.

FIGURE 7. Net assimilation rate of sweet corn plants grown with organic or chemical fertilizers with or without microbial inoculant application at different stages (Experiment I).

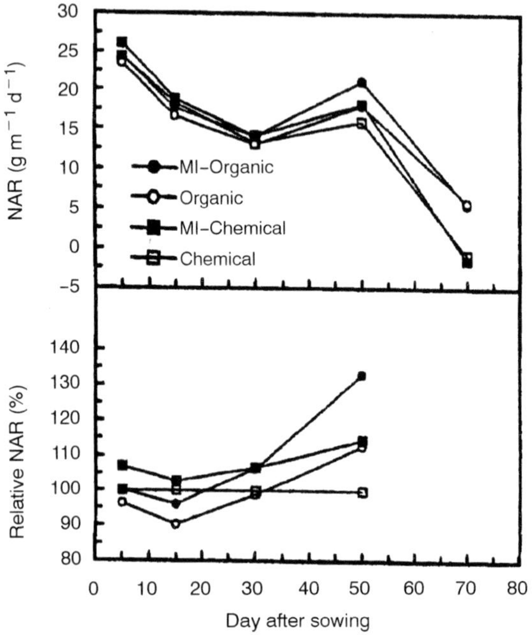

Senescence Shown by P_C in Leaves at Low Positions

The decrement of P_C as senescence developed was much larger in leaves of chemical-fertilized plant than in leaves of organic-fertilized plants (Figure 19 and Figure 20). The positive effect of EM treatment was apparent for plants under both chemical and organic fertilizations at all stages except for the early growth stage of plants fertilized with anaerobically-fermented organic materials. The EM treatment also lessened the photosynthetic declines in senescent leaves.

Photosynthetic Hysteresis

The model for photosynthetic hysteresis is shown in Figure 21. The photosynthetic hysteresis showed the sensitivity of the photosynthetic processes to lighting changes. The detailed descriptions can be found in the section of

FIGURE 8. Net assimilation rate of sweet corn plants grown with organic or chemical fertilizers with or without microbial inoculant application at different stages (Experiment II).

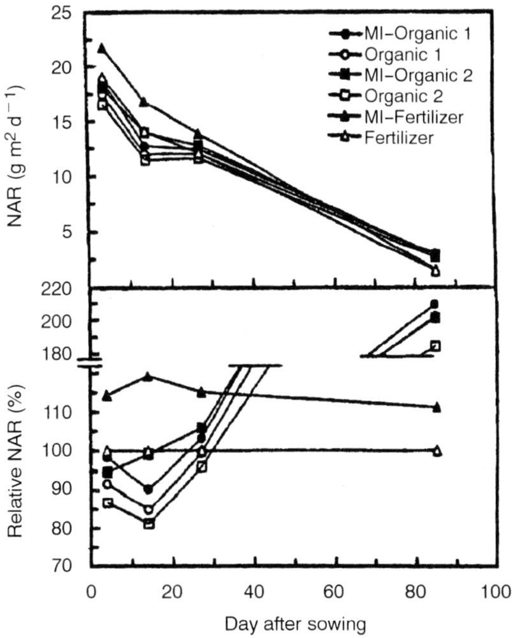

TABLE 1. Dry mass and yield of sweet corn plants grown under organic and chemical fertilizations with or without microbial inoculant (i.e., EM) application.

Parameter	Organic 1		Organic 2		Chemicals	
	EM+	EM−	EM+	EM−	EM+	EM−
Experiment I						
Total dry mass	86a	77b			54c	52c
Ear dry mass	39a	35b			12c	4d
Harvest index (%)	45a	46a			22b	7c
Experiment II						
Total dry mass	67b	63c	77a	70b	75a	68b
Ear dry mass	29a	25b	28a	24b	26ab	21c
Harvest index (%)	30a	28	27	25	26	23

Data in the same column sharing the same letter are not significantly different at $P = 0.05$.

FIGURE 9. Root number and length of sweet corn grown with organic or chemicals with or without microbes application at different stages (Experiment I).

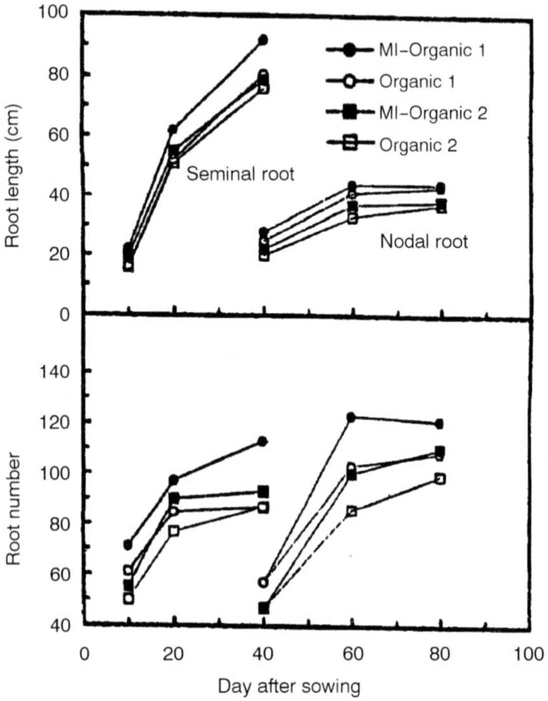

Materials and Methods. Actually, the photosynthetic hysteresis was shown by the relative area surrounded by the two light response curves by changing light intensity from low to high and then from high to low. As shown in Figure 21, photosynthetic hysteresis was larger in chemical-fertilized than in organic-fertilized plants or larger in plants of non-EM plot than in plants of EM treatments. This suggested that, at the later growth stages, senescence in aspects of photosynthetic activity developed more in chemical-fertilized than in organic-fertilized plants and in plants of non-EM plots than in plants of EM-treated plots.

Nutrient Availability and Uptake

Figure 22 and Figure 23 shows the dynamic changes of nitrate nitrogen and available phosphorus for Experiment I. Figure 24 shows the results of nitrogen for Experiment II. The lower left part of Figure 24 shows the avail-

FIGURE 10. Root number and length of sweet corn grown with organic or chemicals with or without microbes application at different stages (Experiment II).

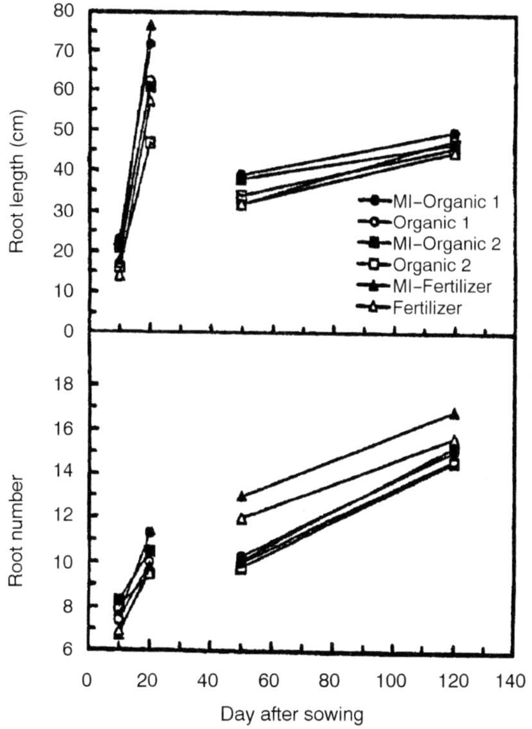

FIGURE 11. Root to total dry mass ratio of sweet corn plants grown with organic or chemical fertilizers with or without microbial inoculant application at different growth stages (Experiment I).

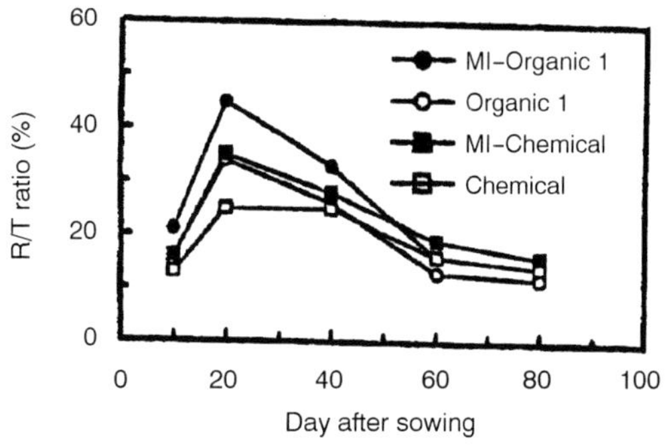

FIGURE 12. Root dry mass and root to total dry mass ratio of sweet corn plants grown with organic or chemical fertilizers with or without microbial inoculant application at different growth stages (Experiment II).

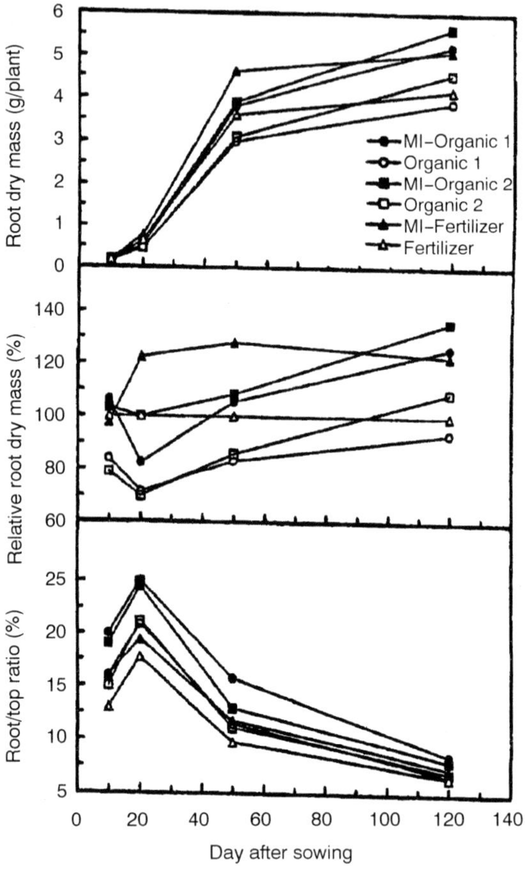

able nitrogen concentration in the soil. The upper left part shows the dynamic changes of the total nitrogen in the soil and that absorbed by the plant. The right part of the graph shows the percentage values of available nitrogen relative to the non-EM-chemical control. At the beginning, available nitrogen was much lower in organic-fertilized pots than in the chemical-fertilized pots. This was the reason why plant growth and dry matter production were lower at early growth stage in organic-fertilized pots than that in chemical-fertilized pots. As the growth stage progressed, both available nitrogen in soil and the

TABLE 2. Characteristics of root system in sweet corn plants grown under different fertilizations.

Parameter	Organic 1		Organic 2		Chemicals	
	EM+	EM−	EM+	EM−	EM+	EM−
Root dry mass (g/plant)	5.24 c	3.91 d	5.66 a	4.54 b	5.11 b	4.17 c
Root/Top ratio (%)	8.51	6.65	7.98	6.98	7.33	6.49
Seminal root number	11.3 a	10.0 b	10.5 b	9.5 c	9.8 c	9.5 c
Nodal root number	15.0 bc	14.7 bc	15.3 bc	14.6 c	16.9 a	15.7 b
Longest seminal root (m)	0.72 a	0.63 b	0.61 bc	0.54 c	0.77 a	0.57 bc
Longest nodal root (m)	0.50 a	0.48 ab	0.47 ab	0.46 ab	0.47 ab	0.45 b
Total seminal root length (m)	3.81 b	3.77 b	3.94 a	3.88 c	3.68 d	3.54 e
Total nodal root length (m)	3.13 b	2.94 c	3.24 a	3.07 bc	3.29 a	3.06 bc
Total root length (m)	916 a	815 b	955 a	836 b	744 c	657 d
Root branching index	130.9	120.4	132.1	119.2	105.7	98.6

Data in the same column sharing the same letter are not significantly different at P = 0.05.

FIGURE 13. The model curves of the root respiration rate in response to temperature changes for sweet corn plants fertilized with aerobic organic fertilizer (Bokashi 2) and plants with chemical fertilizers.

The exponential model: $R = R_0 \exp(\alpha T)$. R_0 is the theoretical respiration rate at 0°C; α is constant; T is the root tissue temperature; Q_{10} is increment of respiration rate in response to temperature change from 20 to 30°C.

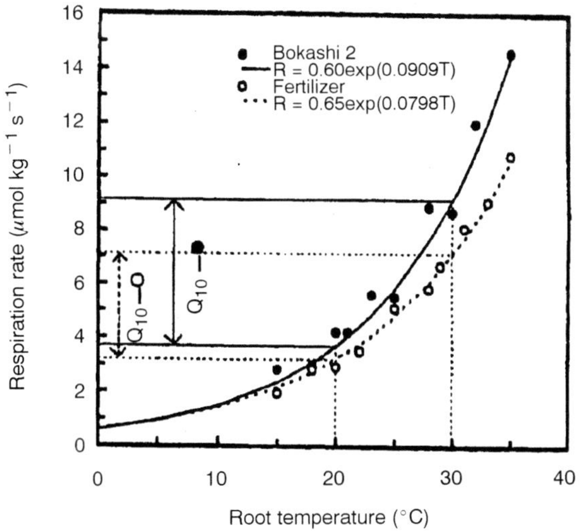

TABLE 3. Root respiration activities, shown by the parameters of temperature response curve given by $R = R_0 e^{\alpha T}$, of sweet corn plants grown under different fertilizations.

Treatment	R_0	R_{25}	Q_{10}	α
	μ mol kg^{-1} s^{-1}		$°C^{-1}$	
Bokashi 1	0.57 b	5.2 c	4.8 c	8.87 b
EM-Bokashi 1	0.58 b	5.6 b	5.8 b	9.42 a
Bokashi 2	0.60 b	5.5 b	6.0 b	9.09 b
EM-Bokashi 2	0.61 ab	6.9 a	6.8 a	9.79 a
Fertilizer	0.65 a	4.0 e	4.8 c	7.98 d
EM-Fertilizer	0.65 a	4.6 d	5.4 b	8.53 bc

R_0 is the theoretical respiration rate at 0°C; α is constant; R_{25} is the respiration rate at 25°C; Q_{10} is increment of respiration rate in response to temperature change from 20 to 30°C.

FIGURE 14. The response curve of photosynthetic rate to the increasing photosynthetic photon flux fitted with an improved exponential model. The exponential equation is as follows:

$$P_N = P_C(1 - e^{-Ki}) - R_D \quad (6)$$

where P_C is photosynthetic capacity; R_D, respiration rate; K, the half time constant and the maximum quantum yield (Y_Q) is shown as $Y_Q = KP_C$; and i, the PPF.

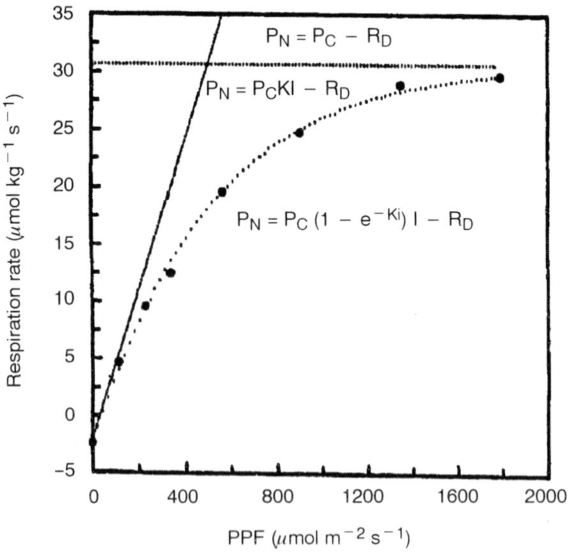

FIGURE 15. The response curves of photosynthetic rate to the increasing photosynthetic photon flux for sweet corn plants grown under organic and chemical fertilizations with or without microbial inoculant applications at the early growth stage (left graph, 25 days after sowing) and later stage (right graph, 80 days after sowing).

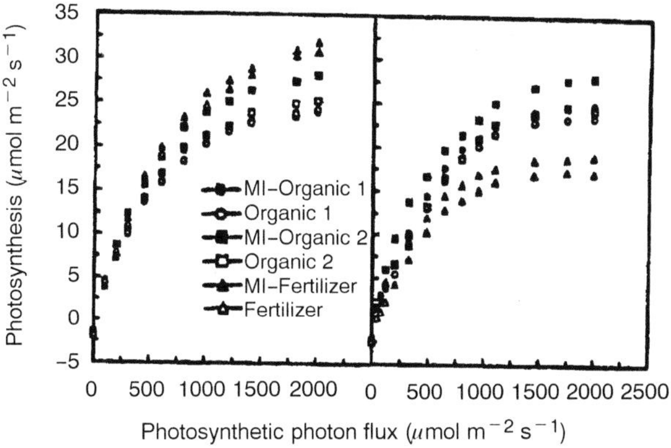

FIGURE 16. Photosynthetic capacity (P_C) at different growth stages in sweet corn plants grown under organic and chemical fertilizations with or without microbial inoculant applications.

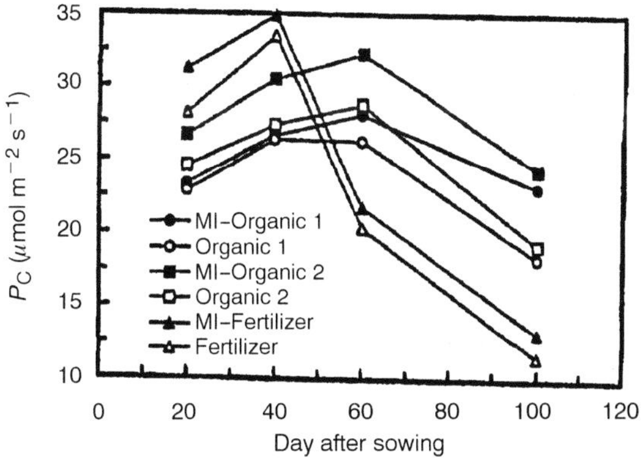

total nitrogen available and absorbed by plants reached a higher level than those for the chemical fertilizer treatment. This also explains why photosynthesis and plant growth became higher in the organic-fertilized plants at later stages. Figure 25 shows the dynamic changes of available phosphorus in the same manner as in Figure 24. The available phosphorus in soil of chemical

FIGURE 17. The maximum quantum yield (Y_Q) at different growth stages in sweet corn plants grown under organic and chemical fertilizations with or without microbial inoculant applications.

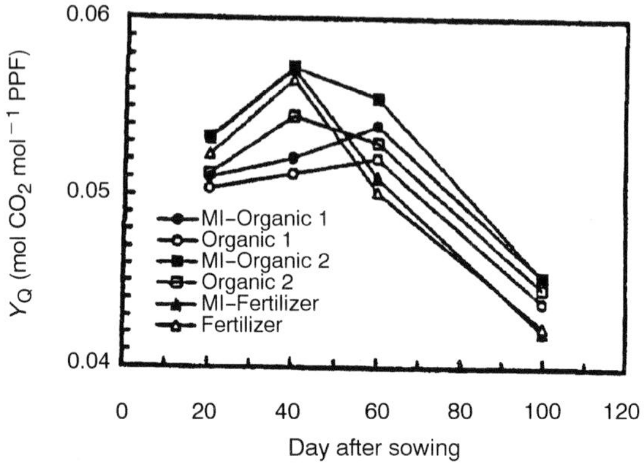

FIGURE 18. The dark respiration rate (R_D) at different growth stages in sweet corn plants grown under organic and chemical fertilizations with or without microbial inoculant applications.

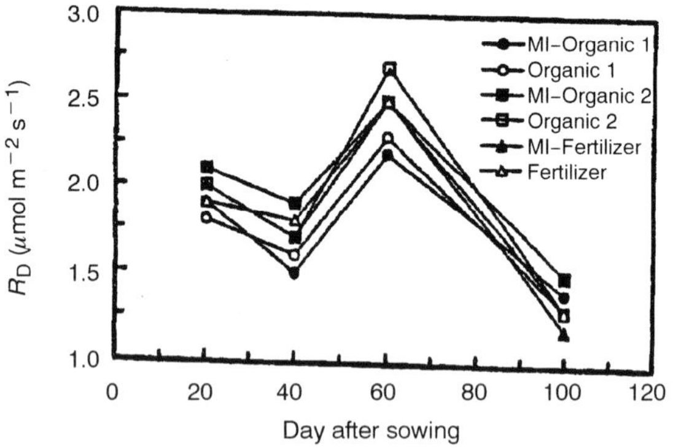

treatments decreased as the growth stage progressed. However, it showed almost no decreases up to the end in the organic fertilizer treatment. The total phosphorus absorbed by the plant plus that available in the soil reached a similar level for both organic and chemical fertilizers. The total available soil phosphorus was slightly higher in EM-treated plots.

FIGURE 19. Photosynthetic capacity (P_C) and the relative value in leaves at different positions from the uppermost of sweet corn plants grown under organic and chemical fertilizations with or without microbial inoculant applications.

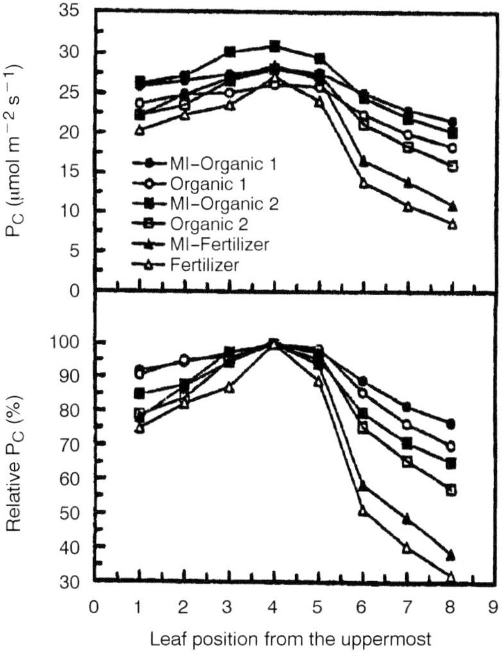

FIGURE 20. A model of photosynthetic hysteresis in responses to changing PPF from low to high (solid) and from high to low (open).

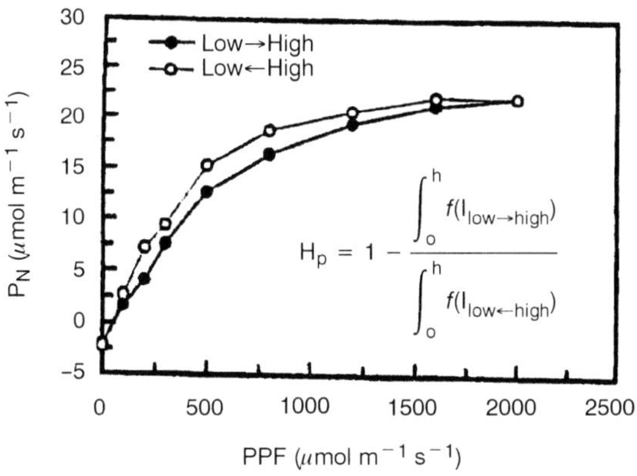

FIGURE 21. Photosynthetic hysteresis in sweet corn plants grown under organic and chemical fertilizations.

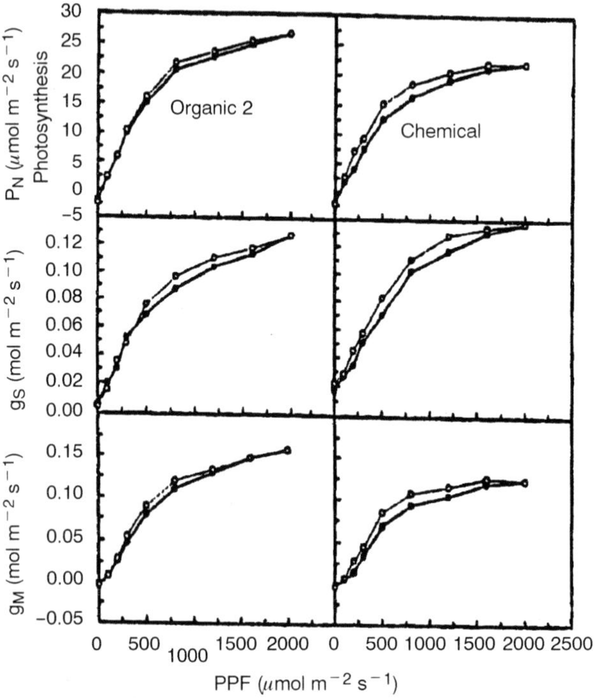

FIGURE 22. Dynamic changes in nitrate nitrogen in the soils for sweet corn plants fertilized with organic materials and chemicals.

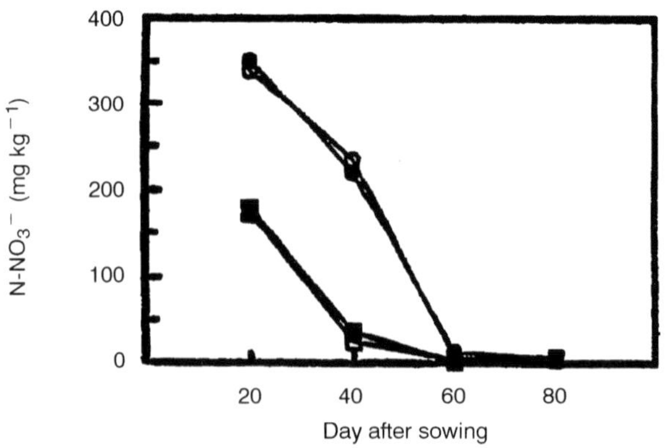

FIGURE 23. Dynamic changes in nitrate nitrogen and available phosphorus in the soils for sweet corn plants fertilized with organic materials and chemicals.

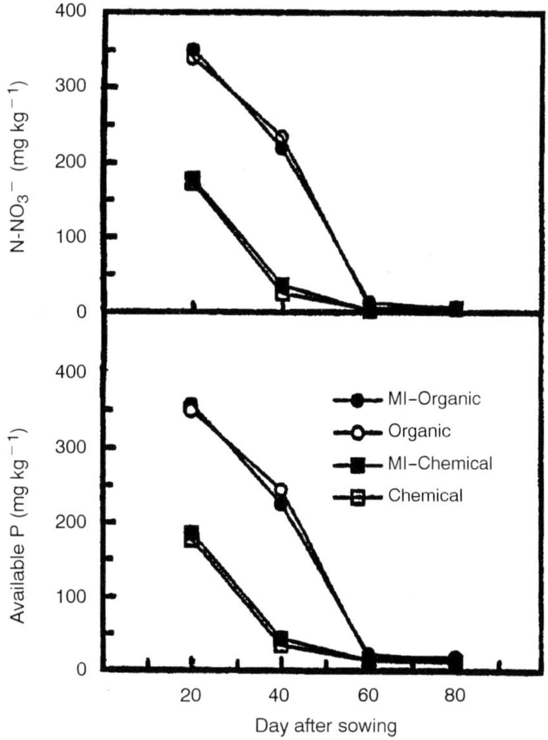

DISCUSSION

At an early stage, growth and dry mass of corn plants were lower for organic-fertilized plants than for chemical-fertilized ones. This might be attributed to the initial low nutrient availability from the organic materials. However, organic-fertilized plants showed a higher root/total dry mass ratio than the chemical-fertilized plants. The physiological activity of roots shown by the respiration rate was also higher in organic-fertilized plants than chemical-fertilized plants. The mechanisms might include both adaptation and nutrient balance. The condition of low available nutrients imposes a stress on the plants, which forces plants to develop more, longer and more active roots scavenging for nutrients. On the other hand, the organic fertilizer contains more kinds of nutrients which makes the nutrients more balanced than the

FIGURE 24. Dynamic changes in nitrogen in the soils and the total in the soil and plant of sweet corn fertilized with organic materials and chemicals.

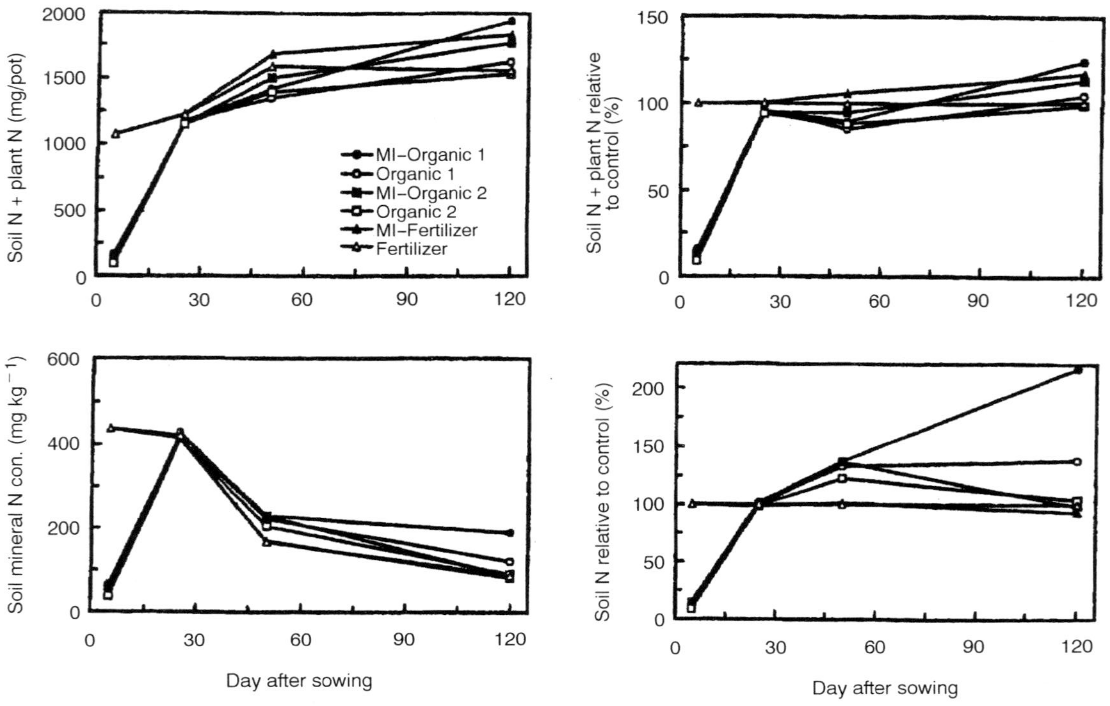

FIGURE 25. Dynamic changes in available phosphorus in the soils and the total in the soil and plant of sweet corn fertilized with organic materials and chemicals.

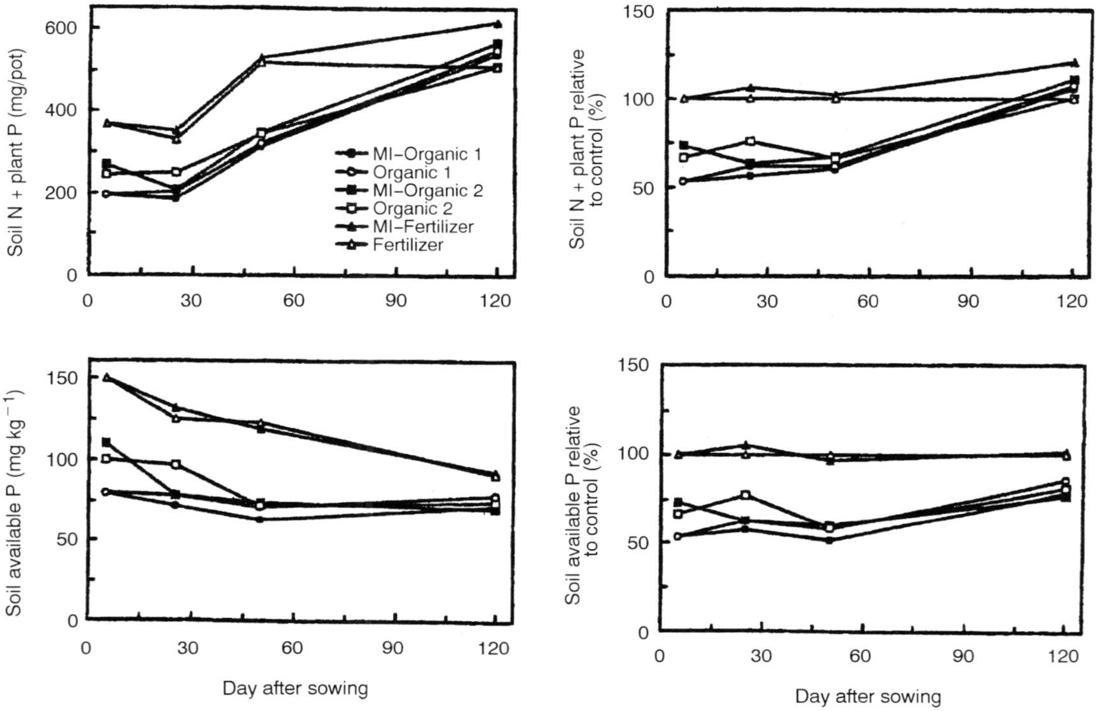

chemical fertilizer that contains only N-P-K. These are just speculative hypotheses and we do not have experimental data to support these suggestions. Because the organic-fertilized plants developed a good root system with more and longer seminal and nodal roots, they possessed a greater potential to encounter and uptake nutrients which became available at the middle and later growth stages. Therefore, the relative growth rate and net assimilation rate based on unit of leaf area was much higher in organic-fertilized than chemical-fertilized plants. The data on photosynthesis supported the results of growth analysis. The photosynthetic capacity was lower in organic-fertilized plants at the early stage, but surpassed that of chemical-fertilized plants after the middle growth stages. Although the organic-fertilized plants could not develop a leaf area greater than or similar to that of chemical-fertilized plants up to the harvest stage, organic-fertilized plants produced a final grain yield higher than or similar to that of chemical-fertilized plants because of the higher photosynthetic rate. The photosynthetic capacity at the later growth stages was higher in the organic-fertilized plants than in chemical-fertilized plants, especially in leaves at lower positions. This meant that the senescence was delayed in organic-fertilized plants. Analysis of the photosynthetic hysteresis also supported this hypothesis. The net photosynthetic rate increased in an exponential manner as PPF increased from low (0) to high (2000 μmol m^{-2} s^{-1}). However, when the PPF was reversed from high to low for the same leaf, the photosynthetic rate was higher at most PPF points during the reversing PPF changes than during the first low-to-high changes. This phenomenon is called hysteresis (Poulovassilis, 1962; Xu et al., 1994). Hysteresis occurs especially when no sufficient time is allowed for completion of the equilibrium at one PPF point (Jones, 1983a). Some authors suggest that photosynthetic hysteresis is caused by stomatal response (Warrit et al., 1980). Usually, in the senescent or water stressed leaves, it is more difficult for the stomata to open fully than those in normal plants in response to PPF changes (Xu et al., 1994). In the present study, organic-fertilized plants show a smaller photosynthetic hysteresis than chemical-fertilized plants. This suggested that, at the later growth stage, there were fewer constraints in leaves of organic-fertilized plants than in leaves of chemical-fertilized plants. Ultimately, the above mentioned results might be associated with both sustainable nutrition of the organic fertilizer and the more active root system where not only the nutrients are absorbed, but some kinds of phytohormones are produced. If the problem of low availability of nutrients in organic materials is solved, organic crop cultivation will lead to a similar or higher productivity compared with chemical farming.

It has been known that long-term fertilization with organic materials improve soil physical, chemical and biological properties (Hillel, 1980). Growth and activity of the root system are promoted by the improved soil properties (Jones, 1983b). To date, there have been many studies on organic farming and use of organic fertilizers (USDA, 1980; Lockeretz and Kohl, 1981; Harwood, 1984; Vogtmann, 1984). However, the mechanism of the effect of EM used in this study is not clear in many aspects. Therefore, the effects of EM applications on plant growth and physiology were examined in a detailed way. The microbial inoculant used in the present study, i.e., Effective Microorganisms or EM, contained a group of beneficial microorganisms. EM applications with both organic and chemical fertilizers promoted plant growth at all growth stages and increased grain yield as a consequence. This was attributed to increased photosynthetic capacity and nutrient availability. The root quantity shown by root number and root length and the root quality shown by the respiration rate were increased by EM applications. Senescence in terms of photosynthetic performance was delayed by EM applications. The exact mechanisms of how EM elicits these effects are not known. The applications and use of microbial inoculants have often preceded a thorough scientific investigation. Many observations of growth promotions, yield increases, and quality improvements are reported by the farmers without strict scientific controls. However, research on the individual species of beneficial microorganisms contained in EM used in the present study have been thoroughly investigated and conducted over times. Some phytohormones and the derivatives are synthesized by soil microbes including some species contained in EM used in this study (Arshad and Frankenberger Jr., 1992). Barea et al. (1976) found that among 50 bacteria isolated from the rhizosphere of various plants, 86, 58, and 90% produce auxins, gibberellins, and kinetin-like substances, respectively. Kampert et al. (1975) reported that 55% of bacteria and 86% of fungi isolated from the rhizosphere of *Pinus silvestris* could produce gibberellins and the derivatives. *Actinomycetes* and *Streptomycetes* produce auxins and similar substances (Purushothaman et al., 1974; Mahmoud et al., 1984), gibberellins (Arshad and Frankenberger Jr., 1992), and cytokinins (Bermudez de Castro et al., 1977; Henson and Wheeler, 1977). Some fungi like *Aspergillus niger* produce gibberellins (El-Bahrawy, 1983). The promotion of root development and activity by EM applications might be due to the effects of plant growth regulators produced by inoculated microbes. However, we do not have available data to support this hypothesis. Some of the plant growth regulators produced by microorganisms contained in the microbial inoculant used in the present study (i.e., EM) are presented in Table 4. Further studies are necessary to examine the mechanistic basis for the effects of EM on plant growth. EM has been used in agriculture, especially in organic farming systems in advance of fundamental researches. Therefore, we need to elucidate the problems remaining in the technology of EM applications.

TABLE 4. Plant growth regulator (PGR) producing microorganisms contained in the microbial inoculant used in the present study.

Microbe	PGR produced	Researcher	Document
Actinomycetes spp.	IAA, ICA	Larsen et al. 1962	Phyiol Plant 15: 552-566.
	IAA, ICA	Kaunat 1969	Zentrbl Bakteriol Abt II 123: 501-515.
	IAA	Purushothaman et al. 1974	Curr Sci 43: 413-414.
	IAA	Clark 1974	Microbios 11A:29-35.
	IAA	Barea et al. 1976	J Appl Bacteriol 40: 129-134.
	GLS	Panosyan et al. 1963	*Proc Symp Rel Soil Micro Plant Root*, Prague, p. 241-244.
	GLS	Kampert et al. 1975	Acta Microbiol Pol 7: 157-166.
	Cytokinins	Bermudez de Castro et al. 1977	*Recent Developments in Nitrogen Fixation*, Academic, London, p. 539-550.
	Cytokinins	Henson & Wheeler 1977	Z Pflanzenphysiol 84: 179-182.
Streptomyces spp.	IAA, IPyA	Mahmoud et al. 1984	Zentrbl Mikrobiol 139: 227-232
Aspergillus spp.	IAA, GLS, GA	El-bahrawy 1983	New Phytol 94: 401-407.
	IAA, GLS, GA	Dvornikova et al. 1968	Microbiol 37: 190-193.
Mucor hiemalis	Ethylene	Lynch 1974	J Gen Microbiol 83: 407-411.
S. cerevisiae	Ethylene	Thomas and Spencer, 1977	Can J Microbiol 23: 1669-1974.

REFERENCES

Arshad, M., and W. T. Frankenberger Jr. (1992). Microbial production of plant growth regulators. In *Soil Microbial Ecology*, ed. F. B. Metting Jr. New York: Marcel Dekker, Inc, pp. 307-348.

Barea, J. M., M. I. Navarro and E. Montoya. (1976). Production of plant growth regulators by rhizosphere phosphate-solubilizing bacteria. *Journal of Applied Bacteriology* 40: 129-134.

Bermudez de Castro, F. A. Canizo, A. Costa, C. Miguel and C. Rodriguez-Barrueco. (1977). Cytokinins and nodulation of the non-legumes *Alnus glutinosa* and *Myri-*

ca gale. In *Recent Developments in Nitrogen Fixation*, eds. W. Newton, J. R. Postgate and C. Rodriguez. London: Academic Press, pp. 539-550.

Clark, A. G. (1974). Indole acetic acid production by *Agrobacterium* and *Rhizobium* species. *Microbiology* 11A: 29-35.

Dvornikova, T. P., G. K. Scryabin and N. N. Suvorov. (1968). Tryptamine conversion by *Aspergillus niger*. *Microbiology* 37: 190-193.

El-Bahrawy, S. A. (1983). Associative effect of mixed cultures of *Azotobacter* and different rhizosphere fungi determined by gas chromatography-mass spectrometry. *New Phytologists* 94: 401-407.

Harwood, R. R. (1984). Organic farming research at the Rodale Research Center. In *Organic Farming: Current Technology and Its Role in a Sustainable Agriculture*, eds. D. F. Bezdicek, J. F. Power, D. R. Keeney and M. J. Wright. Madison: American Society of Agronomy, pp. 1-18.

Henson, I. E. and C. T. Wheeler. (1977). Hormones in plants bearing nitrogen-fixing root nodules: Cytokinins in roots and root nodules of some non-leguminous plants. *Journal of Plant Physiology* 84: 179-782.

Higa, T. (1994) *The Completest Data of EM Encyclopedia*. Tokyo: Sogo-Unicom, pp. 1-385 (In Japanese).

Hillel, D. (1980). *Fundamentals of Soil Physics*. New York: Academic Press, pp. 195-222.

Ishitsuka, J. (1985). Measurements of minerals and nitrogen compounds. In *Experiment Methods of Crop Physiology*, eds. Y. Kitajo and J. Isjitsuka. Tokyo: Association of. Agricultural Technology, pp. 281-316.

Jones, H. G. (1983a). Estimation of an effective soil water potential at the root surface of transpiring plants. *Plant, Cell Environment* 6: 671-674.

Jones, H. G. (1983b). *Plant and Microclimate–A Quantitative Approach to Environmental Plant Physiology*. London: Cambridge University Press, pp. 1-323.

Kampert, M., E. Strzelczyk and A. Pokojska. (1975). Production of gibberellin-like substances by bacteria and fungi isolated from the roots of pine seedlings (*Pinus sylvesreis* L.). *Acta Microbiologica*. 7: 157-166.

Kaunat, H. (1969). Bdung von indolederiaten durch rhizosphärenszifisch Bakterien und Aktinomyzeten. *Zentrbl. Bakteriol. Abt. II* 123: 501-515.

Larsen, P., A. Harbo, S. Klungsöyr and T. Aaseim. (1962). On the biogenesis of some indole compound in *Acetobecter xylinum*. *Physiologia Plantarum* 15: 552-565.

Lockeretz, W., G. Shearer and D. H. Kohl. (1981). Organic farming in the corn belt. *Science* 211: 540-547.

Lynch, J. M. (1974). Mode of ethylene formation by *Mucor hiemalis*. *Journal of Genetic Microbiology* 83: 407-411.

Mahmoud, S. A. Z., M. I. Ramadan, F. M. Thabet and T. Khater (1984). Production of plant growth promoting substances by rhizosphere microorganisms. *Zentrbl. Mikrobiol*. 139: 227-232.

Morita, S., S. Thongpae, J. Abe, T. Nakamoto and K. Yamazaki. (1992). Root branching in maize. I. "Branching index" and methods for measuring root length. *Japanese Journal of Crop Science* 61: 101-106.

Nakaseko, K. (1985). Measurements of plant productivity. In *Experiment Methods of*

Crop Physiology, eds. Y. Hokujo and I. Isjitsuka. Association of Agricultural Technology, Tokyo, pp. 232-254.

Okada, M. (1987). *The True Health*. Church World Messianity, U.S.A. 184 p.

Panosyan, A. K., Z. V. Marshavina, R. S. Arutunyan and S. G. Aslanyan. (1963). The nature of physiologically active substances of actinomycetes and the effect of their metabolites on plant growth. In *Plant Microbes Relationship–Proceedings of the Symposium on Relations Between Soil Microorganisms and Plant Root*, eds. J. Macura and V. Vancura. Prague. pp. 241-244.

Poulovassilis, A. (1962). Hysteresis of pore water, an application of the concept of independent domain. *Soil Science* 93: 405-412.

Purushothaman, D., T. Marimuthu, C. V. Venkataramanan, and R. Kesavan. (1974). Role of actinomycetes in the biosynthesis of indole acetic acid in soil. *Current Science* 43: 413-414.

Thomas, K. C. and M. Spencer. (1977). L-Methionine as an ethylene precursor in *Saccharomyces cerevisiae. Canadian Journal of Microbiology* 23: 1669-1674.

Vogtmann, H. (1984) Organic farming practices and research in Europe. In *Organic Farming: Current Technology and Its Role in a Sustainable Agriculture*, eds. D. F. Bezdicek, J. F. Power, D. R. Keeney and M. J. Wright. Madison: American Society of Agronomy. pp. 19-36.

Warrit, B., J. J. Landsberg and M. R. Thorp. (1980). Responses of apple leaf stomata to environmental factors. *Plant, Cell and Environment* 3: 13-22.

Xu, H. L., L. Gauthier and A. Gosselin. (1994). Responses of the photosynthetic rate to photon flux density in tomato plants affected by high electrical conductivity of nutrient solution and low water content in substrate. *Photosynthetica* 30:279-286.

Use of Effective Microorganisms to Suppress Malodors of Poultry Manure

Weijiong Li
Yongzhen Ni

SUMMARY. A serious problem and major constraint to the development of more efficient and sustainable animal and poultry production system in China is the generation of malodors from manure in rearing facilities and from handling, storage and use of these wastes after removal, particularly in the urban and suburban sectors. Because of their high malodor potential, these wastes are not effectively recycled on agricultural land. A cooperative research study was conducted with the International Nature Farming Research Center (INFRC), Atami, Japan to evaluate the use of a microbial inoculant, Effective Microorganisms or EM, to suppress malodors of poultry manure. Feeding trials consisted of adding EM either to drinking water and feed, or to both water and feed, and comparing the results with non-EM controls. EM markedly reduced the malodor level of the poultry manure, associated mainly with a dramatic decrease in the ammonia (NH_3) levels (i.e., 42 to 70% lower than the controls). The amino acid content of the feed increased 28% after it was inoculated and fermented with EM. EM also improved the growth and disease resistance of the birds and net returns in the marketplace. These results indicate that EM use in poultry operations has great potential for suppressing malodors of manure, improving sustainable production, and protecting the environment, all on a cost-effective basis. *[Article copies available for a fee from The Haworth Document Delivery Service: 1-800-342-9678. E-mail address: getinfo@haworthpressinc.com <Website: http://www.HaworthPress.com>]*

KEYWORDS. Animal waste, effective microorganisms, EM, malodor, microbial inoculant, poultry manure, waste management

---

Weijiong Li and Yongzhen Ni are Professors, College of Resource and Environment, China Agricultural University, Beijing 100094, China.

[Haworth co-indexing entry note]: "Use of Effective Microorganisms to Suppress Malodors of Poultry Manure." Li, Weijiong, and Yongzhen Ni. Co-published simultaneously in *Journal of Crop Production* (Food Products Press, an imprint of The Haworth Press, Inc.) Vol. 3, No. 1 (#5), 2000, pp. 215-221; and: *Nature Farming and Microbial Applications* (ed: Hui-lian Xu, James F. Parr, and Hiroshi Umemura) Food Products Press, an imprint of The Haworth Press, Inc., 2000, pp. 215-221. Single or multiple copies of this article are available for a fee from The Haworth Document Delivery Service [1-800-342-9678, 9:00 a.m. - 5:00 p.m. (EST). E-mail address: getinfo@haworthpressinc.com].

INTRODUCTION

It is estimated that at the present time there are about 2.3 billion chickens being reared intensively in China. This number of birds produces an annual output of 84 million tons of fresh manure or 21 million tons on a dry weight basis. This output combined with the manure of cattle, swine, ducks, geese and other animals amounts to an annual production of several billion tons, much of which contributes to severe problems of environmental pollution, and human and animal health. Improper handling and storage of poultry manure from rearing facilities has generated malodors, attracted flies, and adversely impaired air quality, especially in the urban and suburban sectors. Impacted citizens continue to submit complaints and grievances about the problem, particularly in the spring and fall months when malodors are most intense. In response, the poultry industry has explored alternative methods and techniques for processing and utilizing the manure to alleviate malodors including methane production through anaerobic digestion, high temperature drying, and heat spraying methodology. Unfortunately, these technologies are energy-consuming ventures that require long-term large capital investments. Thus, the problem of malodors of manure from rearing facilities remains unsolved and continues to intensify.

In seeking alternatives and solutions to the problem of malodors we became aware of a new technology called "Effective Microorganisms" or EM, a microbial inoculant that has been developed by professor Teruo Higa, University of the Ryukyus, Okinawa, Japan. EM is a mixed culture of naturally-occurring beneficial microorganisms (i.e., predomiumthy lactic acid bacteria, photosynthetic bacteria, yeast, actinomycetes, and certain fungi). EM has reportedly improved soil quality; the growth, yield and quality of crops; and suppressed the generation of malodors from swine and poultry manure (Higa, 1993; Li et al., 1998). In view of this, we established a cooperative research project with the International Nature Farming Research Center (INFRC) to determine whether EM technology could help to resolve the malodor problem. In doing so, we applied the principle of micro-ecological engineering combined with biological technology (Ma and Li, 1987). This paper presents the results of this project.

MATERIALS AND METHODS

Preparation and Presentation of Test Materials

Treatment of poultry feed with EM. Poultry feed was inoculated with EM and subject to anaerobic fermentation to increase the population and activity of EM microorganisms. Fermentation also increases the availability of feed nutrients and their utilization after ingestion to enhance animal growth and weight gain (Li and Ni, 1995a, b). The period and temperature of fermentation are considerably influenced by ambient conditions and seasons. For example, in warm summer months complete anaerobic fermentation may require only 3 to 5 days, while in cold winter months it may take twice as long, possibly 7 to 8 days. Anaerobic condition also suppresses the growth of molds or other contaminating organisms. Feed that is successfully fermented will have a pleasant "sweet and sour" odor and large populations of EM-inoculated microorganisms, often reaching levels of 1×10^8 organisms per gram of material. Nevertheless, the extract quantitative index of EM fermentation of various feeds is in need of further research and refinement.

Treatment of drinking water with EM. Presenting EM to poultry at all stages of growth and maturity is relatively simple, since liquid EM culture is easily miscible with water, and dilutions can be selected to provide the desired intake depending on age/size of the bird, the drinking water device, and the water metering system. Even so, some poultry farmers prefer not to add EM through the drinking water because of these and other variables that can influence the rate of EM intake per bird.

Experimental design and analysis. Table 1 shows the main treatment designations for the study including the group species and number of birds to determine the effect of EM applied to drinking water or feed, or both water and feed on malodor levels of poultry manure. A non-EM control treatment was established for each of the three main treatments with the same number of birds, same kind and amount of common feed and drinking water (no EM added), and all other conditions the same.

All treatments were applied to separate rooms or compartments and tests were conducted for 30 days. Air samples were collected throughout and ammonia concentration determined by the Kjeldahl Method. Amino acid composition and concentration of the feed were determined before and after inoculation and anaerobic fermentation with EM. An economic analysis was conducted to determine the effect of EM on rate of weight gain, market weight, feed conversion efficiency and the meat:feed ratio.

TABLE 1. Main treatment designations, including group species and number of birds, to determine the effect of EM applied to drinking water or feed, or both water and feed on malodor levels of poultry manure

Treatment	Group/species	Number	EM Applied
1	Ruman table poultry	400	EM drinking water + common feed
2	Dike laying poultry	500	EM feed + common water
3	AA laying poultry	50	EM feed + EM drinking water

RESULTS

The main constituents of animal and poultry wastes that induce malodors are ammonia, hydrogen sulfide, mercaptan and methylmercaptan. Because ammonia usually occurs in higher concentrations in the gaseous environments of animal and poultry rearing facilities, and because it is known to have adverse effects on human and animal health, the ammonia concentration in these facilities is often used as an environmental index. In the present study, the effects of adding EM to drinking water (Treatment 1), to feed (Treatment 2), and to both drinking water and feed (Treatment 3) on the ammonia concentrations from poultry manure and percentage reduction due to EM compared with non-EM controls are reported in Table 2. The greatest reduction in ammonia concentration was 69.7% and occurred when EM was added to both drinking water and feed. The second and third best reductions in ammonia concentration were 54.25% and 41.12% and occurred with treatment 2 (EM added to feed) and treatment 1 (EM added in drinking water), respectively.

The test was conducted for 30 days after which ammonia (NH_3) concentrations were monitored continuously for the number of days indicated.

The dramatic reduction in ammonia concentration, especially where EM was inoculated into feed and fermented, suggests that ammonia was utilized in some biosynthetic pathway. An analysis of the composition and concentration of amino acid before and after inoculation and fermentation with EM showed that the concentration of amino acids in EM-fermented feed had increased 28% (Table 3). This suggests that EM microorganisms accelerate nutrient transformations and availability, enhancing the rate of uptake and utilization by the animal which, in turn raises the nutrient value of the feed and increases the rate of gain and production of animal products. At the same time, EM microorganisms likely suppress the growth and activity of putrefactive microorganisms in the intestine that produce malodorous nitrogen compounds, thereby reducing the concentration of ammonia in their manure.

The action of EM increases the utilization and transfer rate of nitrogenous

TABLE 2. Effect of EM treatments on ammonia concentrations from poultry manure and percentage reduction due to EM compared with non-EM controls

Treatment (No.)	Sampling time (Days)	NH_3 (ppm) Controls (No EM)	Main treatment EM	NH_3 reduction by EM (%)
1	3	8.95	5.18	42.12
2	6	16.13	7.38	54.25
3	3	87.6	26.5	69.7

compounds from the feed to the growth and production phase of animal and poultry. For example, egg production by some laying hen facilities has increased by 13% from use of EM, which has also increased the length of the laying period. Table 4 shows an economic analysis of the effect of EM (in feed) on broiler production. EM increased the rate of weight gain, increased meat production, increased the feed utilization efficiency and increased the meat:feed ratio compared with the non-EM control group. These results show that EM technology can suppress malodors of poultry manure without a great

TABLE 3. The analyzed results of the amino acids concentration in the EM fermented foodstuff of poultry

Amino Acid	A. After treatment (%)	B. Before treatment (%)	A/B
Aspartic	2.94	2.48	1.19
Threonine	1.25	1.05	1.19
Serine	1.88	1.63	1.15
Glutamate	6.17	3.60	1.71
Glycine	1.74	1.39	1.25
Alanine	1.71	1.37	1.25
Cystine	0.45	0.42	1.07
Valine	1.17	1.01	1.16
Methionine	0.63	0.48	1.31
Isoleucine	0.97	0.78	1.24
Leucine	2.16	1.81	1.19
Tyrosine	0.93	0.85	1.09
Phenylalanine	1.38	1.16	1.19
Lysine	1.02	0.84	1.21
Histidine	0.72	0.57	1.26
Arginine	1.89	1.68	1.13
Proline	3.96	3.16	1.25
Tryptophan	–	–	–

TABLE 4. Economic analysis of EM for feeding table poultry (Yuan)

Treatment	Chicken cost	Feed cost	EM cost	Revenue from chicken sale	Net gain	Relative gain	Meat:feed ratio
EM	100	207.74	10.0	427.00	109.26	125.18	2.06
CK	100	227.73	0.0	415.01	87.28	100.00	1.82

All values in the table are in terms of Chinese Yuan. Exchange rate is 1 US$ = 8.3 Chinese Yuan.

investment of money and equipment, and likely would improve the management and efficiency of poultry farms.

DISCUSSION

In recent years, our applied efforts in the poultry rearing industry have indicated that EM fed to birds can effectively suppress malodors of the manure, improve the growth of animals and enhance their disease resistance, and thereby improve the local environment surrounding the rearing facility. EM applied in feed and drinking water improves the feed conversion rate and other production parameters as well as social and economic benefits (Cai and Huo, 1993). Thus, EM provides a multiple function of micro-ecological applications that are related to its predominant microbial types and numbers, i.e., lactic acid bacteria, photosynthetic bacteria, yeast, actinomycetes and certain fungi.

The ability of EM to suppress malodors of animal manure, either through direct or indirect mechanisms or modes-of-action require further study and evaluation. According to our research and observations, the following mechanisms maybe involved: (1) EM is a mixed culture of many species of naturally-occurring, beneficial microorganisms (Li and Ni, 1995a, b) some of which can transform NH_4^+-N to NO_3-N, thereby decreasing the potential for N-volatilization and increasing the potential for nitrogen fixation (through photosynthetic bacteria); (2) EM contains beneficial microorganisms which come to reside in the animals intestines as feed and drinking water are utilized. The EM microorganisms suppress the growth and activity of the indigenous putrefactive types that cause malodors in the manure and transform proteins and amino acids into NH_3-N and NH_4^+-N; (3) Thus, the EM in the intestines reduce the ammonia levels in the manure and blood (Lu, 1992; Kang, 1988). These three combined actions of EM may transfer protein in the feed into effective and available nutrients, thereby increasing the feed utilization rate and suppressing malodor production by indigenous intestinal microorganisms, alleviating the problem of malodors from manure both at the rearing farm and the surrounding urban/suburban environment. This suggested action of EM is similar to the "anti-oxidation theory" proposed by Dr. Teruo Higa, the innovator of EM technology (Higa, 1993).

Feeding EM allows animals to utilize beneficial microorganisms that enhance their health and well-being and restore their micro-ecological balance. This helps the farmer to avoid the malpractice of using commercial antibiotics, which can disrupt the micro-ecological balance and reduce the animal's natural recuperative capacity. Thus, EM allows the animal to produce high quality products for the marketplace, which increases the farmer's profitability and net returns. All of this means that the farmer can better conserve energy and resources; recycle animal wastes and manure to improve soil productivity and avoid malodors throughout (i.e., from rearing facilities to the external environment); protect the natural environment; and reduce the usage of chemical fertilizers, pesticide and antibiotics.

REFERENCES

Cai Huiyi and Huo Qiguang. (1993). The research and applied progress of the feeding microbial additive, *Foodstuff Industry* (Beijing) 14:7-12.

Higa, T. (1993). Effective Microorganisms and their role in Kyusei Nature Farming and sustainable agriculture. In *Proceedings of the Fourth International Conference on the technology of Effective Microorganisms*, Saraturi, Thailand, pp. 1-6.

Kang Bai. (1988). *Micro-Ecology.* Beijing: Dalian Publishing Press, pp. 114-271.

Li Weijong and Ni Yongzhen. (1995a). The research of EM (Effective Microorganisms). *Journal of Ecology* (Beijing) 14:58-62.

Li Weijong and Ni Yongzhen. (1995b). Application of microbiological engineering in ecological animal husbandry. *Journal of Ecology* (Beijing) 14:6-9.

Li Weijong, Ni Yongzhen and H. Umemura. (1998). Effect Microorganisms for sustainable animal production in China. In *Proceedings of the Fourth International Conference on Kyusei nature Farming*, eds. J. F. Parr, S. B. Hornick and M. E. Simpson. Washington, DC: U. S. Department of Agriculture, pp. 171-173.

Lu Tingfang. (1992). The effects of feeding microorganisms directly. *Foreign Animal Husbandry (Pig and Poultry)* 1:10-13.

Ma Shijun and Li Songhua. (1987). *Chinese Agricultural Ecological Engineering*. Beijing: Science Publication Press, pp. 1-10.

Effects of a Microbial Inoculant, Organic Fertilizer and Chemical Fertilizer on Water Stress Resistance of Sweet Corn

Hui-lian Xu

SUMMARY. A glasshouse pot study was conducted to determine the effect of a microbial inoculant (with commercials as EM), organic fertilizers and chemical fertilizer on water stress resistance of sweet corn plants (*Zea mays* L. cv. Honey Bantam). The microbial inoculant is known as Effective Microorganisms or EM and is a mixed culture of 80 species of naturally-occurring beneficial microorganisms with predominant numbers of the following types and genera: (1) Lactic acid bacteria (*Lactobacillus*); (2) Photosynthetic bacteria (*Rhodopseudomonas*): (3) Yeasts (*Saccharomyces*); (4) Actinomycetes (*Strptomyces*); and (5) Fungi (*Aspergillus*). Five weeks after sowing and treatment applications, the plants were subjected to soil water deficits, with photosynthetic rate (P_N) maintenance as an indicator; water stress resistance (R_{WS}) of plants was calculated as tolerance (T_{WS}) or avoidance (A_{WS}). For example, plants treated with organic fertilizer or EM showed higher R_{WS}, i.e., they maintained a higher photosynthetic rate under soil water deficits than those treated with chemical fertilizers and without EM. The relatively high R_{WS} was attributed largely to water stress avoidance (A_{WS}), whereby plants maintained a relatively high leaf water potential (Ψ_L) and a relatively high P_N as a consequence. However, water stress tolerance (T_{WS}), i.e., P_N maintenance ability

Hui-lian Xu is Senior Crop Scientist, International Nature Farming Research Center, 5632 Hata, Nagano 390-1401, Japan (E-mail: huilian@janis.or.jp).

The author thanks Shingo Ishiko, Xiaoju Wang, Jihua Wang, Masao Fujita, Teruo Higa and Hiroshi Umemura for their technical assistance and advice in this research.

[Haworth co-indexing entry note]: "Effects of a Microbial Inoculant, Organic Fertilizer and Chemical Fertilizer on Water Stress Resistance of Sweet Corn." Xu, Hui-lian. Co-published simultaneously in *Journal of Crop Production* (Food Products Press, an imprint of The Haworth Press, Inc.) Vol. 3, No. 1 (#5), 2000, pp. 223-233; and: *Nature Farming and Microbial Applications* (ed: Hui-lian Xu, James F. Parr, and Hiroshi Umemura) Food Products Press, an imprint of The Haworth Press, Inc., 2000, pp. 223-233. Single or multiple copies of this article are available for a fee from The Haworth Document Delivery Service [1-800-342-9678, 9:00 a.m. - 5:00 p.m. (EST). E-mail address: getinfo@haworthpressinc.com].

(PMA) at a given Ψ_L, did not differ between organic- and chemical-fertilized plants or between EM-treated and untreated plants. The enhanced water stress avoidance by plants treated and grown with organic fertilizer and the microbial inoculant is likely due to the more extensive development and higher respiratory activity of the roots of these plants compared with the other treatments.

KEYWORDS. Drought resistance, Effective Microorganisms, microbial inoculant, organic farming, organic fertilizer, nature farming, photosynthesis, water stress

INTRODUCTION

Recent concerns about cost increases, food safety and quality, and environmental protection have prompted plant scientists to reevaluate the current practices in chemical agriculture. Alternative production methods have been proposed in horticulture and agriculture to reduce production costs, improve food safety and quality and reduce environmental degradation (Harwood, 1984; Lockeretz and Kohl, 1981; USDA, 1980; Vogtmann, 1984). The alternatives are generally referred to as "organic farming" practices, although other definitions such as sustainable agriculture and nature farming could be appropriate in special cases (Higa, 1994; Fukuoka, 1985). Supplemental organic matter can increase the soil water holding capacity and thereby minimize water deficits during drought (Hillel, 1980; Huck, 1982; Letey, 1977). Recently, the use of a microbial inoculant widely known as Effective Microorganisms or EM was introduced to the nature farming system in Japan (Higa, 1994). This inoculant is a mixed culture of about 80 species of naturally-occurring beneficial microorganisms and can enhance the fermentation of compost and organic fertilizers. It has also been reported that EM additions to the organic materials or directly to the soil can promote root development and the growth and yield of crops (Higa, 1994; Li and Ni, 1996). However, there have been no reports to suggest that EM can improve drought resistance of crop plants. Recently, some reports have indicated that EM applications to crops may be more beneficial in semiarid regions than in humid areas (Li and Ni, 1996). This suggests that EM application could be associated with soil water relations. Therefore, a study was conducted to determine whether water stress resistance of sweet corn could be significantly affected when plants were treated and grown with EM as a microbial inoculant applied with and without organic or chemical fertilizer.

Usually, in research on plants surviving a dry environment, the term "drought resistance" is used (Levitt, 1980). However, drought is a meteorological term, and is commonly defined as a period without significant rainfall and/or with a low air humidity. In a controlled pot experiment, where water stress is artificially imposed on plants, it is better to substitute the term "water stress resistance" for "drought resistance" as proposed by Ludlow et al. (1983). In ecological research, drought or water stress resistance is defined as the ability of plants to survive dry environmental conditions; while it refers to the economic yield performance under dry conditions in agronomic research (Aspinal et al., 1965; Fischer and Maurer 1978). In the present research, the maintenance of photosynthesis under soil water deficit conditions was used as an indicator of water stress resistance because photosynthesis is one of the most fundamental physiological processes associated with both survival and yield under drought conditions. Levitt (1980) has separated drought resistance into drought tolerance and avoidance for mechanism analysis. Plants may use tolerance and/or avoidance mechanisms to resist water stress induced by low soil moisture. Even when plants resist a similar soil water deficit, the mechanisms might be different. Some plants maintain a high Ψ_L and as a consequence a high growth rate or physiological activity, while some maintain a high growth rate or physiological activity even if Ψ_L decreases in response to the soil water deficit. Therefore, in this study, water stress resistance was separated into tolerance and avoidance, and the mechanisms for effects of a microbial inoculant (commercial name, EM) applied with organic or chemical fertilizer on plant water relations were investigated and discussed.

MATERIALS AND METHODS

Plant Materials

Sweet corn (*Zea mays* L. cv. Honey Bantam) plants were grown in plastic pots each with a soil surface area of 0.02 m^2 and a height of 0.25 m. One plant was grown in each pot under glasshouse conditions from June to October, 1996. The daily temperature and air humidity were not controlled and fluctuated with ambient conditions. The photoperiod was 12 h in June and 9 h in October.

Soil

A fine textured Andosol soil was taken from a field where a soybean crop was previously cultivated. The N-P-K levels in the soil were 3.4-0.025-0.44 g kg^{-1} with a C:N ratio of 13. The field capacity or capillary capacity of the soil was 80% on a gravimetric basis.

Fertilizers

Chemical fertilizers used in this study were ammonium sulfate, superphosphate and potassium sulfate. Anaerobically- and aerobically-fermented organic materials were used as the organic fertilizers and designated as Organic 1 and Organic 2, respectively. These were produced with oilseed mill sludge (10 kg), rice bran (30 kg), fish-processing by-product (10 kg) with molasses (80 ml) and the microbial inoculant (EM-1, 800 ml) inoculated and fermented under anaerobic or aerobic conditions, respectively. The water content was adjusted to 300 g kg^{-1}. The total N-P-K concentrations of both organic fertilizers were 58, 30 and 2 g kg^{-1}, respectively.

Microbial Inoculant

The microbial inoculant used in the study was EM-1 produced by the Effective Microorganisms Institute, International Nature Farming Research Center, Atami, Japan. The microbial inoculant consisted of about 80 species of naturally-occurring, beneficial microorganisms that can co-exist under liquid culture conditions. The predominant species in the microbial inoculant are: (1) Lactic acid bacteria: *Lactobacillus, Lactobacillus casei and Streptococcus lactis*; (2) Photosynthetic bacteria: *Rhodopseudomonas palustris* and *Rhodobacter sphaeroides*; (3) Yeasts: *Saccharomyces cerevisiae* and *Candida utilis*; (4) Actinomycetes: *Streptomyces albus* and *Streptomyces griseus*; (5) Fungi: *Aspergillus oryzae* and *Mucor hiemalis*; and (6) Others: some microorganisms that are naturally-occurring and introduced into the microbial inoculant in the manufacturing process, which can survive in the microbial inoculant liquid at pH 3.5 and below. The density of most of the above mentioned microbes are in the range of 1×10^6 to 1×10^8 per ml.

Fertilizer Treatments

Each of the pots was filled with 3000 g of an Andosol soil (water content 30%, w/w). Six fertilizer treatments were implemented as follows: (1) EM + Organic 1: anaerobically-fermented organic materials (80 g), in which EM was added to the organic materials and then fermented; (2) Organic 1: anaerobically-fermented organic materials (80 g); (3) EM + Organic 2: aerobically-fermented organic materials (80 g), in which EM was added before fermentation; (4) Organic 2: aerobically-fermented organic materials (80 g); (5) EM + Chemical Fertilizer: chemical fertilizers (with the same N-P-K levels as in organic fertilizers, ammonium sulfate 5.3 g, long-period coated urea 2.8 g, superphosphate 13 g and potassium sulfate 4.95 g, with EM 80 ml applied into soil the same time before sowing; and (6) Chemical Fertilizer: the same fertilizers as in (5) as the control.

Soil Water Treatment

Five weeks after sowing, the soil water treatments started. Pots were separated into six groups. On the first day, all pots in the six groups were irrigated to saturation and on the second day water supply was stopped for pots in Group 1 but pots in the other five groups were still irrigated to saturation. The third day, irrigation of pots in Group 2 was stopped and others were watered to saturation again. Finally, on the sixth day, the pots in the six groups had different water contents. Because the soil water content was not exactly the same even within the same group, every group was separated into two plots according to the soil water content. Therefore, the total twelve soil water contents were obtained, and consequently plants in different groups experienced different water regimes. Three plants from each soil water regime were used for measurements of photosynthesis and leaf water potential.

Measurements of Photosynthesis

Six days after the soil water depletion started, the photosynthetic rate in the seventh leaf of plants with different soil water conditions was measured using an open gas exchange system (LI-6400, Li-Cor, Inc., Lincoln, Nebraska, USA). Leaf temperature was constant at 23°C and the relative humidity was maintained at 65% in the assimilation chamber during the photosynthetic rate measurement. Photosynthetic light response curves were obtained by measuring the photosynthetic rate at different photosynthetic photon fluxes under an LED light source. The maximum photosynthetic rate at 2000 μmol $m^{-2} s^{-1}$ was recorded as the photosynthetic activity.

Measurement of Leaf Water Potential

Leaf water potential (Ψ_L) was measured using a pressure chamber (Model 3000, Soilmoisture Equipment Corp. Santa Barbara, California, USA) on the same day as the photosynthetic rate was measured. Because the leaf blade of corn is much broader and longer than those of other cereal crops, a particular procedure was adopted. The top half part of the leaf blade was cut, aligned on the adhesive surface of a 60-cm wide Scotch tape, and then folded along the main vein. The leaf blade was then enclosed in the Scotch tape to prevent evaporative water loss and the pressure gas penetration into the tissue. The prepared leaf blade sample was set in a pressure chamber for measurement of Ψ_L. The tip of the leaf blade with the tape was cut and the cut end was left 1 cm above the rubber stop. Other processes remained the same in using the pressure chamber technique by Turner (1988).

Measurement of Soil Water Potential

Because total soil water potential could not be easily measured *in situ* under undisturbed conditions, soil matric water potential was measured and used to indicate the relative levels of soil water potential. Soil matric water potential was measured using tensiometers (Hirose-Lika, Inc. Tokyo, Japan). Three tensiometers were placed with the sensor heads in different places within the soil volume of the sampled pots. One was inserted at the center of the soil volume; the second one to the lower left of the first one and the third one to the upper right of the first one. The average value of the three tensiometer readings was recorded as the soil matric water potential. Pots were weighed at the time when photosynthesis and leaf conductance were measured. The soil water content was determined on a gravimetric basis. The correlation ($y = 7011.9\,e^{-0.0767x}$; $r^2 = 0.83$; $n = 28$) between soil water matric water potential (y) and soil water content (x) was obtained.

Determinations of Water Stress Resistance, Tolerance and Avoidance

Water stress resistance (R_{WS}) was defined as the ability of photosynthetic maintenance under low soil water potential (Ψ_{soil}) and calculated as

$$R_{WS} = \Psi_{soil[0]} - \Psi_{soil[HD]}$$

where $\Psi_{soil[0]}$ is the Ψ_{soil} without water stress on plants and close to zero in the present experiment; and $\Psi_{soil[HD]}$ is the Ψ_{soil} at which the photosynthetic rate was half depressed. Water stress tolerance was defined as the ability of photosynthetic maintenance under low leaf water potential (Ψ_L) and calculated as

$$T_{WS} = \Psi_{L[0]} - \Psi_{L[HD]}$$

where $\Psi_{L[0]}$ is Ψ_L without water stress and is about -0.2 MPa in the present experiment, and $\Psi_{L[HD]}$ is Ψ_L at which P_N was half depressed. The half depression point of the P_N was obtained from the relationship curves between P_N and Ψsoil or Ψ_L. Water stress avoidance (A_{WS}) was defined as the ability of Ψ_L maintenance under low Ψ_{soil} and calculated as the ratio of R_{WS} to T_{WS} or R_{WS}/T_{WS}.

Measurement of Root Respiration

Tip parts of the root with a length of 5 cm were cut and collected as root samples (each with 1 g of fresh mass) for respiration measurements. The fresh root sample was inserted in the assimilation chamber and measured under a relative humidity of 85% at 15, 20, 25, 30 and 35°C using a gas exchange system for photosynthetic measurement (LI-6400). After measurement, the root sample was dried and the respiration rate was expressed on a dry mass basis.

RESULTS

As shown in Table 1, the maximum photosynthetic rate (P_{max}) under adequate moisture conditions was higher in chemical-fertilized sweet corn plants than in organic-fertilized plants. This was most likely due to low initial availability of inorganic nitrogen nutrients in organic-fertilized soil as reported elsewhere (Kato et al., 1997). P_{max} was higher in plants treated with EM than plants without EM treatments. This was attributed to the promotion of nutrient availability by EM application. It was also found that plants grown with organic fertilizers and/or with EM developed more roots than those grown with chemical fertilizer or without EM treatment (Table 1).

Plants grown with organic fertilizers or with EM treatment maintained higher P_N as Ψ_{soil} decreased than those grown with chemical fertilizer or without EM treatment. This means that plants grown with organic materials and/or with EM develop higher water stress resistance compared with those

TABLE 1. Photosynthetic water stress resistance (R_{WS}), tolerance (T_{WS}) and avoidance (A_{WS}) as well as leaf water potential at 70 kPa of soil matric water potential ($\Psi_{L[70]}$) and the maximum photosynthetic rate (P_{max}), root dark respiration at 25°C (D_{25}) and root/top dry mass ratio (R/T) of sweet corn plants with different fertilizations.

Fertilization	R_{WS} (kPa)	T_{WS} (MPa)	A_{WS} (Pa Pa^{-1})	$\Psi_{L[70]}$ (MPa)	P_{max} (μmol m^{-2} s^{-1})	D_{25}	DM_R (g Plant^{-1})	R/T (%)
EM-Organic 1	129.9 a	0.56 a	0.23 a	−0.61 a	22.3 c	5.2 c	1.83	27 a
Organic 1	88.5 b	0.57 a	0.16 b	−0.76 c	22.2 c	5.8 b	1.57	24 b
EM-Organic 2	120.3 a	0.60 a	0.20 a	−0.67 b	25.0 b	6.0 b	1.99	28 a
Organic 2	82.0 b	0.59 a	0.14 b	−0.79 c	22.6 c	6.8 a	1.68	24 b
EM-Chemical	88.5 b	0.60 a	0.15 b	−0.74 c	27.1 a	4.8 c	2.30	20 c
Chemical	60.3 c	0.56 a	0.11 c	−0.89 d	24.5 b	5.4 d	1.90	18 d

fertilized with chemicals and without EM. This result can be visually understood from the comparison between plants fertilized with Organic 2, i.e., aerobically-fermented organic materials, and those fertilized with chemicals in Figure 1A, where P_N was maintained higher in plants with EM than those without EM applied as Ψ_{soil} decreased. If water stress resistance was separated into tolerance and avoidance, high water stress resistance induced by organic fertilizers and/or EM application was mainly attributed to the water stress avoidance mechanism instead of water tolerance at a given Ψ_L. In other words, plants grown with organic fertilizers and/or EM maintained a higher Ψ_L and consequently maintained P_N as Ψ_{soil} decreased. This result can be explained with the values of Ψ_L at 70 kPa of Ψ_{soil} ($\Psi_{L[70]}$) (Table 1). At a given Ψ_L, photosynthetic maintenance was not different between treatments of fertilizers or between with and without EM applications. Consequently, there was no difference in water stress tolerance between treatments of fertilizers or between with and without EM applications. This can be understood from Figure 1 B and C, where Ψ_L was maintained higher in plants fertilized with Organic 2 than those fertilized with chemicals as Ψ_{soil} decreased but there was no difference in responses to Ψ_L changes. The promotions of water stress avoidance by organic fertilizers and/or EM application might be associated with the roots, which were found to be more developed and higher in their respiratory activity than those grown with chemical fertilizer and without EM (Table 1).

DISCUSSION

Results showed that sweet corn plants grown with organic fertilizers and/or with EM developed a larger root:shoot ratio with higher respiratory activi-

FIGURE 1. Photosynthetic response to decrease in soil matric water potential (A), to decreases in leaf water potential (B), and the response of leaf water potential to decreases in soil matric water potential (C).

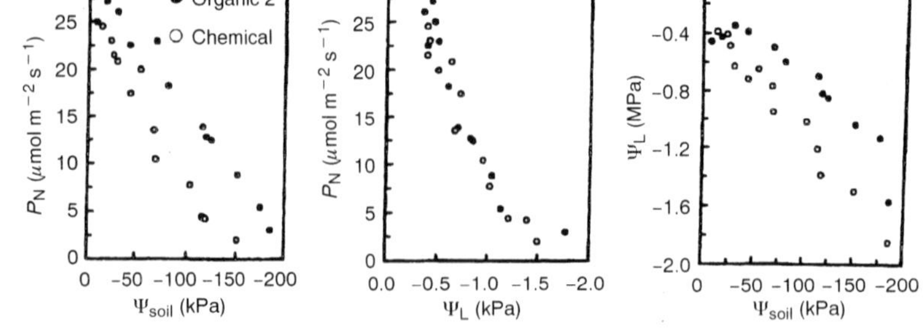

ty compared with the plants grown with chemical fertilizer and without EM treatment. It is logical that a plant with a larger root:shoot ratio can resist drought due to the greater potential for water uptake relative to transpiration. Moreover, the water retention ability was tested using the excised leaves to eliminate the root function and effect on water uptake. Interestingly, plants with a high root:shoot ratio and a high water stress resistance as indicated by photosynthetic maintenance possess a high water retention ability in their excised leaves (Xu et al., 1997). This suggests that in the present experiment a high water stress resistance is attributed not only to the large root system, but also to the leaf water retention ability. Both the strengthened water uptake by the more developed root and diminished water loss from the leaves would likely contribute to the maintenance of Ψ_L under soil water deficit conditions. These results are consistent with the water stress analysis. Actually, no matter how low the Ψ_{soil} is, the plant is water stressed only when Ψ_L decreases to a critical level. If the plant Ψ_L does not decrease or decreases little under low Ψ_{soil}, it can be considered as a water stress avoider. However, if Ψ_L decreases under low Ψ_{soil} but the plant can maintain a high P_N, it indicates the function of a water stress tolerance mechanism. In this study, water stress resistance was determined five weeks after sowing. Sweet corn plants grown with organic fertilizers produced less biomass than those grown with chemical fertilizers at the early growth stage. This was likely due to the low initial availability of inorganic forms of nitrogen and phosphorus from organic fertilizers. Therefore, early, nutrient availability and supply to plants from chemical fertilizer were able to sustain greater growth and biomass production than from organic fertilizers. This can be seen from the maximum photosynthetic rate under well-watered conditions, which is lower for plants grown with organic fertilizers than those grown with chemical fertilizer. Interestingly, the morphological characteristics are different between these two fertilizer treatments. In addition to the more developed root system, leaves of plants grown with organic fertilizers appeared lighter and more glaucous in color with more cuticular waxes on/in the surface than those grown with chemical fertilizer. These characteristics might contribute to the prevention of water loss from the leaf. Therefore, the relatively high water stress resistance of sweet corn plants grown with organic fertilizer might be attributed more to the improved and adapted morphological characteristics rather than differences in physiological characteristics. These results are consistent with those reported by Levitt (1980) and Ludlow et al. (1983), that the water stress avoidance mechanism is mainly associated with morphological characteristics, while water stress tolerance is mainly associated with physiological characteristics.

There was little difference in morphological characteristics between plants treated with and without EM, except for a more developed root system in EM treated plants. The mechanisms for EM effect is not entirely clear in many respects. For example, it is not known whether the effect is directly from the microbes themselves, or from substances produced by the activities of these organisms. Nevertheless, the effects of EM might be different from those of environmental conditions such as water, light and humidity as well as nutritional factors, which can cause changes not only in the internal status but also in external appearances. Therefore, further studies are needed to elucidate the mechanistic basis for the effects of EM on plant characteristics associated with drought resistance.

REFERENCES

Aspinal, D., P.B. Nicols and H.L. May. (1965). The effect of soil moisture on the growth of barley. *Australian Journal of Agricultural Research* 15:729-734.

Fischer, R.A. and R. Maurer. (1978). Drought resistance in spring wheat cultivars. I. Grain yield response. *Australian Journal of Agricultural Research* 29:897-902.

Fukuoka, M. (1985). *Nature Farming*. Tokyo: Sunshusha, 419 p.

Harwood, R.R. (1984). Organic farming research at the Rodale Research Center. In *Organic Farming: Current Technology and Its Role in a Sustainable Agriculture*, eds. D.F. Bezdicek, J.F. Power, D.R Keeney, and M.J. Wright. Madison: American Society of Agronomy, pp. 1-18.

Higa, T. (1994). *The Completest Data of EM Encyclopedia*. Tokyo: Sogo-Unicom (in Japanese). 385 p.

Hillel, D. (1980). *Fundamentals of Soil Physics*. New York: Academic Press, pp. 195-222.

Huck, M. G. (1982). Water flux in the soil-root continuum. In *Roots, Nutrient and Water Flux, and Plant Growth*, eds. S. A. Baber, D. R. Bouldin, D. M. Kral and S. L Hawkins. Madison: American Society of Agronomy, pp. 47-64.

Kato, S., K. Yamada, M. Fujita, H. L. Xu, K. Katase and H. Umemura. (1997). Effects of organic materials and microbial inoculant on soil and plant nutrition. *Abstracts of Annual Meeting of Japanese Society of Soil Science and Plant Nutrition* 43: 163.

Letey, J. (1977). Physical properties of soils. In *Soils for Management of Organic Wastes and Waste Waters*, American Society of Agronomy. Madison: American Society of Agronomy, pp.101-114.

Levitt, J. (1980). *Responses of Plants to Environmental Stresses. II. Water, Radiation, Salt, and Other Stresses*. New York: Academic Press, pp. 25-178.

Li, W.-J. and Y.-Z. Ni. (1996). Researches on application of microbial inoculant in crop production. In: *Research and Application of EM technology*. Beijing: China Agricultural University Press, pp. 42-87.

Lockeretz, W. and D. H. Kohl. (1981). Organic farming in the cornbelt. *Science* 211: 540-547.

Ludlow, M. M., A. P. C. Chu, R. J. Clements and R. G. Kerslake. (1983). Adaptation of species of *Centrosema* to water stress. *Australian Journal of Plant Physiology* 10:449-470.

Turner, N.C. (1988). Measurement of plant water status by the pressure chamber technique. *Irrigation Science* 9: 289-308.

U.S. Department of Agriculture. (1980). *Report and Recommendations on Organic Farming. A Special Report Prepared for the Secretary of Agriculture.* Washington DC: U.S. Government Printing Office. 94 p.

Vogtmann, H. (1984). Organic farming practices and research in Europe. In *Roots, Nutrient and Water Flux, and Plant Growth*, eds. S.A. Baber, D.R. Bouldin, D.M. Kral and S.L. Hawkins. Madison: American Society of Agronomy, pp. 19-36.

Xu, H.L., R. Ajiki, C. Sakakibara and H. Umemura. (1998). Corn leaf water retention as affected by organic fertilizations and effective microbes application. *Pedosphere* 8:1-8.

Effect of a Microbial Inoculant on Stomatal Response of Maize Leaves

Hui-lian Xu
Xiaoju Wang
Jihua Wang

SUMMARY. Laboratory tests were conducted to determine the effect of a microbial inoculant on the stomatal response of maize leaves (*Zea mays* L.). The microbial inoculant investigated is known as Effective Microorganisms or EM and consists of a mixed culture of naturally occurring, beneficial microorganisms. Research has shown that EM applied to soils and plants can improve soil properties and enhance the growth, yield and quality of crops. The exact mechanisms or modes-of-action of how EM cultures elicit beneficial effects on plant growth and metabolism is not known. However, it is likely that some of these cultures can synthesize phytohormones (i.e., auxins and others) or growth regulators that stimulate plant growth. Consequently, the effects of EM and partial illumination on stomatal response of intact and excised maize leaves were evaluated. Potted plants were dried to the wilting point and rehydrated with either a 1:100 dilution of EM and water or water alone applied to the soil. Sudden illumination of plants maintained in the dark showed that the leaf stomata of the EM-treated plants opened more rapidly than water-treated control plants. When leaves were excised and subjected to dehydration, the stomata closed

Hui-lian Xu is Senior Crop Scientist and Xiaoju Wang is Research Soil Scientist, International Nature Farming Research Center, 5632 Hata, Nagano 390-1401, Japan. Jihua Wang is Crop Scientist, Beijing Academy of Agriculture, Beijing, China.

Address of correspondence to: Hui-lian Xu at the above address (E-mail. huilian@janis.or.jp).

[Haworth co-indexing entry note]: "Effect of a Microbial Inoculant on Stomatal Response of Maize Leaves." Xu, Hui-lian, Xiaoju Wang, and Jihua Wang. Co-published simultaneously in *Journal of Crop Production* (Food Products Press, an imprint of The Haworth Press, Inc.) Vol. 3, No. 1 (#5), 2000, pp. 235-243; and: *Nature Farming and Microbial Applications* (ed: Hui-lian Xu, James F. Parr, and Hiroshi Umemura) Food Products Press, an imprint of The Haworth Press, Inc., 2000, pp. 235-243. Single or multiple copies of this article are available for a fee from The Haworth Document Delivery Service [1-800-342-9678, 9:00 a.m. - 5:00 p.m. (EST). E-mail address: getinfo@haworthpressinc.com].

more slowly (i.e., remained open longer) for the EM-treated plants compared with the water-treated control plants. There was no effect of EM on cuticular conductance in any of the experiments. The results of this study indicate that EM cultures contains bioactive substances that can significantly affect leaf stomatal response. *[Article copies available for a fee from The Haworth Document Delivery Service: 1-800-342-9678. E-mail address: getinfo@haworthpressinc.com <Website: http://www.HaworthPress.com>]*

KEYWORDS. Cuticular conductance, effective microorganisms, EM, microbial inoculant, stomatal conductance, *Zea mays*, excised leaves

INTRODUCTION

There is increasing evidence that mixed cultures of naturally occurring, beneficial microorganisms applied to soils and plants can improve soil quality and the growth, yield and quality of crops. One such microbial inoculant is known as Effective Microorganisms or EM and has been developed by Professor Teruo Higa, University of the Ryukyus, Okinawa, Japan (Higa and Parr, 1994). A number of theories have been proposed as to the modes-of-action of EM on plant growth and metabolism (Higa and Wididiana, 1991). However, the exact mechanisms of how beneficial effects are derived by either (a) direct effects of microorganisms on the plant or (b) indirect effects of microbially-synthesized substances (e.g., phytohormones and growth regulators) are largely unknown. In this regard, Xu et al. (1998) reported that EM significantly increased the growth and grain yield of maize by promoting root development and activity that was largely auxin-mediated. Others have also reported that a number of microbes can synthesize phytohormones and physiologically-active compounds (Arshad and Frankenberger, 1992; Kampert et al., 1975; Panosyan et al., 1963). However, these reports did not differentiate between direct effects of microorganisms on plants, or indirect effects of microbially-synthesized phytohormones. Therefore, the purpose of this study was to determine whether the liquid stock solution of EM cultures contained biologically-active substances that could affect stomatal responses of intact or excised maize leaves to partial illumination and dehydration.

MATERIALS AND METHODS

Plant Materials

Sweet corn plants (*Zea mays* L. cv. Honey Bantam) were grown in plastic pots each with a soil surface area of 0.02 m^2 and a height of 0.25 m. Pots were filled with a fine textured Andosol. The N-P-K levels in the soil were 3.4, 0.025 and 0.44 g kg^{-1}, respectively, with a C:N ratio of 13. The field capacity or capillary capacity of the soil was 80% on a gravimetric basis. Chemical fertilizers used in this experiment were ammonium sulfate (5.3 g pot^{-1}), long period coated urea (2.8 g pot^{-1}), superphosphate (5 g pot^{-1}) and potassium sulfate (3.5 g pot^{-1}). One plant was grown in each pot under glasshouse conditions. The daily temperature and air humidity were not controlled and fluctuated with ambient conditions.

Pot Treatment

When the 7th leaf was fully expanded, water supply to the pots was stopped. When the leaves began to wilt, EM was diluted 1:100 with water and supplied to the pots until the plants were rehydrated (i.e., had regained maximum turgor).

Partial Illumination

Partial illumination 2000 μmol m^{-2} s^{-1} was applied to the seventh leaf using an LED light source in the leaf chamber. Other parts of the whole plant were maintained in dark. Stomatal conductance was measured using a LI-6400 instrument The stomatal response to partial illumination (Figure 1) was analyzed using a sigmoid model as follows:

$$g_s = g_{max}\ \{1/[1 + (1 - \beta t)\ e^{-\alpha(t-\tau)}]\} + g_{cc}(1 + \beta t),$$

where, g_{max} and g_{cc} are the maximum stomatal conductance and cuticular conductance; β and α are the cuticular and stomatal response constants; t, time from beginning of illumination; τ, half saturation constant.

Stomatal Closure in Excised Leaves

Stomatal conductance of the 7th leaf of the fully turgid plants was measured under 2000 μmol m^{-2} s^{-1} until the maximum level was reached. The leaf was then excised from the plants and measurement continued. The stomatal response to leaf dehydration (Figure 2) was analyzed using an exponential model as follows:

$$g_s = [g'_{max} - g_{res}(1 - \beta' t)]\ e^{-\alpha'(t-\tau')} + g_{res}(1 - \beta' t)$$

where g'_{max} is the maximum stomatal conductance before excision; g_{res} is residual conductance after the stomata are closed; β' and α' are the cuticular and stomatal response constants; τ' is the time prior to rapid closure of stomata.

FIGURE 1. A sigmoid model for the response of stomatal conductance (g_S) to partial illumination on the leaves of sweet corn plant in the dark.

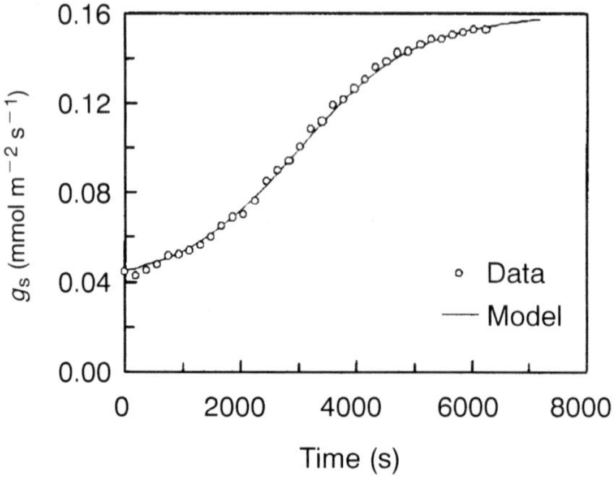

FIGURE 2. A model for the decline of stomatal conductance (g_S) in an excised sweet corn leaf blade.

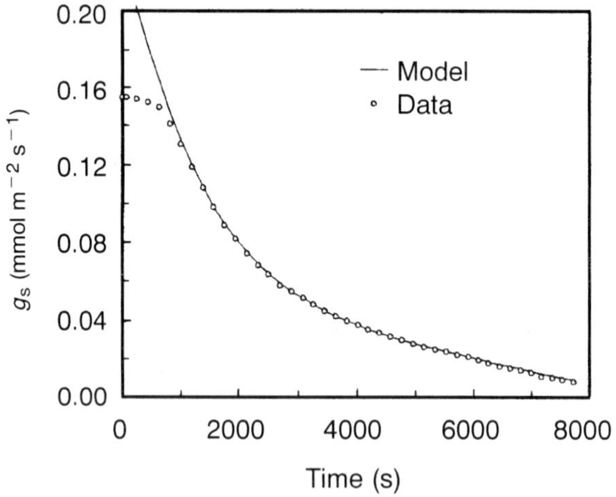

RESULTS AND DISCUSSION

Stomatal Response to Illumination

Preliminary experiments have shown the response of stomatal conductance (g_S) to partial illumination of sweet corn leaves that had equilibrated in the dark. The rapid increase in g_s upon illumination and the close agreement between the sigmoid model and experimental data are shown in Figure 1. The decline in stomatal conductance (g_S) that occurs when a leaf is excised from a corn plant is shown in Figure 2, which indicates close agreement between the model and experimental data.

When the leaf was partially illuminated in the dark, the stomata started to open slowly. In this case, the stomata in the EM-treated leaves opened faster than those of the control plants (Figure 3). The parameters analyzed from the model are presented in Table 1. The variable g_{max} shows the maximum g_S.

FIGURE 3. Comparison of the sigmoid curves for the response of stomatal conductance (g_S) to partial illumination on the leaf blade of sweet corn plants treated with EM liquid and water in the dark.

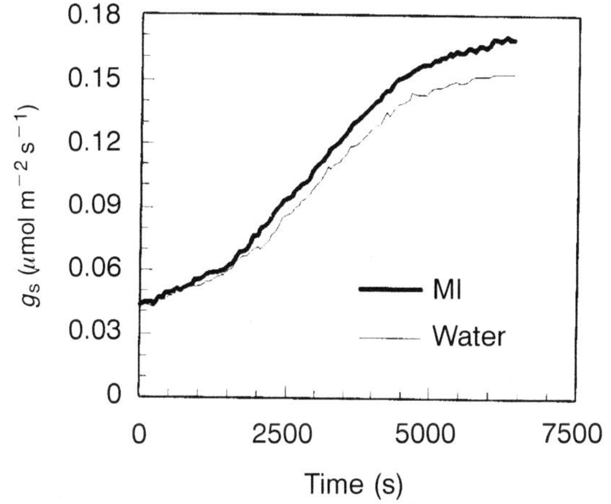

TABLE 1. Parameters of the sigmoid curve of leaf stomatal response to illumination in the dark for intact leaves of sweet corn plants treated with a microbial inoculant (EM) compared with water-treated and untreated control plants.

Treatment	g_{max}	g_{cc}	a	β	τ
	(mmol m^{-2} s^{-1})		(10^{-3} s^{-1})	(10^{-5} s^{-1})	(s)
EM-absorbed	0.118 a	0.0398 a	−1.10 a	4.01 a	2795 a
Water-absorbed	0.110 a	0.0401 a	−1.03 a	2.10 b	2985 b
Control (untreated)	0.113 a	0.0396 a	−1.05 a	2.19 b	2953 b

g_S, stomatal conductance; g_{max}, the maximum g_S; g_{cc}, cuticular or residual conductance in dark, a, stomatal response constant; β, residual conductance response constant; τ, time at which stomatal conductance reaches half of the maximum.

The value of g_{max} was higher for EM treatment than for water-treated or untreated plants because EM increased the stomatal opening. The variable g_{cc} shows the cuticular or residual conductance in the dark. There was no significant difference between treatments in this parameter as related to cuticular characteristics. The coefficient α shows the slope of the fast opening phase of the stomatal response curve, and β shows the initial slope of the conductance response curve. The value of β was higher in the EM-treated plants than in water-treated or untreated control plants. This means that the EM treatment increased stomatal opening at the beginning of illumination. Moreover, there was also a difference in α between treatments. The value of α showed the stomatal response property and was larger for the EM-treated plants than for the control plants. The coefficient τ shows the time at which stomatal conductance reaches half of the maximum. The value of τ was higher in the EM-treated plants than in the water-treated or untreated control plants. This suggests that stomatal conductance was maximized earlier in EM-treated plants than in the water-treated or untreated control plants.

Decline of Stomatal Conductance in the Excised Leaf Blade

When the leaf was excised from the plant, stomata started to close soon in response to leaf dehydration. In this case, stomata in the EM-treated leaves remained open longer than those in the control plants (Figure 4). However, when stomatal closure reached a fast phase, stomata in the EM-treated leaves closed faster than those in leaves of control plant.

FIGURE 4. Comparison of the modeling curves for the decline of stomatal conductance (g_s) in excised leaf blades of sweet corn plants treated with EM liquid and water.

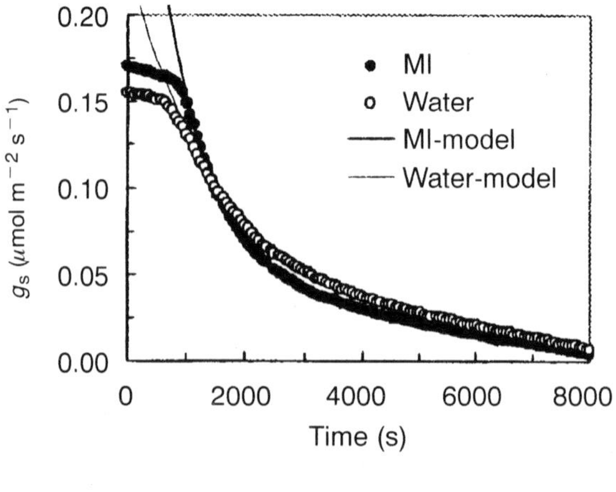

The parameters analyzed from the model are presented in Table 2. The variable g'_{max} shows the initial total g_s before the leaf is excised. As mentioned earlier, EM treatment increased stomatal opening and consequently g'_{max} was higher in EM-treated plants than in water treated or untreated plants. The variable g_{res} shows the cuticular or residual conductance when stomata are roughly closed. There was no statistically significant difference in g_{res} found between these treatments.

The coefficient α' shows the stomatal response constant. The value of α' was higher for the EM-treated plants than in water-treated or untreated control plants. The variable g_{res} shows the cuticular or residual conductance. This indicated that the stomata in leaves of the EM-treated plants closed more rapidly in response to water loss from the excised leaves compared with the controls. There was no difference in the residual conductance response constant (β') between treatments. Obviously, the short-term treatment with a microbial inoculant would not change the morphological structure of the leaf cuticle. The coefficient τ' indicates the time required for rapid stomatal closure to begin. The value of τ' was higher for the EM-treated plants than for water-treated or untreated control plants. This suggests that the leaf stomata in the EM-treated plants could remain open longer under leaf water-deficit conditions than those of the control plants.

Throughout these experiments, there was no significant effect of EM on cuticular conductance. Overall, the results indicate that EM contains substances that can affect stomatal response. Although we do not have direct evidence to support this conclusion, the various species that comprise EM used in the present study have been studied extensively for decades. A number of them are known to synthesize phytohormones, growth regulators and other biologically-active substances (Arshad and Frankenberger, 1992). Barea et al. (1976) found that among 50 bacteria isolated from the rhizosphere

TABLE 2. Parameters of the stomatal closure curve for excised leaves of sweet corn plants treated with a microbial inoculant (EM) compared with water-treated and untreated control plants.

Treatment	g'_{max}	g_{res}	α'	β'	τ'
	(mmol m^{-2} s^{-1})		(10^{-3} s^{-1})	(10^{-5} s^{-1})	(s)
EM-absorbed	0.169 a	0.0505 a	−1.24 a	7.4 a	873 a
Water-absorbed	0.158 b	0.0575 a	−1.13 b	7.9 a	546 b
Control (Untreated)	0.153 b	0.0593 a	−1.16 b	7.1 a	596 b

g'_{max}, the initial total g_s before the leaf is excised; g_{res}, cuticular or residual conductance when stomata are roughly closed. α', stomatal response constant; β', residual conductance response constant; τ', time at which stomatal conductance get into sharp closing course.

of various plants, 86, 58, and 90% produce auxins, gibberellins, and kinetin-like substances, respectively. Kampert et al. (1975) reported that 55% of bacteria and 86% of fungi isolated from the rhizosphere of *Pinus silvestris* could produce gibberellins and their derivatives. *Actinomyces* and *Streptomyces* produce auxins and similar substances (Purushothaman et al., 1974; Mahmoud et al., 1984), gibberellins (Arshad and Frankenberger, 1992), and cytokinins (Bermudez de Castro et al., 1977; Henson and Wheeler, 1977). Some fungi like *Aspergillus niger* also produce gibberellins (El-Bahrawy, 1983). The promotion of stomatal responses by intact and excised leaves of EM-treated plants is likely due to the effects of plant growth regulators existing in the liquid phase of the cultures. Further studies are needed to determine the exact mechanisms and modes-of-action whereby these bioactive compounds can affect leaf stomatal responses, growth and metabolism of plants.

REFERENCES

Arshad M. and W.T. Frankenberger Jr. (1992). Microbial production of plant growth regulators. In *Soil Microbial Ecology*, ed. F.B. Metting. New York: Marcel Dekker, pp. 307-348.

Barea J.M., M.I. Navarro and E. Ontoya. (1976). Production of plant growth regulators by rhizosphere phosphate-solubilizing bacteria. *Journal of Applied Bacteriology* 40: 129-134.

Bermudez de Castro F., A. Canizo, A. Costa, C. Miguel and C. Rodriguez-Barrueco. (1977). Cytokinins and nodulation of the non-legumes *Alnus glutinosa* and *Myrica gale*. In *Recent Developments in Nitrogen Fixation*, eds. W. Newton, J.R. Postgate and C. Rodriguez. London: Academic Press, pp. 539-550.

El-Bahrawy, S.A. (1983). Associative effect of mixed cultures of *Azotobacter* and different rhizosphere fungi with *Rhizobium japonicum* on nodulation and symbiotic nitrogen fixation of soybean. *Zentrbl. Mikrobiol.* 138: 443-449.

Henson, I.E. and C.T. Wheeler. (1977). Hormones in plants bearing nitrogen-fixing root nodules: Cytokinins in roots and root nodules of some non-leguminous plants. *Zeitscrift für Pflanzenphysiol* 84: 179-782.

Higa, T. and G.N. Wididiana. (1991). Concept and theories of effective microorganisms. In *Proceedings of the First International Conference on Kyusei Nature Farming*, eds. J.F. Parr. Washington, DC: U.S. Department of Agriculture, pp. 118-1124.

Higa, T. and J.F. Parr. (1994). *Beneficial and Effective Microorganisms for a Sustainable Agriculture and Environment*. Atami, Japan: International Nature Farming Research Center, 16 p.

Kampert, M., E. Strzelczyk and A. Pokojska. (1975). Production of gibberellin-like substances by bacteria and fungi isolated from the roots of pine seedlings (*Pinus sylvesreis* L.). *Acta Microbiologia* 7: 157-166.

Mahmoud, S.A.Z., M.I. Ramadan, F.M. Thabet and T. Khater. (1984). Production of plant growth promoting substances by rhizosphere microorganisms. *Zentrbl Mikrobiol* 139: 227-232.

Panosyan, A.K., Z.V. Marshavina, R.S. Arutunyan and S.G. Aslanyan. (1963). The nature of physiologically active substances of actinomycetes and the effect of their metabolites on plant growth. In *Plant Microbes Relationship*, eds. J. Macura and V. Vancura. Prague, Czech: Symposium of the Relationship between Soil Microorganisms and Plant Roots, pp. 241-244.

Purushothaman, D., T. Marimuthu, C.V. Venkataramanan and R. Kesavan. (1974). Role of actinomycetes in the biosynthesis of indole acetic acid in soil. *Current Science* 43: 413-414.

Xu, H.L., N. Ajiki, X.J. Wang, C. Sakakibara and H. Umemura. (1998). Corn leaf water retention as affected by organic fertilizations and effective microbes applications. *Pedosphere* 8: 1-8.

Modeling Photosynthesis Decline of Excised Leaves of Sweet Corn Plants Grown with Organic and Chemical Fertilization

Hui-lian Xu
Xiaoju Wang
Jihua Wang

SUMMARY. The transpiration or photosynthesis decline curve of excised plant leaves has been used to predict water stress tolerance ability. This study analyzed the characteristics of the declining photosynthetic rate in excised leaves of sweet corn (*Zea mays* L.) plants after treatment with effective microorganisms (EM), a mixed culture of naturally occurring beneficial microorganisms, organic fertilizer (Bokashi) and chemical fertilizer. The net photosynthetic rates (P_N) of intact leaves were measured using LI-6400 gas exchange system under 2000 μmol m^{-2} s^{-1} PPF. When P_N reached the maximum (P_{max}), the leaf was excised and P_N was continuously monitored. A declining P_N curve with

Hui-lian Xu is Senior Crop Scientist and Xiaoju Wang is Research Soil Scientist, International Nature Farming Research Center, 5632 Hata, Nagano 390-1401, Japan.

Jihua Wang is Crop Scientist, Beijing Academy of Agricultural Sciences, Beijing, China.

Address correspondence to: Hui-lian Xu at the above address (E-mail: huilian@janis.or.jp).

The authors thank K. Katase, M. Fujita, K. Yamada, S. Kato and S. Ishiko at the Agricultural Experiment Station of International Nature Farming Research Center for their technical assistance in this research.

[Haworth co-indexing entry note]: "Modeling Photosynthesis Decline of Excised Leaves of Sweet Corn Plants Grown with Organic and Chemical Fertilization." Xu, Hui-lian, Xiaoju Wang, and Jihua Wang. Co-published simultaneously in *Journal of Crop Production* (Food Products Press, an imprint of The Haworth Press, Inc.) Vol. 3, No. 1 (#5), 2000, pp. 245-253; and: *Nature Farming and Microbial Applications* (ed: Hui-lian Xu, James F. Parr, and Hiroshi Umemura) Food Products Press, an imprint of The Haworth Press, Inc., 2000, pp. 245-253. Single or multiple copies of this article are available for a fee from The Haworth Document Delivery Service [1-800-342-9678, 9:00 a.m. - 5:00 p.m. (EST). E-mail address: getinfo@haworthpressinc.com].

time is obtained and modeled according to $P_N = [P_{max} - P_{res}(1 - \beta t)] e^{-\alpha(t-\tau)} + P_{res}(1 - \beta t)$, where t is the time elapsed after the leaf was excised; τ is the time point when the stomata begin to close; P_{res} is the residual P_N when stomatal closure completed; and α and β are constants related to stomatal and residual conductances respectively. Photosynthesis maintenance ability (PMA) is expressed as $1/(\alpha + \beta)$. Leaves of plants fertilized with organic materials and/or treated with EM showed a higher PMA and P_N than the controls after the leaves were excised. *[Article copies available for a fee from The Haworth Document Delivery Service: 1-800-342-9678. E-mail address: getinfo@haworthpressinc.com <Website: http://www.HaworthPress.com>]*

KEYWORDS. Drought resistance, effective microorganisms, excised leaves, modeling, organic fertilizer, photosynthesis, water stress

INTRODUCTION

Water retention ability of excised leaves has been used to predict water stress resistance in plants (Clarke and Thomas, 1982; Quisenberry et al., 1982). Water loss from a leaf is mainly determined by the stomatal transpiration. When leaves are excised, the water supply is completely stopped, stomata will close and photosynthetic rate and transpiration rate will decline accordingly. As time elapses after the leaf is excised, photosynthesis will continue provided that sufficient leaf tissue water is retained. Even if the tissue water decreases to the same level, a leaf with higher physiological activities might maintain a higher photosynthetic rate than a leaf with lower physiological activities. Therefore, in the case of an excised leaf, maintenance of photosynthesis may be more important than that of transpiration or water content itself. When stomata close, gas exchange can still occur through incompletely closed stomata and cuticular pores. This process may be affected by the stomatal sensitivity on one hand, and by characteristics of the cuticular layer on the other. The cuticular layer is affected by the wax deposition on and/or into the leaf surface (Richards et al., 1986). Cuticular wax deposition is usually induced under conditions of high light intensity (Baker, 1974), high temperature (Haas, 1977), low humidity (Whitecross and Armstrong, 1972) and soil water deficit (Svenningsson and Liljenberg, 1986). It is reported that the application of organic materials inoculated with EM promotes root development and consequently increases water stress resistance of sweet corn plants (Xu et al., 1997, 1998). However, it is not clear whether the declining characteristics of photosynthesis in excised leaves are consistent with its water stress resistance *in situ*. Therefore, we measured the time-rate of change of photosynthesis of excised leaves through the dehydration phase and analyzed the photosynthesis decline curve using an exponential model. The effects of EM, organic and chemical fertilizer application on the photosynthetic maintenance ability in excised leaves of sweet corn plants were observed and analyzed.

MATERIALS AND METHODS

Plant Materials and Treatments

Seeds of sweet corn (*Zea mays* L. cv. Honey-Bantam) were sown in Wagner pots each filled with fine textured Andosol and one plant remaining in each pot after thinning. The total nitrogen and available phosphorus of the soil were 3.4 g kg^{-1} and 0.025 g kg^{-1} with a C:N ratio of 13. The field capacity was 80% on a gravimetric basis. Organic materials such as oilseed mill sludge (cake), rice husk and bran and fish-processing by-product were fermented with EM under anaerobic or aerobic conditions to produce two experimental organic fertilizers. The total N, available P, and K concentrations in both anaerobically- and aerobically-fermented organic materials were 58 g kg^{-1}, 30 g kg^{-1} and 2 g kg^{-1}, respectively. Ammonium sulfate, superphosphate and potassium sulfate were applied in the chemical fertilizer treatment with the N-P-K level adjusted to that of organic fertilizer treatments. The microbial inoculant used in this study is widely known as Effective Microorganism (EM) and is a mixed culture of naturally-occurring beneficial microorganisms. Predominantly the following types and genera: (1) Lactic acid bacteria (*Lactobacillus*, *Streptococcus* and *Pediococcus*); (2) Yeasts (*Acharaomyces* and *Candida*); (3) Photosynthetic bacteria (*Rhodopseudomonas*, *Rhodospirillum*, *Chromatium* and *Chlorobium*), (4) Actinomyces (*Streptomyces*, *Propionibacterium*, *Nocardia* and *Micromonospora*); and (5) Certain fungi (*Aspergillus* and *Mucor*). Six treatments were designed as follows: (1) EM-Organic 1–anaerobically-fermented organic materials (80 g), in which EM was added to the materials before fermentation; (2) Organic 1–anaerobically fermented organic materials (80 g); (3) EM-Organic 2–aerobically-fermented organic materials (80 g), in which EM was added before fermented; (4) Organic 2–aerobically-fermented organic materials (80 g); (5) EM-Fertilizer–chemical fertilizers (ammonium sulfate 5.3 g, long-period coated urea 2.8 g, superphosphate 13 g and potassium sulfate 5.0 g), with EM, 80 mL, applied to soil at the same time before sowing; (6) Fertilizer–the same fertilizers as in 5.

Measurement and Modeling of Photosynthesis

Photosynthetic measurement was made five weeks after sowing seeds using a gas exchange system (LI-6400, LI-COR, Inc., Lincoln, Nebraska,

USA). Photosynthetic rates of the just fully expanded 7th leaf was measured under 2000 μmol m^{-2} s^{-1} until the maximum level was reached, then the leaf was excised off the plants and measured continuously. The response of the photosynthetic rate to leaf dehydration was analyzed using an exponential model as follows:

$$P_N = [P_{max} - P_{res}(1 - \beta t)]\, e^{-\alpha(t-\tau)} + P_{res}(1 - \beta t),$$

where, P_{max} is the maximum photosynthetic rate before the leaf was excised; P_{res}, photosynthetic rate after stomata are closed; β and α, cuticular and stomatal response constants; τ, time until the stomata begin to close rapidly. Photosynthetic maintenance ability (PMA) was expressed as PMA = $1/(\alpha + \beta)$. The model curve is shown in Figure 1.

RESULTS AND DISCUSSION

The Photosynthesis Decline Curve

Photosynthetic rate (P_N) in the leaf of sweet corn reached the maximum about fifteen minutes after the illumination begins. When the leaf was ex-

FIGURE 1. A model of photosynthetic declining curve of an excised leaf. See Table 1 for the abbreviations.

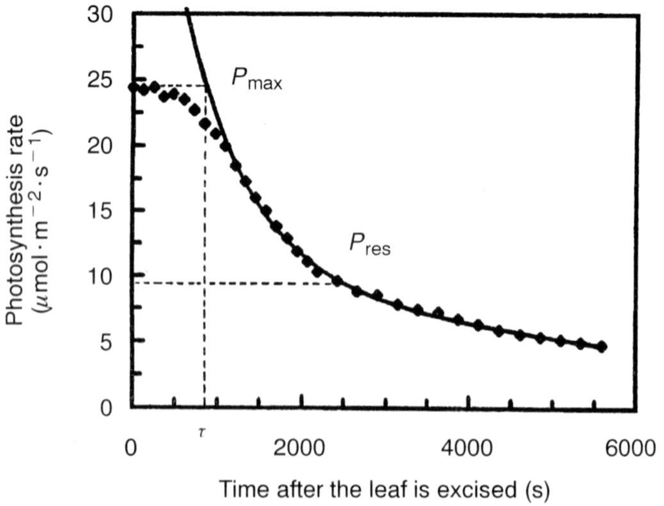

cised off the plant, P_N soon began to slowly decline in response to leaf dehydration. However, P_N did not decrease sharply until sometime later. This was because the leaf water did not immediately decease to the critical level. During a short time P_N maintained an almost constant level with a curve of slight decline and gentle slope. In the modeling equation, τ designates this duration. In due course, P_N starts to decrease sharply in a negative exponential manner and the curve shifts to the second phase. This occurred because the leaf tissue was dehydrated to the critical water level, where the stomata began to close rapidly. It is a known physiological phenomenon that sensitive stomata close in response to leaf dehydration in negative exponential manner. The sensitivity might be affected by the physiological activity and internal water status of the leaf. For example, it may be associated with endogenous hormones and osmotic concentration in the tissue. The stomatal sensitivity is shown by constant α in the model equation.

The extent of the sharp decrease must also include the influence of decreased cuticular conductance. The influence of the cuticular conductance is linear in response to leaf dehydration and is shown by the constant β in the model equation. When the stomata are almost closed, P_N is mainly controlled by the cuticular conductance and the effect of the incompletely closed stomata. Consequently, P_N at this time is defined as the residual photosynthetic rate (P_{res}). Beyond this point, the curve shifted to the third phase, which was mainly controlled by cuticular conductance.

The transpiration rate and leaf conductance also showed a curve similar to that for P_N after the leaf was excised from the plant. Transpiration and leaf conductance can also be analyzed using the model for photosynthesis as mentioned earlier. Actually, some researchers have tried using water retention ability of excised leaves as an indication of water stress resistance (Clarke and Thomas, 1982; Quisenberry et al., 1982), in response to increasing test samples for drought resistant genotype selections. This method was used from the time when perfect gas exchange measurement system was not yet available. Today, it is more appropriate to use the photosynthesis decline curve rather than the water content or transpiration decline curve because drought resistance of a plant is not only related to tissue water maintenance, but also associated with dehydration tolerance when tissue water decreases. Relationships involving the leaf physiological and morphological characteristics are discussed in studies using the water retention curve of excised leaves. Plants with high water stress resistance or plants grown under such stress conditions as soil water deficit, low humidity, high irradiation and high salinity can adjust their internal solute concentration to retain tissue water and/or accumulate more waxes on the leaf surface to prevent water loss from transpiration (Richards et al., 1986).

In an excised leaf, photosynthetic maintenance is largely determined by water loss that is controlled by stomatal and cuticular transpiration, because water uptake by the root system has been terminated. If other factors associated with stomata are known, evaporation from the leaf tissue may be, at least in part, related to the solute concentration in cells because the chemical activity of the water in solution is reduced by osmotically active molecules (Jones, 1983). Thus, a leaf with a high solute concentration may retain water stronger and longer at a low solute concentration. If a high tissue water content is maintained, photosynthesis can be logically maintained at a relatively high rate. Moreover, a leaf with substantial deposits of waxes deposited in or on the surface may prevent water loss from the cuticular layer. In an earlier study, Xu et al. (1998) investigated the effect of the same treatments used in the present study (i.e., EM, organic fertilizer and chemical fertilizer) on water relations of corn leaves. Their results showed that the leaves of sweet corn grown with organic fertilizer fermented with EM looked more glaucous in color, had more waxes on the surface, showed a higher water retention in the excised leaves and a higher photosynthetic water stress resistance under water deficits, than leaves of plants treated with chemical fertilizer.

Effects of Organic Fertilizations and Effective Microbes Inoculum

The maximum photosynthetic rate (P_{max}) before the leaf excision was slightly higher in leaves of plants treated with organic fertilizer and EM than in leaves of chemical-fertilized plants and plants without EM application. After the leaf was excised, photosynthesis in leaves of EM-treated or organic-fertilized plants continued at a higher rate for a longer duration than the control (Figure 2). The higher photosynthetic maintenance ability (PMA) at the first and second phases of the curve for leaves of EM-treated and organic-fertilized plants is reflected by the greater value of τ and the smaller value of α (Table 1). The P_{res} was slightly higher in leaves of EM-treated plants. The value of β was significantly smaller for leaves of organic-fertilized plants than for leaves of chemical fertilized plants but just slightly smaller for leaves of EM-treated plants than their control. The PMA, shown as $1/(\alpha + \beta)$, was higher in leaves of organic-fertilized and EM-treated plants than in leaves of chemical-fertilized plants and plants without EM application.

It is reported that organic fertilizers applied with EM promote root growth and activity and increase water stress resistance of sweet corn plants, compared with chemical fertilizers or the controls without EM application (Xu et al., 1997, 1998). It is logical that a plant with a larger root system can resist water stress more than one with a smaller root system because water is absorbed through the root system. However, a remaining question is whether the higher water stress resistance is attributed mainly to the larger root. If the water supply from the root is completely stopped, as with an excised leaf, can the leaf maintain a higher tissue water status to support photosynthesis, and

FIGURE 2. The comparison of the photosynthetic declining curves between leaves of chemical fertilized (solid) and organic fertilized (open) plants.

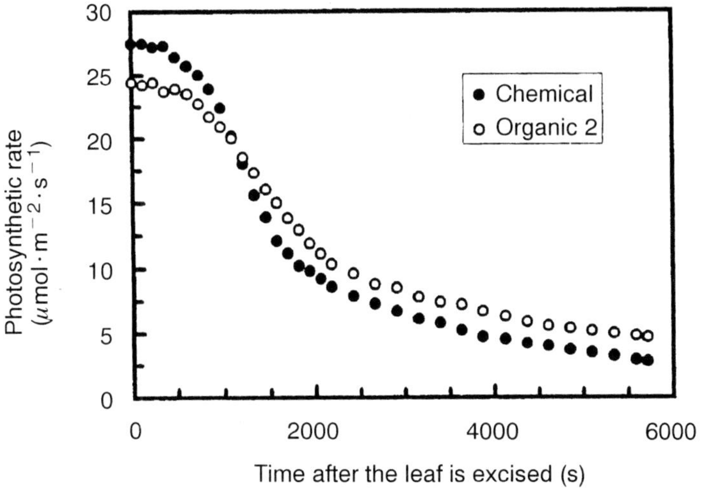

TABLE 1. Parameters showing the photosynthetic declining characteristics for the excised leaves of sweet corn with different fertilizations.

Fertilization	P_{max}	P_{res}	α	β	τ	PMA
	(μmol m^{-2} s^{-1})		(10^{-4} s^{-1})		(s)	
EM-Organic 1	23.7	9.8	11.4	0.91	886	8123
Organic 1	22.9	9.4	12.7	0.95	823	7326
EM-Organic 2	25.9	10.1	11.9	0.89	904	7819
Organic 2	23.4	9.7	13.2	0.93	871	7077
EM-Chemical	29.3	9.1	14.3	1.09	814	6498
Chemical	26.5	8.9	16.5	1.21	778	5647

P_{max}, the maximum photosynthetic rate before the leaf is excised; P_{res}, the residual photosynthetic rate when stomata are almost closed; α, stomatal response constant; β, residual conductance constant; τ, time point when stomata start to close fast; PMA, photosynthetic maintenance ability.

would this provide greater tolerance or resistance to dehydration than the control leaf? This may be a leading question. Therefore, the photosynthetic maintenance ability (PMA) of excised leaves was tested exclusive of root interactions. Earlier studies have assessed tissue water retention in the excised leaves (Xu et al., 1997) and the water stress resistance under *in situ* conditions (Xu et al., 1997). Results indicate that plants with a high root:top ratio and a high water stress resistance possess a high water retention ability and a high photosynthetic maintenance ability in their excised leaves. This suggests that in the present experiment a high water stress resistance is attributed not only to the large root system but also to the high leaf water retention ability and the high tolerance or resistance of photosynthesis against the leaf dehydration. Thus, it is possible to indicate water stress resistance with the photosynthetic maintenance ability of excised leaves as one of the criteria. As to the effects of organic fertilizers, there have been many studies reported (Harwood, 1984; 1981; USDA, 1980; Vogtmann, 1984). The role of EM used in this study was reported earlier by Higa (1994). However, in the case of EM, practical have greatly proceeded the fundamental mechanistic research on this microbial inoculant. Although there are many documents describing the functions and effects of photosynthetic bacteria (Kobayashi, 1980) and beneficial microorganisms such as the actinomycetes and certain fungi (Arshad and Frankenberger, Jr., 1992), the exact mechanism or mode-of-action of many of the species of microorganisms that comprise EM cultures are yet to be elucidated. Considerable basic and applied research on EM and other such microbial inoculants is needed for the immediate future.

REFERENCES

Arshad M. and W. T. Frankenberger, Jr. (1992). Microbial production of plant growth regulators. In *Soil Microbial Ecology*, ed. F. B. Metting, Jr. New York: Marcel Dekker, Inc., pp. 307-347.

Baker E. A. (1974). The influence of environment on leaf wax development in *Brassica oleracea* var. gemmifera. *New Phytologists* 73:955-966.

Clarke J. M. and N. M. Thomas. (1982). Evaluation of techniques for drought resistance in wheat. *Crop Science* 22: 503-506.

Haas K. (1997). Einfluβ von temperatur und Blattalter auf das das cuticularxachs von *Hedera helix*. *Biochemi. Pflanz.* 171:25-31.

Harwood R. R. (1984). Organic farming research at the Rodale Research Center. In *Organic Farming: Current Technology and Its Role in a Sustainable Agriculture*, eds. D. F. Bezdicek, J. F. Power, D. R. Keeney and M. J. Wright, Madison: American Society of Agronomy, pp. 1-18.

Higa T. (1994). *The Completest Data of EM Encyclopedia*, Tokyo: Sogo-Unicom, 385 p.

Jones H. G. (1983). *Plant and Microclimate–A Quantitative Approach to Environmental Plant Physiology*. London and New York: Cambridge Univ. Press, 323 p.

Kobayashi, M. (1980). Waste water treatments using photosynthetic microbes. In *Soil Microbes*, ed. S. Furusaka, Tokyo: Hakuyusha, Inc., pp. 421-433.

Quisenberry J. E., B. Roark B. and B. L. McMichael. (1982). Use of transpiration decline curves to identify drought-tolerant cotton germplasm. *Crop Science* 22: 918-922.

Richards R. A., H. M. Rawson and D. A. Johnson. (1986). Glaucousness in wheat: its development and effect on water-use efficiency, gas exchange and photosynthetic tissue temperatures. *Australian Journal of Plant Physiology* 13: 465-473.

U.S. Department of Agriculture. (1980). *Report and Recommendations on Organic Farming, A Special Report Prepared for the Secretary of Agriculture.* Washington, DC: U.S. Govern. Print. Office, 1545 p.

Vogtmann H. (1984). Organic farming practices and research in Europe. In *Organic Farming: Current Technology and Its Role in a Sustainable Agriculture*, eds. D. F. Bezdicek, J. F. Power, D. R. Keeney and M. J. Wright. Madison: American Society of Agronomy, pp. 19-36.

Svenningsson M. and C. Liljenberg. (1986). Changes in cuticular transpiration rate and cuticular lipids of oat (*Avena sativa*) seedlings induced by water stress. *Physiologia Plantarum* 66: 9-14.

Whitecross M. I. and D. J. Armstrong. (1972). Environmental effects on epicuticular waxes of *Brassica napus*. *Australian Journal of Botany* 20: 87-95.

Xu H. L., S. Kato, K. Yamada, M. Fujita, K. Katase and H. Umemura. (1997). Soil-root interface water potential in sweet corn plants affected by organic fertilization and effective microbes application. *Japanese Journal of Crop Science* 66 (Extra 1): 110-111.

Xu H. L., N. Ajiki, X. Wang, C. Sakakibara. and H. Umemura. (1998). Corn leaf water retention as affected by organic fertilizations and effective microbe applications. *Pedosphere* 8:1-8.

Properties and Applications of an Organic Fertilizer Inoculated with Effective Microorganisms

Kengo Yamada
Hui-lian Xu

SUMMARY. Research studies were conducted to elucidate the chemical, physical and microbiological properties of an organic fertilizer that was inoculated and fermented with a microbial inoculant (Effective Microorganisms or EM). The quality estimation methods employed addressed the mechanistic basis for beneficial effects of soil improvement and crop yield. Effective Microorganisms or EM was utilized as the microbial inoculant that is a mixed culture of beneficial microorganisms. Tests showed that the fermented organic fertilizer contained large populations of propagated *Lactobacillus* spp. Actinomycetes, photosynthetic bacteria and yeasts; high concentrations of intermediate compounds such as organic acids and amino acids; 0.1% of mineral nitrogen mainly in the ammonium (NH_4^+) form, and 1.0% of available phosphorus; and a C:N ratio of 10. The quality of the fermented organic fertilizer depends on the initial water content; addition of molasses as a carbon and energy source; and the microbial inoculant. The medium pH appears to be reliable fermentation quality criterion for producing this organic fertilizer. Beneficial effects of the fermented organic fertilizer on soil fertility and crop growth will likely depend upon the organic fraction, direct effects of the introduced microorganisms, and indirect

Kengo Yamada is Research Agronomist, Hui-lian Xu is Senior Crop Scientist, International Nature Farming Research Center, 5632 Hata, Nagano 390-1401, Japan.

Address correspondence to: Hui-lian Xu at the above address (E-mail: huilian@janis.or.jp).

[Haworth co-indexing entry note]: "Properties and Applications of an Organic Fertilizer Inoculated with Effective Microorganisms." Yamada, Kengo, and Hui-lian Xu. Co-published simultaneously in *Journal of Crop Production* (Food Products Press, an imprint of The Haworth Press, Inc.) Vol. 3, No. 1 (#5), 2000, pp. 255-268; and: *Nature Farming and Microbial Applications* (ed: Hui-lian Xu, James F. Parr, and Hiroshi Umemura) Food Products Press, an imprint of The Haworth Press, Inc., 2000, pp. 255-268. Single or multiple copies of this article are available for a fee from The Haworth Document Delivery Service [1-800-342-9678, 9:00 a.m. - 5:00 p.m. (EST). E-mail address: getinfo@haworthpressinc.com].

effects of microbially-synthesized metabolites (e.g., phytohormones and growth regulators). *[Article copies available for a fee from The Haworth Document Delivery Service: 1-800-342-9678. E-mail address: getinfo@haworthpressinc.com <Website: http://www.HaworthPress.com>]*

KEYWORDS. Effective Microorganisms, EM, EM bokashi, fermentation, microbial inoculant, nature farming, organic farming, plant nutrients

INTRODUCTION

The concept of nature farming was first introduced in 1935 by Mokichi Okada, a Japanese naturalist and philosopher (Anonymous, 1993). While nature farming is somewhat similar to organic farming, e.g., both advocate the non-use of synthetic chemicals, there are conceptual and ideological differences. The use of Effective Microorganisms or EM has become an important part of nature farming (Arakawa, 1985; Suzuki, 1985; Higa, 1998). EM consists of mixed cultures of naturally-occurring, beneficial microorganisms applied as inoculants to soil and plants which are widely documented to improve soil quality and the growth and yield of crops (Higa and Parr, 1994; Iwahori and Nakagawara, 1996; Iwaishi, 1994; Suzuki, 1985). Although EM is comprised of a large number of microbial species, the predominant populations include lactic acid bacteria, yeasts, actinomycetes and photosynthetic bacteria. Because most of the microorganisms in EM cultures are heterotrophic, i.e., they require organic sources of carbon and nitrogen, EM has been most effective when applied in combination with organic amendments to provide carbon, nitrogen and energy.

Consequently, there has been considerable interest in applying EM as a component of organic fertilizers. One such product is EM bokashi in which a mixture of rice bran, oil mill sludge or cake and fish meal is inoculated with EM and fermented, often under poorly-defined conditions. While researchers have often shown EM bokashi to be effective in improving soil quality and crop growth, some results have not shown consistent beneficial effects (Kato et al., 1997; Noparatraraporn, 1996). The reasons for these discrepancies can likely be attributed to (a) the wide range in type and quality of organic materials used to produce bokashi, (b) fluctuations in environmental conditions, (c) variable conditions of fermentation, and (d) differences in practical application technology. Moreover, research is needed to determine the mechanisms or modes-of-action on how EM bokashi actually elicits beneficial effects on soil quality and on crop growth and yield.

Therefore, the purpose of this paper was to assess the properties of EM bokashi produced under standardized conditions as well as by farmers themselves, to evaluate methods for estimating product quality; and to determine the mechanistic basis for the effects of EM bokashi on soil improvement and crop production.

MATERIALS AND METHODS

Experiment 1: Aerobic Fermentation of Organic Materials with EM Added

EM bokashi was prepared by adding molasses (8 ml), water (800 ml) and EM-1 (8 ml) to the mixed materials of rice bran (3.5 kg) and rice husk (2.0 kg), rapeseed oil mill cake (1.5 kg), and fish meal (1.0 kg) in a closed container. The treatment was repeated three times with the mixed non-EM materials as a control. The microbiological and chemical properties were examined 7, 21, 42 and 84 days after the beginning of fermentation. The numbers of aerobic and anaerobic microorganisms, fungi, aerobic dye tolerant bacteria, *Lactobacillus* spp. and yeast were evaluated using the dilution plate method with media of YG, VL, rose bengal, crystal violet added YG, GYP agarose, and YM, respectively (Kanbe, 1990; Koto, 1992; Uchimura and Okada, 1992). EC and pH were measured with glass electrodes (CM-20E and F-7AD, TOA Electrics Ltd., Tokyo, Japan). The anaerobes and *Lactobacillus* spp. were cultured in a nitrogen gas exchange incubator (BNR-110, TABAI ESPEC Corp., Tokyo). C:N ratios were determined with a carbon-nitrogen analyzer (MT-700 Yanaco Analytical Industries Ltd., Kyoto, Japan). Total and available phosphorus as well as NO_3^- and NH_4^+ were extracted with vapor distillation methods (Bremner, 1965). Organic acids such as lactic, acetic and butyric were measured by a high performance liquid chromatography.

Experiment 2: The Quality of Bokashi as Affected by EM and Molasses Additions and Water Content

EM bokashi was prepared similarly to that in Experiment 1 by adding water (800 ml), molasses (8 ml) and EM 1 (8 ml) to mixed materials of rice bran (2.7 kg), rapeseed oil mill cake (1.0 kg) and fish meal (1.0 kg). Treatment varied with or without additions of EM and molasses, and with water content (20% and 30%). Because material properties were observed within 21 days in Experiment 1, chemical properties and microbial numbers were evaluated on days 7 and 21 in Experiment 2. Items determined and methods used were the same as in Experiment 1.

Experiment 3: The Standards of EM Bokashi Prepared by Farmers

Although EM bokashi is commercially-marketable, most bokashi products are produced and used by farmers themselves. Therefore, chemical analyses were conducted for 9 different EM bokashi products made by farmers. EC, pH, C:N ratios, inorganic N and available P were determined as in Experiment 1.

RESULTS

The Characteristics of Bokashi During Fermentationwith EM Added

As shown in Figure 1, the *Lactobacillus* spp. population was only 10^3 CFU g^{-1}, at the beginning of fermentation, but increased to 10^8 CFU g^{-1} after seven days. Yeast populations increased from 10^4 CFU g^{-1} at the beginning to 10^8 CFU g^{-1} after three weeks of fermentation. The fermentation of bokashi with EM resulted in different trends for microbial numbers. *Lactobacillus* and yeast showed higher populations and lasted longer in EM bokashi than in non-EM bokashi. On day 84, the population of *Lactobacillus* was lower than 10^5 CFU g^{-1} and yeast was about 10^2 CFU g^{-1} in EM bokashi. Fungi were never detected at levels higher than 10^4 CFU g^{-1}.

FIGURE 1. Changes in *Lactobacillus* and yeast concentrations during the fermentation period.

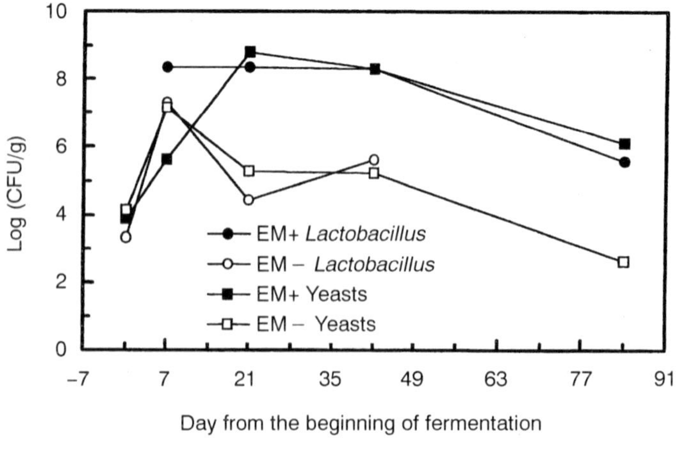

Electrical conductivity was higher and pH was lower in EM bokashi than non-EM bokashi. The ratios of population of actinomycetes to fungi and of bacteria to fungi changed during the fermentation period (Figure 2). However, it is not clear what the trends mean and why the ratios change in these ways.

The changes in concentration of organic acids are shown in Figure 3. Of the three organic acids analyzed, L-lactic acid showed the highest concentration, increasing steadily from day 7 to 21, and remaining high until the end of fermentation. This pattern was amplified by EM addition. An increase in acetic acid concentration was also observed for EM bokashi but not for non-EM bokashi.

Changes in pH, EC, and inorganic nitrogen (NO_3^--N and NH_4^+-N) are shown in Figure 4. In EM bokashi, pH decreased significantly from day 42 and reached 4.5 on day 84. However, pH for non-EM bokashi remained at a high level of 6.0 until day 84. EC increased rapidly until day 21 for EM bokashi and then decreased slowly while a slow and steadily increase in EC was observed for non-EM bokashi. NH_4^+-N concentration was higher than NO_3^--N that was only 1/2 to 1/10 that of NH_4^+-N. On day 42, the total

FIGURE 2. Changes in actinomycetes/fungi and bacteria/fungi ratios during the fermentation period

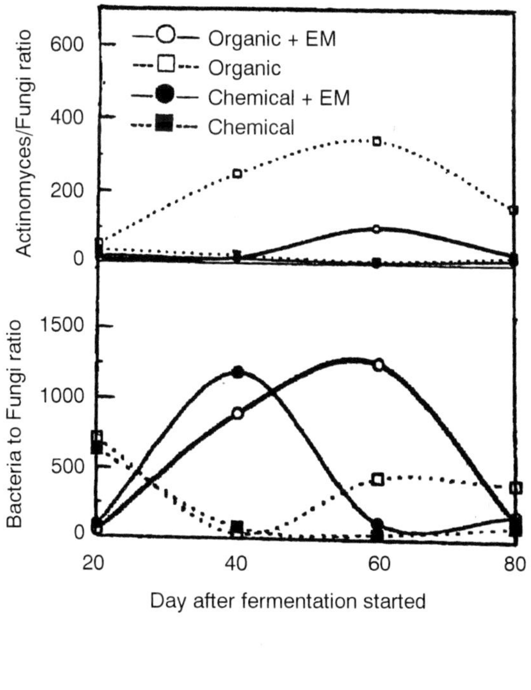

FIGURE 3. Changes in concentration of organic acids during the fermentation period.

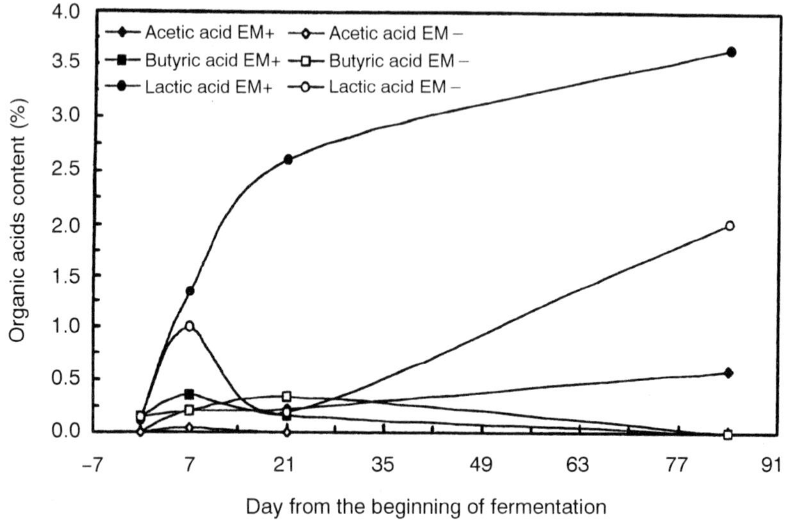

nitrogen content was higher in EM bokashi, but on day 84 it was higher for non-EM bokashi.

Fermentation Quality as Affected by EM, Molasses and the Water Content

In the present study, only when the water content of the materials was maintained at 30% during fermentation was lactic acid production observed (Figure 5). Addition of molasses did not show any effect on lactic acid production. As shown in Figures 5 and 6, pH at the beginning of fermentation was 6.1 in EM bokashi with a water content of 20%, while it was 4.8 for EM bokashi with a 30% water content. If no EM was added, pH was always about 6.0 whether the water content was low or high. EC was 3.2 mS cm^{-1} in bokashi with a 20% water content, but 5.1 with a water content of 30%. EC was 5.0 for EM bokashi without molasses added and only 3.4 for non-EM bokashi.

The population of *Lactobacillus* was 10^7 CFU g^{-1} in bokashi with a 30% water content and EM added (Figure 5). However, the number declined markedly when EM was not added, and was only 10^3 CFU g^{-1} when the water content was 20%. There was no fixed trends observed for the numbers of aerobes and yeast. Moreover, only when the water content was 30% with EM added did L-lactic acid reach a high concentration.

FIGURE 4. Changes in chemical properties during the fermentation period.

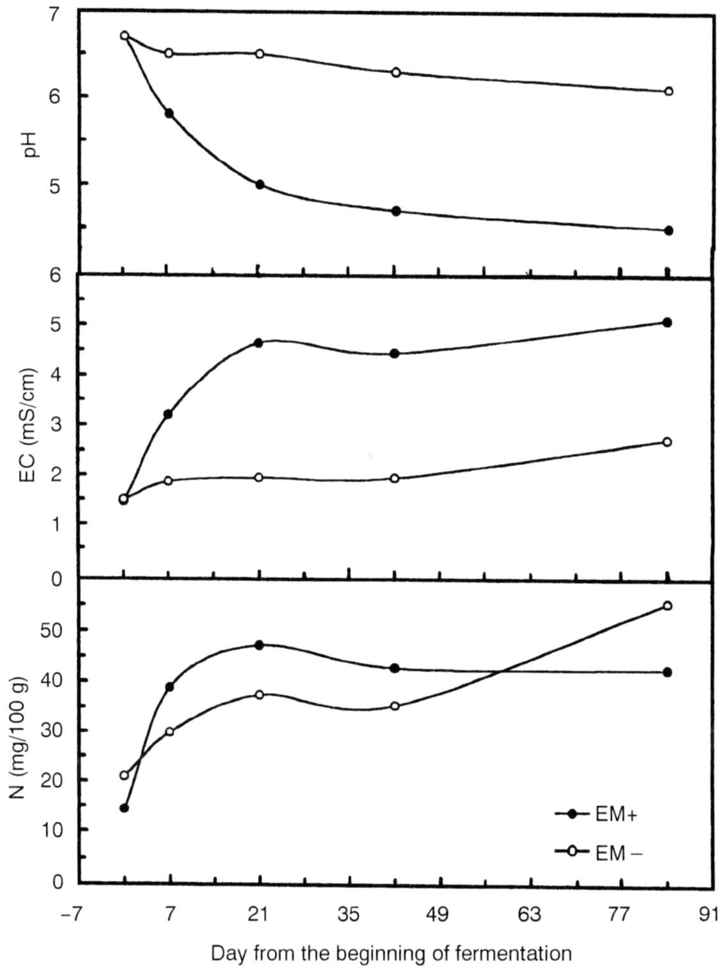

Properties of EM Bokashi Produced by Farmers

The farmer's EM bokashi was made mainly from rice bran that was locally available. The range in quality parameters for 9 different bokashi products produced by farmers is presented in Table 1. The lowest pH was 4.5 and the highest was 6.8 with a mean of 5.5 and standards deviation (SD) of 0.76. The lowest EC was 2.5 mS cm^{-1} and the highest was 6.5 with a mean of 4.9 and

FIGURE 5. Effects of preparing factors on the quality of EM bokashi.

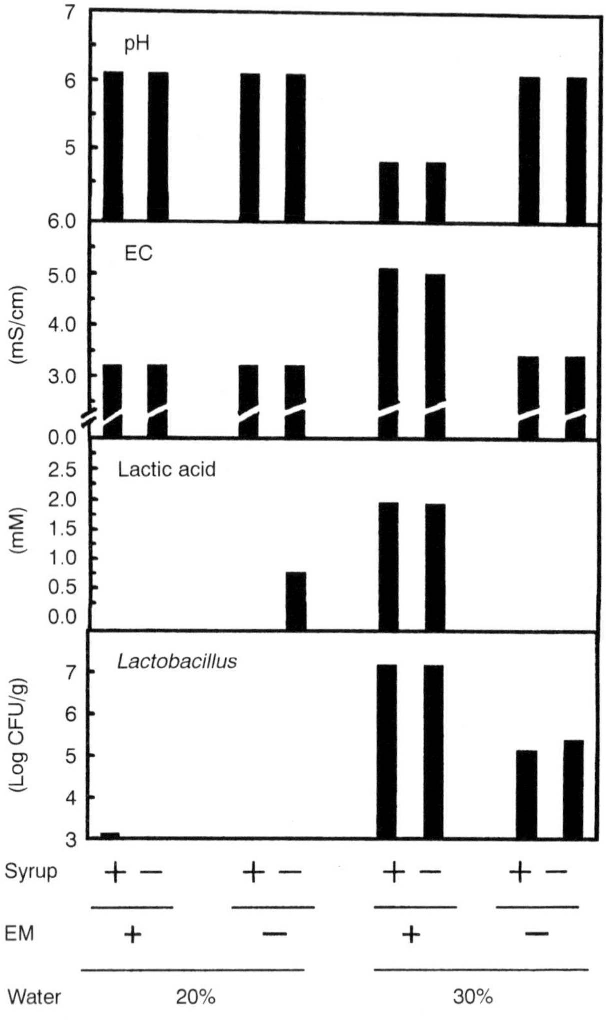

SD of 1.38. The total carbon and total nitrogen contents were 44.5% and 4.5%, respectively, with a C:N ratio of 10.3. The average NH_4^+-N and NO_3^--N contents were 1007 and 85 mg kg^{-1}, respectively, both having large variations. The available phosphorus was as high as 9934 ± 1549 mg kg^{-1}, but variable among the products. Even so, the survey showed that these bokashi products are all good nutrient resources.

FIGURE 6. Effects of water content on pH of EM bokashi.

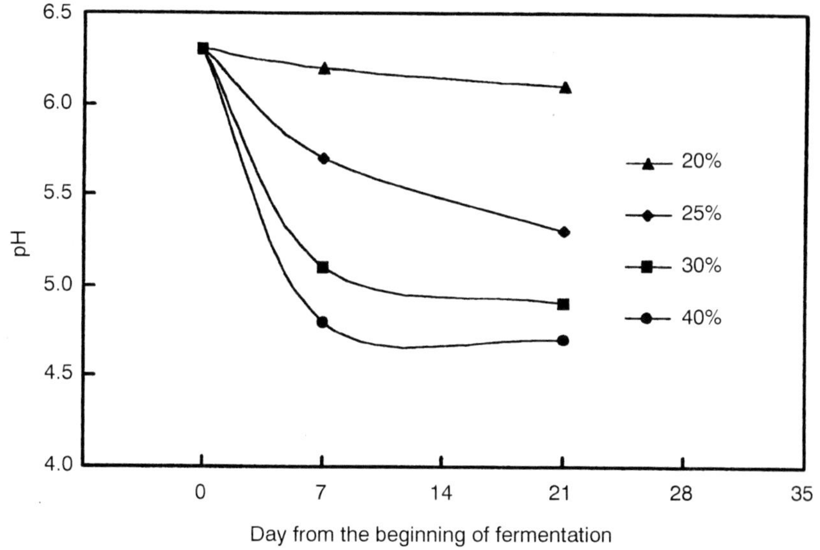

TABLE 1. Quality Parameters (EC [mS cm^{-1}], Electrical Conductivity; Av.-P [g kg^{-1}], Available Phosphorus; TC [g kg^{-1}], Total Carbon; TN [g kg^{-1}], Total Nitrogen; C:N, Carbon to Nitrogen Ratio) of the Commonly Used EM Bokashi.

pH	EC	NH_4^+-N	NO_3^--N	Av.-P	TC	TN	C:N
5.5	4.9	1.007	0.085	9.934	445	45	10.3
±0.7	±1.3	±0.6	±0.076	±1.549	±26.9	±6.1	±2.3

DISCUSSION

The beneficial effects of EM bokashi for improving soil quality and crop production have been widely reported. However, some experimental results have not shown a clear effect of EM because of fluctuations in environmental conditions and a lack of practical technology (Kato et al., 1997; Noparatrara-porn, 1996). The application of EM bokashi is vital for the adoption of EM technology and sustainable crop production in nature farming systems. Considerable research on cultivation technology with EM application has been conducted (Iwahori and Nakagawara, 1996; Iwaishi, 1994; Suzuki, 1985). However, the various properties of EM bokashi and application characteristics have not been elucidated clearly. The farmer's bokashi is aerobically fermented by mixing the organic materials with soil to provide beneficial microorganisms, while EM bokashi is characterized by anaerobic fermentation in enclosed containers with EM added. EM bokashi produced and used in the present study is quite different in materials and fermentation conditions than the common bokashi produced by farmers. Therefore, the material properties of EM bokashi for both chemical and microbiological aspects will likely be different from the common bokashi.

Results of the present research suggest that lactic acid fermentation does occur during the incubation period. *Lactobacillus* propagated rapidly under anaerobic conditions that resulted from the activities of microorganisms at the early stage. The pH decreased as the lactic acid concentration increased. This low pH suppressed the propagation of many other microbes and enabled yeast to reproduce dominantly. Consequently, the intermediate substances like lactic acid, amino acid and others increased due to the activities of *Lactobacillus* and yeast. The principles of lactic acid fermentation technology are extensively utilized by industries that process foods and agricultural products (Kanbe, 1990; Uchimura and Okada, 1992), but rarely used for soil improvement and crop production. However, EM bokashi is now considered as an organic fertilizer that is uniquely different from other organic fertilizers.

As mentioned earlier, the quality of EM bokashi depends on whether or not lactic acid fermentation is predominant. One of the most important conditions is the water content of the materials at the beginning of fermentation. If the water content is too low, the aeration of the materials will increase and activities of anaerobic microbes will be suppressed, resulting in a poor quality bokashi as a consequence. The research presented in this paper found that maintaining the water content at a proper level was critical to producing high quality bokashi. The water content must be maintained at 30% or a little higher to ensure the desired decrease in pH, increase in EC, synthesis of lactic acid, and propagation of *Lactobacillus*. In general, farmers tend to use little water in preparing bokashi that may result in incomplete fermentation and poor product quality.

Even so, under a wide range of conditions, a decrease in pH is usually indicative of lactic acid synthesis and propagation of *Lactobacillus*. Therefore, pH can be a reliable indication and a simple criterion of the quality of EM bokashi. This is supported by results of the present research.

The chemical properties of EM bokashi are characterized by high NH_4^+-N and very low NO_3^--N levels, which result from suppressed aerobic nitrification because of anaerobic conditions. The high available phosphorus content suggests that EM bokashi can be a good nutrient source for plants.

Based on the present research, the following conclusions can be drawn for EM bokashi prepared according to stated directions: (1) EM enhances anaerobic fermentation of organic materials, increases the production of lactic acid and decreases the media pH; (2) EM bokashi is an organic fertilizer that contains 0.1% mineral N, 1% available P and has a C:N ratio of 10; (3) the quality and maturity of EM bokashi can be simply estimated by changes in pH and EC. Although it was not determined, it is likely that photosynthetic bacteria and actinomycetes might also exist in EM bokashi along with *Lactobacillus* spp. and yeast. Therefore, EM bokashi can be considered as a "living fertilizer" or "microbial fertilizer."

Research is needed to elucidate the exact mechanisms and modes-of-action whereby EM bokashi elicits beneficial effects on soils and plants. As mentioned earlier, EM bokashi includes propagated EM microbes; intermediate bioactive products from fermentation and other metabolic processes; and inorganic nutrients and undecomposed organic substances. Obviously, there will be individual effects and interactive effects when EM bokashi is applied to soils and plants. The occurrence and magnitude of these effects may depend on soil conditions. According to a recent study the (Kato et al., 1997), most NH_4^+-N was nitrified within 20 days when EM bokashi was applied to an Andosol soil. Such rapid nitrification was also observed in the field where EM bokashi was applied. The release of available plant nutrients from EM bokashi depends on the activities of ammonium-oxidative microbes and nitrite-oxidative microbes in the soil. Therefore, rapid nitrification does not occur in degraded and infertile soils because these nitrifying microbes do not thrive there. The nutrient availability of EM bokashi is comparable to clover leaves and chicken manure because the C:N ratio is only 10 and the undecomposed materials are quickly mineralized.

Extensive research has been conducted on the effects of bokashi on plant growth, photosynthesis and grain yield compared with chemical fertilizers (Fujita et al., 1997; Xu et al., 1997). The total dry matter of plants produced by chemical fertilizer was clearly higher at the early stage of growth, but lower at the later stages. However, plants nourished with bokashi maintained vigorous growth with greater root mass and activity and a higher rate of photosynthesis until harvest, showing a different growth pattern compared with plants treated with chemical fertilizer (Fujita et al., 1997). The well-developed roots of the bokashi-treated plants would likely play an important role in maintaining a higher rate of growth and photosynthetic activity (Yamada et al., 1997). This may largely be the result of the sustained nutrient supply of bokashi (Kato et al., 1997). However, the possibility still exists that EM contains phytohormones or other biologically-active substances that can delay senescence of plants. Similar phenomena have been observed for other organic materials with low C:N ratios. It was also found that plants nourished with aerobic bokashi showed a higher growth rate, higher photosynthetic activity, and finally higher grain yields than plants treated with anaerobic bokashi (Fujita et al., 1997; Xu et al., 1997). This was also due to the more developed root system of plants treated with aerobic bokashi. Compared with anaerobic bokashi where nitrification is needed after application to the soil, aerobic bokashi contains available NO_3^--N that is immediately available to plants after soil application. In most cases, however, nitrification can occur rapidly after application of anaerobic bokashi and should provide adequate available N for plants. However, in the case of rapid early plant growth, the anaerobic bokashi should be treated aerobically to promote nitrification and a higher level of NO_3^--N before application.

Because EM bokashi was prepared by fermenting organic materials with the EM inoculant, comparisons between non-EM bokashi and EM bokashi were made (Fujita et al., 1997; Xu et al., 1997). In addition to a higher growth rate, and increased photosynthetic activity, the most obvious effect of EM was enhanced root development and root growth (Yamada et al., 1997). The percentage ratio of root dry mass was clearly higher for the EM treatments than for non-EM treatments. It is also possible that phytohormones or other auxin-type growth regulators produced by EM and present in EM bokashi, could have stimulated root activity (Yamada et al., 1997).

From the foregoing discussion one may conclude that the beneficial effects of EM bokashi can be mainly attributed to (a) the sustained release of available plant nutrients from decomposition of organic materials, and (b) biologically-active substances such as phytohormones and growth factors synthesized by the EM cultures or produced as by-products during organic matter decomposition. This concept is illustrated in the schematic diagram of Figure 7. It is well

FIGURE 7. A speculative illustration of the effect of EM bokashi on soil fertility and plant growth.

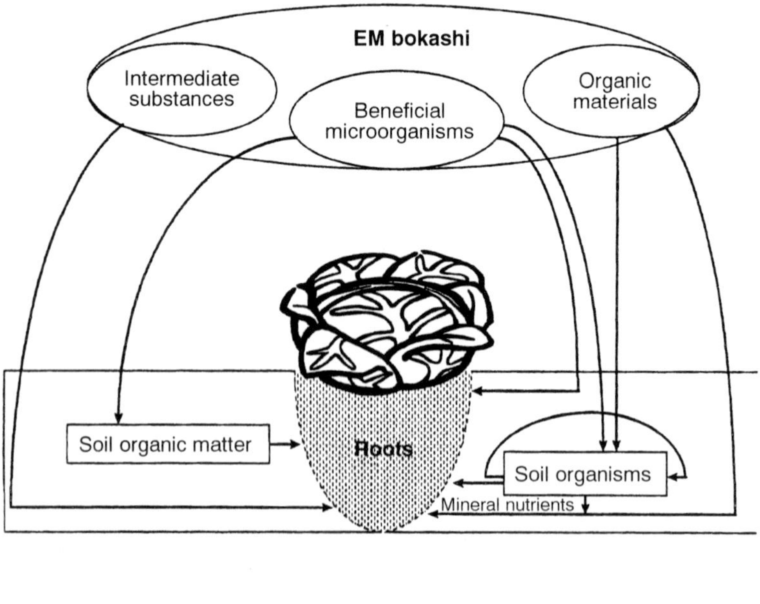

known that plant hormones such as auxins, gibberellins and abscisic acid play important roles in root growth and development (Schneider and Wightman, 1974; Kuraishi, 1983). Moreover, research has shown that many soil microorganisms, i.e., bacteria, fungi and actinomycetes produce a variety of bioactive compounds that can enhance plant growth and metabolism (Arshad and Frankenberger, 1992). Some researchers have also speculated that to a large extent, the beneficial effects of EM can be attributed to the biosynthesis of antioxidants, although this has yet to be scientifically proven. Consequently, research is needed to determine the mechanisms or modes-of-action as to how EM elicits "growth-promoting" or "growth-stimulating" effects on plant growth and metabolic processes. Finally it was recently reported that EM can enhance soil aggregation. Research is needed to determine the exact conditions under which this occurs and the mechanisms that are involved.

REFERENCES

Anonymous. (1993). About the name Kyuse Nature Farming. In *The Basis of Paradise*, ed. The Editorial Board of Kyusei Dogma. Tokyo, Atami: Kyusei Publisher.

Arakawa, Y. (1985). Effects of EM application on snow thawing as well as snow rotting and powdery mildew problems of crops (Part V). *Proceedings of Symposium of Applied Soil Microbiology*. November 22, 1985, Urazoe, Okinawa, Japan.

Arshad, M. and W.T. Frankenberger. (1992). Microbial production of plant growth regulators. In *Soil Microbial Ecology*, ed. F.B. Metting. New York: Marcel Dekker, pp. 307-348.

Bremner, J.M. (1965). Inorganic forms of nitrogen. In *Methods of Soil Analysis*, ed. C.A. Black. Madison, Wisconsin: American Society of Agronomy, pp. 1179-1237.

Fujita, M., S. Kato, K. Yamada, H.L. Xu, K. Katase and H. Umemura. (1997). Applications of effective microorganisms in nature farming. VIII. Growth and yield of sweet corn as affected by applications of organic fertilizer with effective microorganisms. *Annual Meeting of Japanese Society of Soil Science and Plant Nutrition*, April 2(4, 1997, Shizuoka, *Proceedings* 43:163.

Higa, T. (1998). Effective microorganisms for a more sustainable agriculture, environment and society: potential and prospects. In *Proceedings of the Fourth International Conference on Kyusei Nature Farming*, eds. J.F. Parr and S.B. Hornick. Washington, DC: U.S. Department of Agriculture, pp. 6-7.

Higa, T. and J.F. Parr. (1994). *Beneficial and Effective Microorganisms for a Sustainable Agriculture and Environment*. Atami (Japan): International Nature Farming Research Center, 16 p.

Iwahori, H. and T. Nakagawara. (1996). Studies on EM application in nature farming. V. Applying methods of EM bokashi in vegetable cultures. *Annual Meeting of Japanese Society of Soil Science and Plant Nutrition*, April 1996, Tokyo, Japan.

Iwaishi, S. (1994). Effects of EM bokashi on various paddy-rice varieties. *Annual Meeting of Asia-Pacific Nature Agriculture Network*. October 6, 1994, Seoul, Korea.

Kanbe, M. (1990). Use of lactic acid bacteria. *Journal of Microorganisms* 6:56-65.

Koto, K. (1992). Quantification, isolation, and identification of soil bacteria. In *Experimental Methods of Soil Microbiology*, eds. T. Hittori, M. Nishio, and K. Miyashita. Tokyo: Yokendo, pp. 15-54.

Kato, S., K. Yamada, M. Fujita, H.L. Xu, K. Katase and H. Umemura. (1997). Applications of effective microorganisms in nature farming. IX. Soil fertility and plant nutrient uptake of sweet corn as affected by applications of organic fertilizer with effective microorganisms added. *Annual Meeting of Japanese Society of Soil Science and Plant Nutrition*, April 24, 1997, Sizuoka, *Proceedings* 43:164.

Kuraishi, S. (1983). *Plant Hormones*. Tokyo: University of Tokyo Publisher, pp. 1-142.

Noparatraraporn, N. (1996). Thailand collaborative research on evaluation of EM and EM products, their feasibility testing and effects of their uses on agriculture and environment. Open Symposium: *Present Situations and Prospects of Microorganisms as Agricultural Materials*. August 23, 1996, Tokyo.

Schneider E.A. and F. Wightman. (1974). Metabolism of auxin in higher plants. *Annual Review of Plant Physiology* 25:487-531.

Suzuki, Y. (1985). Effects of effective microorganisms on yield and quality of ginseng herbs. *Symposium of Applied Soil Microbiology*. November 22, 1985, Urazoe, Okinawa.

Uchimura, T. and W. Okada. (1992). Experimental methods for *Lactobacillus*. In *Manual of Lactobacillus Experiments–from Isolation to Identification*, ed. M. Osaki. Tokyo: Asakura Books, pp. 21-125.

Xu, H.L., S. Kato, K. Yamada, M. Fujita, K. Katase and H. Umemura. (1997). Applications of effective microorganisms in nature farming. XI. Photosynthesis of sweet corn as affected by applications of organic fertilizer with effective microorganisms added. *Annual Meeting of Japanese Society of Soil Science and Plant Nutrition*, April 24, 1997, Shizuoka, *Proceedings* 43:164.

Yamada, K., S. Kato, M. Fujita, HL Xu, K. Katase and H. Umemura. (1997). Applications of effective microorganisms in nature farming. X. Root development of sweet corn as affected by applications of organic fertilizer with effective microbes added. *Annual Meeting of Japanese Society of Soil Science and Plant Nutrition*, April 24, 1997, Shizuoka, *Proceedings* 43:163.

Effect of Organic Fertilizer and Effective Microorganisms on Growth, Yield and Quality of Paddy-Rice Varieties

Shinji Iwaishi

SUMMARY. The effect of an organic fertilizer inoculated with Effective Microorganisms (EM) on the growth, yield and quality of 13 paddy-rice varieties varying with maturation period was studied. EM inoculation increased kernel enlargement after the panicle formation stage and also increased ear number and length and kernel number. The yield of brown rice from EM inoculation was higher for the standard fertilizer rate and lower for the higher rate of organic fertilizer. EM inoculation increased the glutinousness and the total quality index of glutinous rice varieties. Under 1993 weather conditions, early and medium-ripening non-glutinous varieties and glutinous varieties were suitable for nature farming with EM-inoculated organic fertilizer. *[Article copies available for a fee from The Haworth Document Delivery Service: 1-800-342-9678. E-mail address: getinfo@haworthpressinc.com <Website: http://www.HaworthPress.com>]*

KEYWORDS. Effective microbes, EM, nature farming, organic fertilizer, quality, rice variety

INTRODUCTION

During the last decades, a microbial inoculant referred to as Effective Microorganisms or EM has been used with considerable success in nature

---

Shinji Iwaishi is Research Agronomist, International Nature Farming Research Center, Hata, Nagano 390-1401, Japan (E-mail: infrc390@janis.or.jp).

[Haworth co-indexing entry note]: "Effect of Organic Fertilizer and Effective Microorganisms on Growth, Yield and Quality of Paddy-Rice Varieties." Iwaishi, Shinji. Co-published simultaneously in *Journal of Crop Production* (Food Products Press, an imprint of The Haworth Press, Inc.) Vol. 3, No. 1 (#5), 2000, pp. 269-273; and: *Nature Farming and Microbial Applications* (ed: Hui-lian Xu, James F. Parr, and Hiroshi Umemura) Food Products Press, an imprint of The Haworth Press, Inc., 2000, pp. 269-273. Single or multiple copies of this article are available for a fee from The Haworth Document Delivery Service [1-800-342-9678, 9:00 a.m. - 5:00 p.m. (EST). E-mail address: getinfo@haworthpressinc.com].

farming and organic farming systems in Japan and throughout the Asia-Pacific region. EM is a mixed culture of naturally-occurring, beneficial microorganisms (predominantly, lactic acid bacteria, photosynthetic bacteria, yeast, actinomycetes and fungi) that has reportedly enhanced soil quality and biodiversity and increased the growth, yield and quality of crops (Higa and Parr, 1994). There is some indication that EM applied in combination with an organic amendment (or organic fertilizer) is more effective than applied alone. This is likely true because most of the microorganisms in EM are classified as heterotropgic, i.e., they require organic forms of carbon and nitrogen for metabolism and biosynthesis. There is considerable interest in inoculating organic fertilizers with EM and allowing a period of fermentation prior to application (Fujita, 1995), thereby improving the quality and effectiveness of organic fertilizers for nature farming systems. Thus, the purpose of this study was to determine the effect of an organic fertilizer with and without EM inoculation on the growth, yield and quality of different paddy rice varieties and to select suitable varieties for nature farming systems.

MATERIALS AND METHODS

Geographic and Climatic Conditions

The experiment was conducted in 1993 at the Agricultural Experiment Station, International Nature Farming Research Center, Matsumoto, Japan (36°N, 101°W, 685 m elevation). The soil was classified as an Andosol derived from volcanic ash. The annual mean temperature was 11.2°C and annual total precipitation was 1,011 mm. The crop season in 1993 was unusually cool with low temperatures, little sunshine and excessive rainfall. Especially in July and August, the temperature stayed far below that of the normal year.

Treatments

Thirteen paddy-rice varieties, 11 of which were non-glutinous and two were glutinous, different in maturation period, were grown with organic fertilizer with or without a microbial inoculant. The organic fertilizer (composition shown in Table 1) was either inoculated with EM and fermented, designated as OF + EM, or uninoculated and fermented by naturally-occurring indigenous microorganisms, designated as OF. The organic fertilizers in black polyethylene bags and fermented anaerobically at room temperature. This fermented product is often referred to as EM bokashi. The two organic fertilizers (OF + EM) and OF were applied to plots at two rates, i.e., a standard rate of 18.7 kg a^{-1} and a higher rate of 27.5 kg a^{-1}. Basal applications of the two fertilizers were made 30 days prior to planting rice and dressing applications were made on June 11 and July 21. The four treatments used throughout the study were:

1. OF + EM applied at 18.7 kg a^{-1},
2. OF applied at 18.7 kg a^{-1},
3. OF + EM applied at 27.5 kg a^{-1},
4. OF applied at 27.5 kg a^{-1}.

Experiment Design

Treatments were applied using a randomized complete block design with four replications. Area of each plot was 18 m^2. The data were analyzed using Fisher's LSD test.

RESULTS AND DISCUSSION

Plant Growth

At the maximum tillering stage, plants treated with the standard or low rate of EM-inoculated organic fertilizer (OF + EM) were taller with fewer tillers compared with organic fertilizer alone (OF). At the high rate of application, plants treated with OF + EM were shorter than those treated with organic fertilizer (OF) alone, but both treatments produced the same number of stems. Leaves in the OF + EM plots were lighter in color than those in the OF plots for both rates of organic fertilizer application. At the maturation stage, plants in the OF + EM plots had more ears and longer culms than the non-EM plots (Table 2). Although there was no difference in plant growth between the

TABLE 1. Composition and application rates of organic fertilizer inoculated and fermented with or without EM as a microbial inoculant.

EM	Rice bran	Fish meal	Rape cake	Molasses MI (1%)	Molasses steamed	Bone meal	Zeolite	Applied rate (kg a^{-1})*
Yes	3.0	2.0	3.0	1.7	0.0	6.0	3.0	18.7
No	3.0	2.0	3.0	0.0	1.7	6.0	3.0	18.7
Yes	6.0	4.0	6.0	2.5	0.0	6.0	3.0	27.5
No	6.0	4.0	6.0	0.0	2.5	6.0	3.0	27.5

*The organic fertilizers were applied at two rates, i.e., 18.7 and 2.57 kg a^{-1}. "Are" is a land measurement used by many countries of the Asia-Pacific region; the abbreviation for "are" is "a", and 1 are or 1 a = 0.01 hectare.

EM-treated plants and non-EM plants at the early growth stage, the increased growth at the maturing stage was likely the result of enhanced plant growth from EM treatment after panicle formation.

Yield and Yield Components

The yield of brown rice at the standard or low rate of organic fertilizer was higher in the EM-treated plots than in the non-EM plots. For all varieties, the yield increase ranged from 8 to 19% (Table 3). At the high rate of organic fertilizer, EM treatment had little effect on the yield of brown rice for the late-ripening, non-glutinous varieties. However, the yield increase due to EM treatment for all other varieties ranges from 2 to 6%. Thus, EM treatment was found to be beneficial on increasing the grain number, ear number and length, and kernel weight, consequently increasing the yield of brown rice.

Quality

The quality of rice was measured by near-infrared spectroscopic analysis. For non-glutinous varieties no correlation was observed between the EM-in-

TABLE 2. Heading date and date of maturity of cultivar group.

	Early	Medium	Late	Glutinous
Heading date	Aug. 12	Aug. 20	Aug. 23	Aug. 15
Maturity date	Sept. 25	Oct. 10	Oct. 23	Sept. 26

Early, early varieties; Medium, medium varieties; Late, late varieties; Glutinous, glutinous varieties.

TABLE 3. Effects of two rates of organic fertilizer inoculated and fermented with or without EM on the growth, yield and quality of rice.

Organic rate	EM	Length (cm) Culm	Ear	Number Ear (m^{-2})	Grain (10^3 m^{-2})	BR yield ($kg\ a^{-1}$)	(%)	1,000 KW(g)	Quality G	T
Low	Yes	70.08 ±3.51	17.05 ±0.33	415.50 ±8.95	29.78 ±1.17	52.63 ±0.45	112 ±2.5	21.38 ±0.26	1.26 ±0.34	0.57 ±0.11
Low	No	68.40 ±2.76	16.33 ±0.29	404.50 ±9.74	26.20 ±1.37	47.13 ±1.33	100 ±0.0	21.10 ±0.24	1.20 ±0.27	0.53 ±0.14
High	Yes	74.63 ±3.72	14.43 ±2.69	475.50 ±18.57	33.30 ±0.62	54.10 ±1.56	103 ±1.3	20.75 ±0.21	1.11 ±0.27	0.42 ±0.13
High	No	73.23 ±3.36	16.85 ±0.22	467.00 ±13.30	31.68 ±0.25	52.83 ±2.00	100 ±0.0	20.75 ±0.25	1.08 ±0.18	0.44 ±0.15

BR, brown rice; %, percentage against the control; 1,000 KW, weight of 1,000 kernels; G, glutinousness; T, total quality index. The quality of rice was measured with near infrared spectroscopic analyzer (NIRECO) for whole grians of polished rice. Low application rate was 18.7 kg a^{-1} and the high application rate was 27.5 kg a^{-1}.

oculation treatment and the total quality index. For glutinous varieties it was found that the glutinousness of rice in the EM-treated plots was higher than the non-EM treated plots.

Desired Varieties

Middle-late-ripening varieties are recommended as suitable for nature farming in this region. However, early-ripening varieties yielded higher with a better quality than the middle-late varieties, which was contrary to our expectation. The reason for this seems to be related with the harsh, unseasonable weather conditions during the cultivation period. The temperature was low with little sunshine which also caused low temperatures ranging from 13 to 15°C in irrigation water of the experimental field. Therefore, under the weather conditions of 1993, early-ripening varieties and medium-ripening varieties, which were considered as cool weather tolerant (Sato and Horisue, 1992), were suitable for nature farming systems among non-glutinous varieties. Furthermore, because of their high yield and good quality, glutinous varieties seem to be suitable for nature farming with EM-inoculated organic fertilizer.

REFERENCES

Fujita, M. (1995). An approach of apple cultivation towards nature farming. In *Proceedings of the Fourth Conference on the Technology of Effective Microorganisms*, ed. J. F. Parr. Saraburi (Thailand): APNAN, pp. 338-355.

Higa, T. and J.F. Parr. (1994). *Beneficial and Effective Microorganisms for a Sustainable Agriculture and Environment.* Atami (Japan): International Nature Farming Research Center, 16 p.

Sato, S. and N. Horisue. (1992). Cool Weather tolerance of rice plants. In *Rice Plant Breeding in Japan*, eds. K. Fushibuchi, T. Kaneda, R. Yamamoto, Y. Akama, and K. Maruyama. Tokyo: Agricultural Technology Association, pp. 338-355.

Effect of Microbial Inoculation on Soil Microorganisms and Earthworm Communities: A Preliminary Study

Zhiping Cao
Weijiong Li
Qinlong Sun
Yongliang Ma
Qin Xu

SUMMARY. A field experiment was conducted in 1991 to assess the effect of Effective Microorganisms (EM) on soil microorganisms and earthworm communities. Three treatments were applied to the experimental plots including compost + EM, compost alone, and chemical fertilizer alone. Soil samples were taken in autumn 1996 and spring 1997 to characterize treatment effects. Soil was analyzed for changes in microbial biomass and earthworm populations and biomass. Preliminary results indicated that (a) EM increased the soil microbial biomass, which was significantly higher in the spring than in autumn, and offset the adverse effects of cold temperatures; and (b) EM enhanced the

Zhiping Cao, Weijiong Li , and Yongliang Ma are Professors, College of Resources and Environmental Science, China Agricultural University, Beijing 100094, China.

Qinlong Sun is Professor, Handan Agricultural School, Hebei, China.

Qin Xu is Professor, Department of Biology, Beijing Educational College, Beijing, China.

Address of correspondence to: Weijiong Li at the above address.

[Haworth co-indexing entry note]: "Effect of Microbial Inoculation on Soil Microorganisms and Earthworm Communities: A Preliminary Study." Cao, Zhiping et al. Co-published simultaneously in *Journal of Crop Production* (Food Products Press, an imprint of The Haworth Press, Inc.) Vol. 3, No. 1 (#5), 2000, pp. 275-283; and: *Nature Farming and Microbial Applications* (ed: Hui-lian Xu, James F. Parr, and Hiroshi Umemura) Food Products Press, an imprint of The Haworth Press, Inc., 2000, pp. 275-283. Single or multiple copies of this article are available for a fee from The Haworth Document Delivery Service [1-800-342-9678, 9:00 a.m. - 5:00 p.m. (EST). E-mail address: getinfo@haworthpressinc.com].

growth and activity of earthworms in the topsoil which was readily apparent in autumn when the soil was in a relatively stable condition, but EM had little effect on earthworm populations and biomass in the spring.

KEYWORDS. Earthworm, microbial inoculation, soil microorganisms

INTRODUCTION

Effective Microorganisms or EM is a mixed culture of naturally-occurring, beneficial microorganisms that has been widely reported to improve soil quality and increase the growth and yield of crops (Li and Ni, 1995). However, there has been very little research on the exact mechanisms of how EM can elicit these beneficial effects. There are several ways that EM can improve crop productivity including: (a) accelerating the decomposition of soil organic matter or organic amendments, which would directly increase the release and availability of plant nutrients, and (b) enhancing the populations of various soil macrobiota (e.g., earthworms) and their activities, which would indirectly improve the decomposition of organic matter and plant nutrient availability. Thus, a study was conducted at the Quzhou field experiment station to determine the role and mechanisms of EM for improving soil quality and crop productivity.

MATERIALS AND METHODS

Field Experiment

A field experiment was begun in 1991 with wheat and maize planted annually as test crops (Sun et al., 1996). The experimental design included 6 treatments and four replications applied to 30 m^2 plots. Three of the treatments involved a study of Effective Microorganisms and are the only results reported in this paper.

Treatment 1 (OM + EM): EM + Compost applied at 15 t ha^{-1}.

Treatment 2 (OM): Compost applied at 15 t ha^{-1}.

Treatment 3 (CF): Chemical fertilizer applied at 450 kg N ha^{-1}.

Sampling Procedure and Analysis

The soil sampling periods were October 1996 (autumn, after corn had been harvested and before planting wheat) and April 1997 (spring). For microbial analysis, each sample was a composite of 8 random sites (100 ml per site) within a treatment plot and taken at a depth of 0 to 20 cm. Three plots were sampled for each treatment. Microbial biomass was determined using a fumigation-incubation procedure. Soil samples for earthworm analysis were taken from an area of 0.5 × 0.5 m^2 at a depth of 0 to 20 cm. Earthworms were separated from the soil by hand.

RESULTS AND DISCUSSION

Changes in Soil Chemical and Physical Properties

Eight indicators were used to characterize treatment effects on soil chemical and physical properties. These included bulk density, cation exchange capacity, organic matter content, total nitrogen, total phosphorus, alkaline hydrolysis nitrogen, active phosphorus and active potassium. As shown in Table 1, there were considerable differences among the treatments. For example, the highest soil organic matter content was found in the EM + Compost treatment which was 14.0% higher than compost alone which, in turn, was 7.5% higher than chemical fertilizer. For this particular parameter, treatments

TABLE 1. Differences in soil chemical and physical properties between fertilizations of EM + compost, compost and chemical fertilizer.

Treatments	Bulk density ($Mg\ m^{-3}$)	Cation exchange capacity ($me\ kg^{-1}$)	Organic matter ($g\ kg^{-1}$)	Total N (%) ($g\ kg^{-1}$)
EM+ compost	1.51 eE	108.9 aA	131.1 aA	0.78 aA
Compost	1.71 cC	99.4 cB	11.4 eC	0.69 bCB
Chemical fertilizer	1.89 cB	93.2 dC	10.6 dC	0.63 cdCD
	Total P ($g\ kg^{-1}$)	Alk. hydrolyzed N ($mg\ kg^{-1}$)	Active P ($mg\ kg^{-1}$)	Active K ($mg\ kg^{-1}$)
EM+ compost	0.74 aA	76.60 aA	42.81 aA	95.73 aA
Compost	0.64 cCB	66.7 bB	36.31 bB	86.30 bB
Chemical fertilizer	0.61 deCD	58.94 cD	27.30 dcDC	75.93 cC

Data sharing the same lowercase letters are not significantly different at P = 0.05 and those sharing the same capital letters are not significantly different at P = 0.01.

followed the order of EM + compost > compost > chemical fertilizer. The cation exchange capacity also declined in the same treatment order with EM + compost 9.5% higher than compost, which, in turn, was 6.7% higher than chemical fertilizer. However, bulk density followed the opposite treatment order of chemical fertilizer > compost > EM + compost. In this case EM + compost was 13.2% lower than compost alone which, in turn, was 10.5% lower than chemical fertilizer. The other five parameters followed the treatment order of EM + compost > compost > chemical fertilizer, which also reflects the overall index order of soil fertility.

Biomass of Soil Microorganisms

The biomass of soil microorganisms in different seasons varied considerably (Figure 1). Usually, the biomass of soil microbial communities reach the highest level in autumn, However, neither the EM + compost nor compost alone showed this trend. The autumn biomass of soil microorganisms for chemical fertilizer was 454 μgC/g, which was 100% higher than in spring. However, for EM + compost and compost alone, the biomass of soil microorganisms was highest in the spring, with values of 917 μgC/g and 709 μgC/g, which were 29.8% and 25.8% higher, respectively, than in autumn.

FIGURE 1

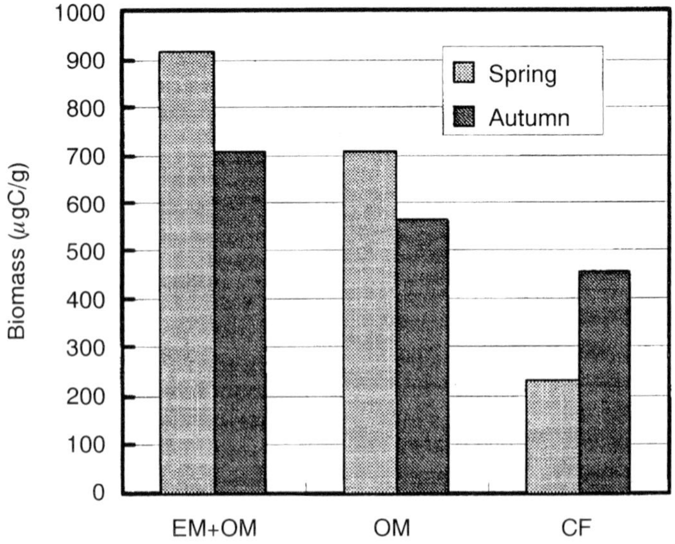

Difference in Biomass Among Treatments

Whether spring or autumn, the biomass of soil microorganisms followed the treatment order of EM + compost > compost > chemical fertilizer. In spring, the increases were highest. In autumn, compost alone was 24.2% higher than chemical fertilizer; EM + compost was 25.4% higher than compost alone; and EM + compost was 55.7% higher than chemical fertilizer. In spring, compost was 205.3% higher than chemical fertilizer; EM + compost was 29.3% and 294.8% higher than compost and chemical fertilizer, respectively.

In both spring and autumn, the soil microbial biomass for EM + compost was higher than compost alone, ranging from 29.3 to 25.4%, which showed that the application of EM increased the soil microbial biomass. EM was applied in October, and reproduced quickly in the following spring. This indicates that EM has a strong ability to resist coldness. Compared with chemical fertilizer, the biomass for EM + compost was 55.7% higher in autumn and 294.8% higher in spring; this was partly due to EM, but mainly the effect of organic matter. Compared with chemical fertilizer, the biomass in the compost treatment was 24.2% higher in the autumn and 205% higher in the spring. This suggests that conventional farming could benefit by application of some organic fertilizer together with chemical fertilizer to increase the soil microbial biomass, especially, in the spring.

In spring, the soil microbial biomass in the EM + compost treatment and compost alone were higher than in autumn. This indicates that soil treated with EM and organic fertilizer was more suitable for beneficial microorganisms to reproduce and develop large numbers that would enhance soil fertility and crop production. Spring is the season when crops develop quickly and demand more available plant nutrients. The enormous reproduction of microorganisms increases the decomposition of organic matter and release of available nutrients to sustain plant growth and yield.

Population and Biomass of Earthworms

Difference between seasons. As shown in Figures 2 and 3, the earthworm populations were higher in autumn than in spring. In autumn, the earthworm populations for EM + compost, compost and chemical fertilizer were 37.6, 44.9 and 46.2% higher than in spring, respectively. In autumn, the earthworm biomass for EM + compost was 30.2% higher than in spring; but the biomass for compost and chemical fertilizer in autumn was 14.0 and 32.2% lower than in spring, respectively. In other words, for the treatment applied with EM, the earthworm biomass reached the highest level in autumn, and for the treatments without EM, the biomass reached the highest level in spring. Spring is the season when EM microorganisms reproduce rapidly and the microbial

FIGURE 2

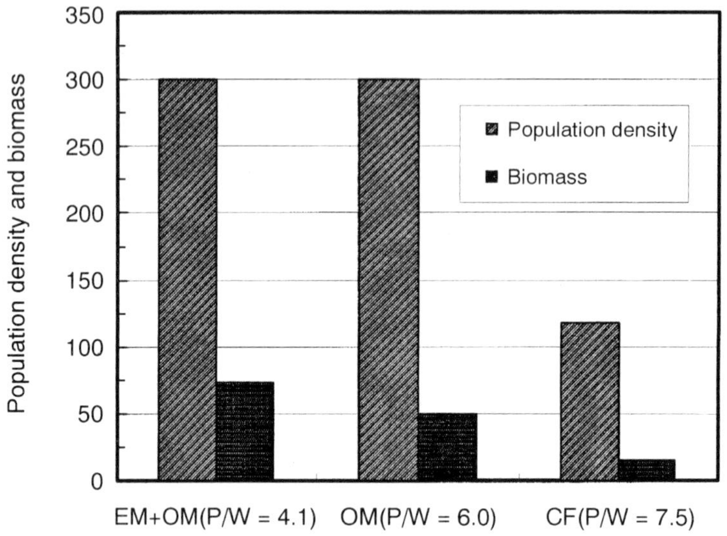

FIGURE 3

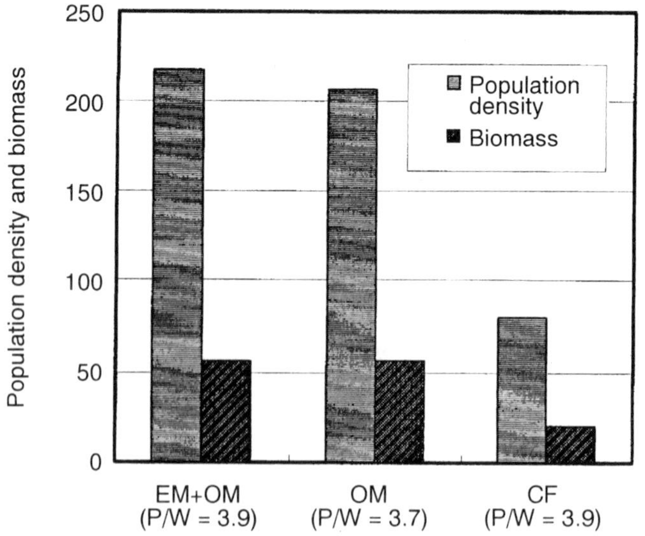

biomass increase greatly. In the spring, the earthworm biomass should also increase according to the trophic interrelation between earthworms and microorganisms, but actually it decreased in the present study. This suggests that there might be some competitive interaction between EM and the earthworm community. The future studies may help to determine whether this is a predictive observation.

Difference between treatments. In autumn, the earthworm biomass for the treatment of EM + compost was 47.7% higher than the compost alone, and 370% higher than for chemical fertilizer. These results indicate that the organic amendment had a greater effect on earthworm biomass than EM. Compost alone increased earthworm biomass more than twice compared with that of chemical fertilizer. Application of EM increased earthworm biomass nearly 50% compared with compost alone, although, this was not as large as the difference between compost and chemical fertilizer. This conclusion also was demonstrated by the earthworm biomass in April 1997. In spring, the biomass for the compost treatment was 174.1% higher than for chemical fertilizer; however, the biomass for EM + compost was nearly equal to that of compost alone.

Ratio of population density to biomass (P/W). The unit for P/W is number/gram. As shown in Figure 2, the P/W for EM + compost, compost and chemical fertilizer was 4.1, 6.0 and 7.5, respectively. These values indicated that in autumn EM could increase the size of earthworms and their weight; while compost could also increase the size of earthworms. However, in spring (Figure 3) the values of P/W for these three treatments were nearly equal, i.e., 3.9, 3.7 and 3.9, respectively. For EM + compost, the values of P/W changed minimally from autumn to spring, but for compost and chemical fertilizer, the values of P/W obviously decreased. Perhaps EM can protect young earthworms from death caused by low winter temperatures.

Structure of earthworm community. The earthworm community in the experimental field consisted of three species: *Aporrectodea trapezoides*, *Eisenia foetida* Savigny and *Drawida japonica*. The dominant species were *Aprorrectodea trapezoides* and *Eisenia foetida* Savigny. Based on the October 1997 sampling, the proportion of various populations are shown in Table 2. In terms of population density, *Eisenia foetida* Savigny was definitely dominant among the three treatments, ranging from 73% to 85%, with lower populations of *Aporrectodea trapezoides* and *Drawida japonica* Japonica. A large P/W value generally indicates a large earthworm size. In autumn, the values of P/W were in an inferior position. However, the individuals of *Aporrectodea trapezoides* were larger than those of *Eisenia foetida* Savigny and *Drawida japonica* Japonica and therefore the biomass proportion of *Aporrectodea trapezoides* was high. With the EM + compost treatment for an example, the population proportion of *Aporrectodea trapezoides* was 17.2%,

TABLE 2. Population and biomass of various earthworm populations under different fertilizer treatments (October 1996)

Parameters	Earthworm Populations	EM+ Compost	Compost	Chemical
Population	*Aporrectodea trapezoides*	53	18	3
density	*Eisenia foetida* Savigny	225	255	100
(number m^{-2})	*Drawida japonica* Japonica	30	27	16
	Total	308	300	119
Proportion of	*Aporrectodea trapezoides*	17	6	2
populations (%)	*Eisenia foetida* Savigny	73	85	84
	Drawida japonica Japonica	10	9	13
	Total	100	100	118
Biomass	*Aporrectodea trapezoides*	43	23	4
(g m^{-2})	*Eisenia foetida* Savigny	25	25	10
	Drawida japonica Japonica	5	5	2
	Total	73	53	16
Proportion of	*Aporrectodea trapezoides*	59	44	24
biomass (%)	*Eisenia foetida* Savigny	34	47	62
	Drawida japonica Japonica	7	9	13
	Total	100	100	99

but the biomass proportion was as high as 58.5%. By contrast, the population proportion of *Eisenia foetida* Savigny was 73.0%, but the biomass proportion was as low as 34.4%. The population density and biomass of *Aporrectodea trapezoides* decreased in the treatment order of EM + compost > compost > chemical fertilizer.

CONCLUSIONS

The following conclusions can be drawn from this study.

- With application of EM, the soil fertility was improved, as shown by soil physical and chemical indicators.
- EM increased the biomass of soil microorganisms, which was higher in spring than in autumn. EM showed the ability to resist coldness if applied in winter. Microbes from EM can reproduce rapidly in the following spring. Meanwhile, the organic amendment (i.e., compost) applied

with EM was important in sustaining the growth and activity of EM species in the spring. Spring is the season when crops grow very rapidly, and the demand for plant nutrients increases accordingly. The combination of EM and an organic fertilizer can increase available nutrients, and keep pace with the overall crop nutrient requirement.

- EM enhanced the growth and reproduction of earthworms in topsoil. This was particularly apparent in autumn when the soil earthworm community was in a relatively stable phase. However, in spring, the effect of EM on earthworm growth and activity was not significant.
- Within the earthworm community of farmland topsoil at Quzhou, *Aporrectodea trapezoides* and *Eisenia foetida* Savigny were the two dominant populations. Different treatments showed that these populations were in a competitive relationship.

The relationship between earthworms and microorganisms is very complex. Microorganisms are a food source for earthworms, and earthworms can change the biomass and activity of microorganisms (Lavelle and Martin, 1992). Wolters and Joergensen (1992) pointed out that earthworms could directly affect microorganisms by reducing their total biomass and affecting their metabolic activity.

Results of this paper can support the conclusion that "earthworms are able to reduce microbial biomass." Especially, in the treatment that applied EM, this phenomenon was very obvious. In autumn, when the earthworm biomass was high, the microbial biomass was low; and in spring, when the microbial biomass increased, the earthworm biomass decreased accordingly. Therefore, it seems that not only earthworms can reduce microbial biomass, but also microbial can reduce earthworm biomass, especially, when non-indigenous microorganisms are introduced. Consequently, our knowledge of the relationship between microorganisms and earthworms is in need of future elucidation.

REFERENCES

Li Wei-jiong and Ni Yougzhen. (1995). Research and application of EM (Effective Microorganisms). *Chinese Journal of Ecology* 14:58-62.

Sun Qinlong, Li Weijiong, Zhang Shukui and Shi Aimin. (1996). Primary report about application of EM technology on crops. In *Research and Application of EM Technology*, eds. W. J. Li and Y. Z. Ni. Beijing: China Agricultural Science Press, pp. 67-72.

Lavelle, P. and A. Martin. (1992). Small-scale and large-scale effects of endogenic earthworms on soil organic matter dynamics in soil of the humid tropics. *Soil Biology and Biochemistry* 24:1491-1498.

Wolters, V. and R. G. Joergensen. (1992). Microbial carbon turnover in beech forest soil worked by *Aporrectodea caliginosa* (Savigny). *Soil Biology and Biochemistry* 24:171-177.

Mycorrhizal Associations and Their Manipulation for Long-Term Agricultural Stability and Productivity

Uma K. Aryal
Hui-lian Xu

SUMMARY. Mycorrhizae refer to an association (largely symbiotic) between plants and fungi that colonize the cortical root tissue of most agricultural crops during the period of active plant growth. The contribution of these symbioses to plant growth and soil fertility maintenance has been well-recognized for past several years. In spite of these benefits to agriculture, at present, the realization of the full potential of these fungi has not yet been reached. It must also be recognized that recent research on the possible application of the mycorrhizal symbiosis in agriculture has revealed many gaps in knowledge of fungal biology and ecology. Scientific knowledge on the role of these fungi in plant development and protection, soil stabilization, aggregate formation and creation of nutrient reserves is still limited. For efficient use and manipulation of these fungal symbioses for long-term agricultural stability and productivity, our understanding of their physiology, function and interactions with existing crops and environmental conditions should be improved. Besides, effects of different agronomic practices, application of chemical fertilizers and pesticides on their ecology and function should be elucidated before their successful utilization in agriculture.

Uma K. Aryal is Visiting Soil Microbiologist, and Hui-lian Xu is Senior Crop Scientist, International Nature Farming Research Center, 5632 Hata, Nagano 390-1401, Japan (E-mail: huilian@janis.or.jp).

[Haworth co-indexing entry note]: "Mycorrhizal Associations and Their Manipulation for Long-Term Agricultural Stability and Productivity." Aryal, Uma K., and Hui-lian Xu. Co-published simultaneously in *Journal of Crop Production* (Food Products Press, an imprint of The Haworth Press, Inc.) Vol. 3, No. 1 (#5), 2000, pp. 285-302; and: *Nature Farming and Microbial Applications* (ed: Hui-lian Xu, James F. Parr, and Hiroshi Umemura) Food Products Press, an imprint of The Haworth Press, Inc., 2000, pp. 285-302. Single or multiple copies of this article are available for a fee from The Haworth Document Delivery Service [1-800-342-9678, 9:00 a.m. - 5:00 p.m. (EST). E-mail address: getinfo@haworthpressinc.com].

This paper presents information on the morphology of different mycorrhizal fungi, their physiology and functions. Methods presently used to produce mycorrhizal inocula, their application in the field, problems to be resolved for their massive exploitation and future research needs have also been described. References have been selected to explain the recent advances in our understanding on these beneficial fungi. *[Article copies available for a fee from The Haworth Document Delivery Service: 1-800-342-9678. E-mail address: getinfo@haworthpressinc.com <Website: http://www.HaworthPress.com>]*

KEYWORDS. Arbuscular mycorrhizae, mycorrhizal symbiosis, ectomycorrhizae, soil aggregation, agronomic practices, inoculation

INTRODUCTION

In the past, soil was treated as a 'resource base' with its function of support subordinate to the production of food and fiber, and on the scale of priorities, the soil had always taken second place (Bethlenfalvay and Schüepp, 1994). In recent years, this perception has been undergoing radical change and scientists have recognized the importance of the soil not only as an agricultural resource base (Stewart et al., 1991), but as a complex, living, and fragile system that must be protected (Reagnold et al., 1990) and managed for its own sake (Pierce and Lal, 1991) to guarantee its long term stability and productivity. At present, agricultural scientists and land managers have realized the importance of soil biota in any production system (Jastrow and Miller, 1991). In response to increased concern for environmental quality, sustainable technologies need to be incorporated into agricultural systems. Proper management of mycorrhizal fungi is an important aspect of such an approach (Bethlenfalvay and Linderman, 1992).

Mycorrhizae refer to an association or symbiosis between plants and fungi that colonize the cortical tissues of roots during periods of active plant growth (Miller and Jastrow, 1994). The root system of a very large number of plants growing under natural conditions does not exist as roots but, as a complex mycorrhizal association resulting from root infection by certain soil fungi (Gianinazzi et al., 1990). Positive growth responses of plants to mycorrhizal infection have been well-documented, particularly on phosphorous (P)-deficient soils (Son and Smith, 1988). This has been attributed to increased P uptake via the fungal symbionts which in turn derive carbon compounds from the autotrophic hosts. In recent years, emphasis has also been put on the importance of mycorrhizal associations to soil structure. Roots and mycorrhizal hyphae are involved in the creation of water-stable soil aggregates via several potential mechanisms (Miller and Jastrow, 1992). Thus, mycorrhizae are important in linking the restoration of vegetation to the establishment of soil processes crucial to the formation of soil structure and to the redevelopment of nutrient cycles.

The purpose of this review paper is to discuss current knowledge on the mycorrhizal symbioses, their contribution to plant growth and development, their application within the context of sustainability in agriculture and finally recent advances and limitation for their efficient manipulation.

TYPES OF MYCORRHIZAE

Five broad groupings of mycorrhizae have come into general use over the years on the basis primarily of morphology and anatomy but also of either host plant taxonomy or fungal taxonomy. These generally accepted groupings are ectomycorrhizae, arbuscular mycorrhizae or endomycorrhizae, ericaceous mycorrhizae, ectendomycorrhizae, and orchidaceous mycorrhizae. All of these groupings do share the commonality of a host plant having a fungal associate in intimate, mutualistic symbiosis where both partners are living and exchange of substances can occur in both directions, i.e., biotrophic association (Lewis, 1973). For our review, only three types of mycorrhizae, ecto-, endo- and ectendomycorrhizae have been described. The ericaceous and orchidaceous mycorrhizae show similar characters to either ectomycorrhizae or endomycorrhizae and therefore have not been described in this review.

Ectomycorrhizal Fungi

The ectomycorrhizal associations exist primarily in the families of Pinaceae, Fagaceae, Betulaceae and Salaceae as well as few genera of other families such as *Eucalyptus, Casuarina*. Several members of the Caesalpiniaceae and Dipterocarpaceae also carry ectomycorrhizae. The fungal partner of the symbiosis in ectomycorriza belongs to Basidiomycetes and Ascomycetes although few members of Endogonaceae also infect roots (Mikola, 1981). Endomycorrhizal infection is initiated from spores or hyphae (collectively referred to as propagules) of the fungal symbionts in the rhizosphere of feeder roots. The fungus completely encloses each feeder rootlet in a sheath or mantle of the fungal hyphae and hyphal branches penetrate the space between the cells of the root cortex (intercellular). The presence of a net like structure called the Hartig net, after Robert Hartig who is considered the father of forest biology, in the cortex, is a key diagnostic feature of this association (Marks and Foster, 1973). These fungi never penetrate the wall into individual cortical cells and also never invade beyond the endodermis (Harley and Smith, 1983). The infection is stimulated by root exudates, and they grow vegetatively over the feeder root surface intercellularly.

Endomycorrhizal Fungi

The Endomycorrhizal fungi commonly referred to as "arbuscular" type are the most wide spread and important root symbionts. They occur in practically all families of angiosperms, gymnosperms, and many pteridophytes and bryophytes. Most of the economically important agronomic grain and forage crops as well as major commercial fruit and nut trees and berries normally form endomycorrhizae.

The fungi form large, conspicuous, thick walled spores on the root surfaces, in the rhizosphere, and sometimes in feeder root tissues. The hyphae enter the cell of the root (intracellular) and are often disintegrated. The infective hyphae may develop specialized absorbing or nutrient-exchanging structures called "arbuscules" in the cortical cells. Arbuscules consist of dense clusters of very fine dichotomously-branched filaments which may occupy the entire lumen of the cell (Kormanik et al., 1977). Vesicles are developed later generally in the middle or outer cortex, and appear as terminal swellings either within or between cells (Nicolson, 1967). No external morphological changes occur in roots infected by endomycorrhizal fungus although with some hosts, a yellow or brown pigmentation has been reported (Kormanik et al., 1977). They rarely invade meristematic tissues. Many conflicting opinions have been expressed about the functions of the arbuscules and vesicles. Generally, vesicles are currently regarded as temporary storage organs, and arbuscules help in the transfer of nutrients from soil into the root system. The fungi that form endomycorrhizae are mainly phycomycetes. They do not produce large aboveground fruiting bodies or wind disseminated spores as do most ectomycorrhizal fungi. Some of them produce large azygospores and chlamydospores on or in roots, while others produce large sporocarps (5 to 10 mm dia.) containing many spores in roots. Spread of these fungi in soil is by root contact, moving water, insects or mammals. In the absence of a host, the spores of these fungi are able to survive for many years in the soil (Smith, 1974).

These fungi are classified on the basis of their spore morphology into five genera: *Glomus*, *Gigaspora*, *Acaulospora*, *Sclerocystis*, and *Endogone*. Many graminaceous plants and legumes are highly susceptible to AM colonization. The AM fungi are intracellular obligate endosymbionts which have not yet been obtained in pure culture (Hayman, 1978; Mosse and Tinker, 1975).

Ectendomycorrhizae

This type of mycorrhizae have the features of both ecto and endomycorrhizae. They have a limited occurrence and are found primarily on the roots of ectomycorrhizal hosts. Very little is known about the association of these species and their importance to plant growth and nutrition (Kormanik et al.,

1977). These mycorrhizae form Hartig-net in the cortex of the root but develop little or no sheath (Smith, 1982).

PHYSIOLOGY AND FUNCTION

Ectomycorrhizae

Ectomycorrhizal fungi do not generally secrete cellulolytic or lignolytic enzymes and they depend on the carbon skeletons supplied by the host plant. These mycorrhizal plants are poor in root hairs and the internal intercellular net work of hyphae (Hartig-net) acts as a liason between the outer mycorrhizal sheath and plant cells in the process of nutrient absorption. The fungal mantle and the Hartig-net serve to offer greater absorptive surface in the root region. Labeled phosphate has been used to demonstrate that nearly 80-90% of the observed phosphate remains in the fungal sheath (Mikola, 1981). The fungal sheath may thus act as a reservoir of nutrients (essentially phosphate) and release them as the occasion demands under adverse conditions. Extensive work has clearly shown the role of the fungal sheath as a nutrient absorbing appendage in ectomycorrhizal associations (Harley, 1969; Mikola, 1981).

One of the physiological functions of ectomycorrhizal association is the protection offered by the symbiosis to feeder root pathogens such as *Phytophthora, Pythium, Rhizoctonia, Fusarium*, etc. Species of *Lactarious, Cartinarius* and *Hygrophorous* produce antifungal substances against pathogens such as *Rhizoctonia solani, Pythium debaryanum*, and *Fusarium oxysporum*. Besides these protective mechanisms, *Boletus variegatus* has been found to produce volatile fungistatic compounds such as isobutanol and isobutyric acid (Marx, 1972; Schonbeck, 1979).

Endomycorrhizae

Endomycorrhizal (AM) fungi are prevalent in most angiosperms, gymnosperms, pteridophytes and bryophytes but absent in Pinaceae, Betulaceae, Ericales, and Orchidaceae. Some species like *Eucalyptus, Leucaena, Casuarina*, etc., have both AM as well as ectomycorrhizae. Dense AM infections are common in most species of Leguminosaceae and Graminae. AM fungi infect most of the economically-important crops. Except under waterlogged conditions, AM fungi are virtually worldwide in temperate, tropical and arctic regions. Highly fertile soils generally show less AM fungal populations and pesticide treatment tends to decrease their numbers (Hayman, 1982).

Even though any AM fungus can infect any host plant species, there can be soils or plant species more suitable for a given species of AM fungus which is also dependent on soil pH, soil fertility and pesticide application. Root colonization and subsequent sporulation are reported to vary depending on the host and fungal species and edaphic factors (Khalil et al., 1992). Varying soil pH may affect the development and functioning of AM (Abbott and Robson, 1985; Hayman and Tavares, 1985) by altering the concentration of many nutrients and toxic ions in soil solutions as well as hydrogen ions (Russel, 1973). The response of AM fungi to soil pH may depend on the species and strains constituting the indigenous AM flora (Robson and Abbott, 1989). The variation in response can also be host-mediated changes of the rhizosphere pH. The nitrate reduction processes of the mycorrhizal host changes the pH of the root exudate, which in turn alters the rhizosphere pH, affecting pH sensitive microorganisms including AM fungi (Smith and Gianinazzi-Pearson, 1988).

Apart from the well-known effect of AM fungal associations on P-uptake (Gray and Gerdmann, 1969) there are reports in the literature showing that Zn uptake is enhanced in peaches (Gilmore, 1971; La Rue et al., 1975), maize, wheat and potatoes (Swaminathan and Verma, 1979). Similarly, the uptake of sulphur is enhanced by AM fungal infection in red clover and maize (Gray and Gerdmann, 1973). Cytokinins and chlorophyll contents of plants have been found to increase by AM association (Allen et al., 1980). There are also reports that AM fungal association with plant roots can help plants to overcome water stress by stomatal regulation in citrus (Levy and Krikun, 1980). AM fungi are determinants in the productivity of their host plants (Verma, 1995). They can help their hosts through periods of stress (Sylvia and Williams, 1992; Sanchez-Dias and Honrubia, 1994) and reduce the incidence of root diseases (Dehne, 1982; Jalali and Jalali, 1991; Hooker et al., 1994). AM fungi are a component of their host root system and facilitate the uptake of several nutrients, particularly P, Cu and Zn (Barea, 1991; Bolan, 1991).

AM fungi are not only a major component of soil fertility but also play a crucial role in the regulation of soil biological activity because of their abundance throughout the uppermost soil layer. AM fungi represent up to 20% of the dry biomass of the mycorrhizae that account for 25% of the biomass of the soil microflora and microfauna combined (Bethlenfalvay et al., 1982; Hamel and Smith, 1991). AM fungi have direct access to plant-fixed carbon and constitute a major input of carbon and energy in soil. They distribute this carbon throughout the soil of the rooting zone (Jakobsen and Rosendahl, 1990; Finlay and Söderström, 1992), for use by soil microbes and animals. AM fungi interact with all types of soil microbes, including invertebrates, plant pathogens, rhizobia, free living N_2-fixers, decomposers and biocontrol agents (Paultiz and Linderman, 1991). In addition, they play an important role in soil aggregation and stability, the corner-stone of agricultural sustainability (Bethlenfalvay, 1992). In a natural ecosystem, up to 27 m of AM mycelium per gram of soil was found (Finlay and Söderström, 1992). The abundant AM fungal hyphae hold soil micro-aggregates (< 0.25 mm diameter) and contribute to their stabilization with microbial gums and other organo-mineral binding agents, and package these micro-aggregates into macro-aggregates (> 0.25 mm diameter) (Miller and Jastrow, 1992). It is generally believed that the AM fungus depends upon the supply of carbohydrates from the host (Ho and Trappe, 1973) but, under poor light and temperature conditions, the fungus may tilt the balance from symbiosis to slight parasitism (Hayman, 1974).

INOCULA PRODUCTION AND USE

In situations where native mycorrhizal inoculum potential is low or ineffective, artificial provision of the appropriate fungi for the plant production system is worth considering. Because crop management strategies can not always favour the maintenance of efficient AM fungal communities, it would be useful for agriculturists to have access to efficient, economical and easy-to-use AM inoculants (Hamel, 1996). The first step in any inoculation program is to obtain an isolate that is both infective, or able to penetrate and spread in the root, and effective or able to enhance growth or stress tolerance of the host.

Isolation and inoculum production of EM and AM fungi present very different problems. Many EM fungi can be cultured on artificial media. Therefore, isolates of EM fungi can be obtained by placing surface disinfected portions of sporocarps or mycorrhizal short roots on an agar growth medium. The resulting fungal biomass can be used directly as inoculum but, for ease of use, inoculum often consists of the fungal material mixed with a carrier or bulking material such as peat. The biotrophic (obligate) nature of AM fungi has traditionally hampered inoculum production and therefore, obtaining isolates of AM fungi is more difficult because they will not grow apart from their host (Sylvia, 1994). Culture of AM fungi on plants growing in disinfected soil has been the most frequently used technique for increasing propagule numbers (Menge, 1984). But, new techniques aimed at the production of AM fungi (Bécard and Fortin, 1988) and formulation of inoculant (Strullu and Plenchette, 1991) could foster the widespread practice of crop inoculation in the near future. Spores can be sieved from soil, surface disinfected and, used to initiate "pot culture" on a susceptible host plant in sterile soil or an artificial plant growth medium (Sylvia, 1994). When the required species multiply on the roots, the authenticity of the species must be checked microscopically. Once this is assured, the entire root system is finely chopped with adhering soil and used as the inoculum for scaling up inoculum production on living roots of test host grown in pots containing sterilized soil. Inoculum is typically produced in a scaled up pot cultures. Onions, sorghum and other grasses are suitable hosts to multiply AM fungal inocula and the plants are grown in 1:1 sand soil mixture to obtain good root growth. Hydrophonic and aeroponic culture systems are also possible, a benefit of these systems is that plants can be grown without a supporting substratum, allowing colonized roots to be sheared into an inoculum of high propagule number. The following points should be considered for inocula production.

A highly susceptible host plant should be used and it should produce root mass quickly and tolerate high light conditions required for the fungus to reproduce rapidly (Sylvia and Jarstfer, 1994). Host plants should be screened to insure that maximum inoculum levels are achieved (Bagyaraj and Manjunath, 1980). Seed propagated hosts are preferable to cuttings since seeds can be more easily disinfected than cuttings (Sylvia and Jarstfer, 1994).

- All components of the culture system should be disinfected prior to initiation of a culture of AM fungi. The objective of soil disinfection is to kill existing AM fungi, pathogenic organisms and weed seeds while preserving a portion of non-pathogenic microbial community (Sylvia and Jarstfer, 1994).
- Containers used for the pot cultures should be shielded from contaminated soil, splashing water and crawling insects. In addition, specific isolates of AM fungi should be kept well separated to reduce cross contamination. Container size should match the potential volumes of the root system within practical space constraints. Larger containers may result in higher spore concentration (Ferguson and Menge, 1982).
- Light quality and intensity and soil moisture and temperature markedly influence root colonization and spore production (Menge, 1984). A moderate irrigation regime (Nelson and Safir, 1982) and a warm growth environment (Schenck and Smith, 1982) support optimal colonization and spore production.
- Fungal responses to P and N fertilization are strain dependent. A nutrient regime low in P but high in N increases colonization of AM fungi (Sylvia and Neal, 1990).
- Some soil-applied biocides greatly reduce or eliminate AM fungi (Trappe et al., 1984; Trappe, 1987), while others increase colonization and sporulation (Sreenivasa and Bagyaraj, 1989). Selected use of pesticides may be useful for better inoculum production, but they are rarely used to protect against or reduce organisms contaminating pot cultures (Menge, 1983).

ROLE OF AM ON PLANT MINERAL NUTRITION, DEVELOPMENT AND PROTECTION

Mineral Nutrition

The primary effect of AM is to improve plant growth by increasing the supply to roots with mineral nutrients in the soil (Gianinazzi et al., 1990). The external hyphal network of AM fungi plays an important role in nutrient uptake, especially for those ions that are not very mobile in soil solution (Miller and Jastrow, 1994). The classical explanation for the mechanism behind a mycorrhizal mediated increase in ion uptake is that the external hyphal network explores soil beyond the root hair zone. Mycorrhizal hyphae may also foster a more integrated nutrient cycle, and occurrence of a hyphal link between two or more plants is a common phenomenon in grassland and forest floor communities (Newman, 1988). Mineral nutrients can be passed between two plants via these hyphal links. Nutrients can also be passed from dying roots directly to living roots via hyphal connections (Newman and Eason, 1989).

Newman (1988) has suggested that mycorrhizal hyphal links promote direct nutrient cycling by avoiding the mineralization process, and increases the system's productivity by keeping more nutrient ions in the biomass pool. Beyond their involvement in nutrient uptake, AM fungal hyphae are also believed to play an important role in the physical enhancement of soil particles to create relatively stable soil aggregates (Miller and Jastrow, 1992). Thus, indirectly, they influence the residence time of nutrients with the system, which in turn, reduces the loss of soil nutrients by erosion (Miller and Jastrow, 1992). The creation of an aggregate soil system accrue organic matter and organically-bound nutrients (Van Veen and Kuikman, 1990). Therefore, AM fungal hyphae are involved with the scavenging and retention of nutrient ions and with the creation of an aggregate system that acts as a control point for accrual and mineralization of organic matter in the soil. At a larger scale, the mycorrhizal association, by its involvement in nutrient accumulation and retention, creates a system that reduces erosion and leaching loss of nutrients.

In addition, highly aggregated soils are likely to have a more stabilized organic matter, diverse biotic communities, and food web structure, which probably is not only more resistant to perturbation but also facilitates tighter nutrient cycling. Hence by contributing to the aggregation process, AM fungi are involved in a feedback mechanism whereby they are both creating and responding to their immediate environment. Mycorrhizally-mediated changes in soil aggregates also contribute to soil structural aspects of the ecosystem, which can affect nutrient cycling through physical properties such as infiltration, aeration, erosion, runoff, and hydraulic conductance (Jastrow and Miller, 1991).

Plant Development and Protection

The influence of AM fungi on plant physiology is not limited to mineral nutrition. Root infection by these fungi can also affect shoot and root development and morphogenesis (Gianinazzi et al., 1990). Results on micro-propagated roses show that number of flowers and branches increased in mycorrhizal plants and also the delay of flowering onset was reduced (Gianinazzi et al., 1990). The pattern of root development is modified in mycorrhizal plants due to changes in meristematic activity. In mycorrhizal plants where root-shoot ratios are often lower, roots become shorter, due to a blocking effect of the fungus on meristem cell division (Barta et al., 1990). In contrast, plants infected by ericoid endomycorrhizal fungi show rapid increases in total and root hair length, possibly due to activation of the meristem, whereas root number is less affected (Barta et al., 1988). Changes in hormone level have been suggested to be responsible for modification in root development and in particular for alterations in the activity of root apices (Trewavas, 1985). AM formation can increase the level of cytokinin in the host plant and changes the level of abscisic acid and gibberellin-like substances (Allen et al., 1980, 1982). *Glomus mossae*, one of the AM fungi has been shown to synthesize phytohormones (Barea and Azon-Aguilar, 1982). Fungus *Pezizella ericae*, forming ericoid mycorrhizae, is able to release IAA and ICA, an intermediate of IAA breakdown (Gasper et al., 1982), in pure culture, and IAA release within host cells might affect root metabolism and development (Barta et al., 1988).

AM infection favors the colonization of roots by other symbiotic microorganisms like *Rhizobium* (Asimi et al., 1980) and often reduce susceptibility, or increases tolerance, of roots to soil-born pathogens like Nematodes, *Phytophthora* spp., *Chalara elegans*, *Fusarium* sp., and *Pythium* sp. (Bondoux and Perrin, 1982; Bagyaraj, 1984). The reasons for increased resistance to pathogens in AM plants are not understood, but a well-established AM infection is a prerequisite for protection (Bartschi et al., 1981). This protection phenomena may be linked to the elicitation in roots infected by AM fungi of phytoalexins and associated isoflavonoids which are usually associated with development of host resistance to microorganisms (Morandi et al., 1984; Bailey and Mansfield, 1982).

PROBLEMS AND FUTURE RESEARCH NEEDS

Several problems still have to be resolved in the selection and production of AM fungi for larger scale inoculation purposes. Little is known of the genetics of these fungi. The availability of molecular techniques such as the polymerase chain reaction (PCR) have now made possible the study of AM fungal genes (Simon, 1995). Geneticists have to answer several important questions such as what kind of genetic variability can be found within a given species of AM, among the spores of a single organisms, or within a single spore and several others (Hamel, 1996). Such information will be very important to improve naturally occurring strains and release superior organisms for testing and commercial use.

Skinners and Bowen (1974) demonstrated transport of P over distance of 12 cm in rhizomorphs, although the importance of this mechanism for P uptake by the plant still has to be demonstrated (Timonen et al., 1996). Excised EM (e.g., Alexander and Hardy, 1981; Dighton, 1991; Lewis, 1973) and EM mycellium from litter (Colpaert and Van Laere, 1996) can produce phosphatases suggesting the ability to access organic forms of P, although this has not yet been confirmed. It may be recognized that recent research on the possible application of the AM symbiosis in agriculture has revealed many gaps in our knowledge of AM fungal biology and ecology. It appears in fact, that before the best use of this symbiosis can be made in those crops which can form effective symbioses under field conditions, the effect of agronomic practices on indigenous communities of AM fungi must be elucidated. Similarly, because of biotrophic nature of AM fungi, maintenance of certain population of weed under a certain threshold, could contribute significantly to the maintenance of overall soil fertility and healthy mycorrhizal population (Hamel, 1996). Future research can be concentrated on the impact of weed control on abundance and diversity of AM fungi. Some studies have also suggested that crop rotation could be used as a tool to manage the indigenous fungal communities in agricultural soils (Johnson et al., 1991). But our understanding of this sequence on AM community is still very limited and more research in future is needed for better management of indigenous population of AM fungi in agricultural production.

Scientists have emphasized the role of AM fungi in plant protection and reduce susceptibility or increased tolerance to soil-born pathogens. But, the reasons for increased resistance or tolerance to pathogens are not well understood. Changes in hormone levels have been suggested to be responsible for modifications in root activity (Trewavas, 1985). But, the importance of such hormones in plant growth, physiology and mycorrhizzal infectivity has yet to be determined.

A wide range of external hyphal lengths has been reported with field studies. The reported differences are probably related to ecosystem properties such as vegetation, soil type, and rooting density. Unfortunately, most of these studies give little information on the hyphal extraction efficiency and counting accuracy of their procedures (Miller and Jastrow, 1994). Therefore, correlation of hyphal abundance with causative factors is problematic. Furthermore, there is a broader issue concerning the identification of AM hyphae. Some studies have based on comparative morphology of the hyphae (Allen and Allen, 1986), while others use hyphal diameter to separate AM hyphae from saprophytic fungal hyphae (Sylvia, 1986). The resolution of this particular problem is not likely until the molecular techniques currently developed to identify AM fungi are refined and become available for routine application (Miller and Jastrow, 1994).

The potential for using AM fungi to make efficient use of phosphate from the soil and fertilizer has become a major topic in recent research on these common fungi (Abbott and Robson, 1982a). But, enhanced growth shown by mycorrhizal plants may obviously depend on a number of environmental factors other than soil phosphorous levels. The main question which is still unresolved is whether the mycorrhizae are able to exploit phosphate in the soil not available to nonmycorrhizal roots or whether they are just more efficient in capturing the available soluble phosphates. The increased uptake by mycorrhizal roots may be due to a more efficient translocation of phosphates coupled with a better exploitation of soluble phosphate in soil by external hyphae, and solubilization of insoluble phosphate by mycorrhizal fungi, although questions still remain. While it is becoming clearer what advantages the plant may derive from AM, the role of plants in fungal metabolism remains relatively unexplored.

CONCLUSIONS

Mycorrhizal systems must be considered as an essential factor for soil fertility and for promoting plant health and productivity. It is well-documented that mycorrhizal fungi can transfer phosphorous to the host and that results in an increase of P uptake by plants colonized by these fungi. The benefits of high fertilization rates on yield are well-established and very few producers are ready to reduce their P-fertilization regime to levels required to allow adequate development of the symbiosis. It must also be recognized that recent research on the possible application of the mycorrhizal symbiosis in agriculture has revealed many gaps in our knowledge of fungal biology and ecology. Therefore, before the best use of these symbioses can be made in crop production, the effect of agronomic practices on indigenous communities of mycorrhizal fungi must be well-elucidated. Efficient fungal isolates could be selected for various desirable traits, multiplied and inoculated onto crops, but the conditions which would allow these organisms to express their full potential will have to be determined.

The impediments in the speedy application of mycorrhizae to improve plant growth are the relatively slow growth of ectomycorrhizal fungi and the very obligate nature of AM fungi which limit large scale production of inocula. If spores of AM fungi can be made to grow and produce large inocula in nutrient media, it is likely that dual inoculation of nitrogen fixing bacteria and AM fungi can be practiced on a wider scale. Advances in molecular technologies will probably facilitate the study of fungal ecology in the near future. A better understanding of the conditions required for utilization of AM inocula will improve inoculation of field crops and lead to increased numbers and quality of fungi in agricultural soils. If nitrogen-fixing genes (*Nif* genes) can be transferred to AM fungi by genetic engineering, the opportunity for biological nitrogen fixation and phosphate mobilization can be combined in one microorganism. The biotechnological possibilities in this area are still in the realm of speculation, but perhaps future research may turn this speculation into a reality.

REFERENCES

Abbott, L. K. and R. D. Robson. (1985). Formation of external hyphae in soil by four species of vesicular-arbuscular mycorrhizal fungi. *New Phytologist* 99: 245-255.

Abbott, K. and R. D. Robson. (1982a). The role of vesicular-arbuscular mycorrhizal fungi in agriculture and the selection of fungi for inoculation. *Australian Journal of Agricultural Research* 33: 389-408.

Alexander, I. J. and K. Hardy. (1981). Surface phosphatase activity of sitka spruce mycorrizas from a serpentine site. *Soil Biology and Biochemistry* 13: 301-305.

Allen, M. F., T. S. Moore, and M. Christensen. (1980). Phytohormone changes in *Bouteloua gracilis* infected by vesicular-arbuscular mycorrhizae. I. Cytokinin increases in the host plant. *Canadian Journal of Botany* 58: 371-374.

Allen, M. F., T. S. Moore, and M. Christensen. (1982). Phytohormone changes in *Bouteloua gracilis* infected by vesicular arbuscular mycorrhizae: II. Altered levels of gibberellin like substances and abscisic acid in the host plant. *Canadian Journal of Botany* 60: 468-471.

Allen, M. F. (1996). The ecology of arbuscular mycorrhizae: a look back into the 20th century and a peak into the 21st century. *Mycological Research* 100(7): 769-782.

Allen, E. B. and M. F. Allen. (1986). Water relations of xeric grasses in the field: Interactions of mycorrhizae and competition. *New Phytologist* 104: 559-571.

Asimi, S., V. Gianinazzi-Pearson, and S. Gianinazzi. (1980). Influence of increasing soil phosphorous levels on interactions between vesicular-arbuscular mycorrhizae and *Rhizobium* in soybeans. *Canadian Journal of Botany* 58: 2200-2205.

Bagyaraj, D. J. and A. Manjunath. (1980). Selection of a suitable host for mass production of vesicular-arbuscular mycorrhizal inoculum. *Plant Soil* 55: 495-498.

Bagyaraj, D. J. 1984. Biological interactions with VA mycorrhizal fungi. In *VA Mycorrhiza*, eds. C. L. Bailey and J. W. Mansfield. CRC Press, Boca Raton, Florida, USA, pp. 131-154.

Bailey, J. A. and J. W. Mansfield. (1982). *Phytoalexins*: Blakie and Son, Glasgow and London, UK. 334 p.

Barea, J. M. and C. Azon-Aguilar. (1982). Production of plant growth-regulating substances by vesicular-arbuscular mycorrhizal fungus *Glomus mosseae*. *Applied Environmental Microbiology* 43: 810-813.

Barea, J. M. (1991). Vesicular-arbuscular mycorrhizae as modifiers of soil fertility. *Advances in soil Science* 15: 1-40.

Bartschi, H., V. Gianinazzi-Pearson, and I. Vegh. (1981). Vesicular-arbuscular mycorrhiza formation and root rot disease (*Phytophthora cinnamoni*) development in *Chamaecyparis lowsoniana*. *Phytopathology* 102: 213-218.

Barta, G., A. Fusconi, A. Trotta, and S. Scannerini. (1990). Morphogenetic modifications induced by the mycorrhizal fungus *Glomus* strain E_3 in the root system of *Allium porrum* L. *New Phytologist* 114: 207-215.

Barta, G., V. Gianinazzi-Pearson, G. Gay, and G. Torri. (1988). Morphogenetic effects of endomycorrhiza formation of the root system of *Calluna vulgaris* (L.) Hull. *Symbiosis* 5: 33-44.

Bécard, G. and J. A. Fortin. (1988). Early events of vesicular-arbuscular mycorrhiza formation on Ri T-DNA transformed roots. *New Phytologist* 108: 211-218.

Bethlenfalvay, G. J. and H. Schüepp (1994). Arbuscular mycorrhizae and agrosystem stability. In *Impact of Arbuscular Mycorrhizas on Sustainable Agriculture and Natural Ecosystem*, eds. S. Gianinazzi and H. Schüepp. Birkhäuser Verlag Basal, Switzerland, pp. 117-131.

Bethlenfalvay, G. J. and R. G. Linderman. (1992). *Mycorrhizae in Sustainable Agriculture*. ASA Special Publication No. 54, Agronomy Society of America, Madison,WI.

Bethlenfalvay, G. J. (1992). Mycorrhizae in the agricultural plant–soil system. *Symbiosis* 14: 413-425.

Bethlenfalvay, G. J., R. S. Pacovsky, and M. S. Brown. (1982). Parasitic and mutualistic associations between a mycorrhizal fungus and soybean: Development of the endophyte. *Phytopathology* 72: 894-897.

Bolan, N. S. (1991). A critical review on the role of mycorrhizal fungi in the uptake of phosphorous by plants. *Plant and Soil* 134: 189-207.

Bondoux, P. and R. Perrin. (1982). Mycorrhizae et protection des plants. *CR Acad. Agric. De France* 15: 1162-1167.

Colpaert, J. V. and A. Van Laere. (1996). A comparison of the extracellular enzyme activities of two ectomycorrhizal and a leaf-saprotrophic basidiomycete colonizing beech leaf litter. *New Phytologist* 133: 133-141.

Dehne, J. (1982). Interaction between vesicular-arbuscular mycorrhizal fungi and plant pathogens. *Phytopathology* 72: 1115-1118.

Digton, J. (1991). Acquisition of nutrients from organic resources by mycorrhizal autotrophic plants. *Experientia* 47: 362-369.

Ferguson, J. J. and J. A. Menge. (1982). Factors that affect production of endomycorrhizal inoculum. *Proceeding of Florida State Horticultural Society* 95: 37-39.

Finlay, R. and B. Söderström. (1992). Mycorrhiza and carbon flow in the soil. In *Mycorrhizal Functioning: An Integrated Plant Fungal Process*, ed. M. F. Allen. Chapman and Hall, New York, London, pp. 134-160.

Gasper, T. H., C. Penel, T. Thorpe, and H. Grepin. (1982). *Peroxidases, A Survey of*

Their Biochemical and Physiological Roles in Higher Plants. Geneva University Publication, Geneva, CH, 234 p.

Gianinazzi, S., A. Trouvelot, and V. Gianinazzi-Pearson. (1990). Role and use of mycorrhizae in horticultural crop production. *Plenary Lecture Presented at the XXIII International Horticultural Congress*, I.S.H.S.-International Society for Horticultural Sciences, Firenze (Italy), August 27-September 1, 1990, pp. 25-30.

Gilmore, A. E. (1971). The influence of endotrophic mycorrhizae on the growth of peach seedlings. *Journal of American Society of Horticultural Sciences* 96: 35-38.

Gray, L. E. and J. W. Gerdemann. (1969). Uptake of phosphorous-32 by vesicular-arbuscular mycorrhizae. *Plant and Soil* 30: 415-422.

Gray, L. E. and J. W. Gerdmann. (1973). Uptake of sulphur-35 by vesicular-arbuscular mycorrhizae. *Plant and Soil* 39: 687-689.

Hamel, C. (1996). Prospects and problems pertaining to the management of arbuscular mycorrhizae in agriculture. *Agriculture, Ecosystem and Environment* 60: 197-210.

Hamel, C. and D. L. Smith. (1991). Interspecific N-transfer and plant development in a mycorrhizal field grown mixture. *Soil Biology and Biochemistry* 23: 661-665.

Harley, J. L. (1969). *The Biology of Mycorrhiza*, 2nd edition. Leonard Hill London.

Harley, J. L. and S. E. Smith. (1983). *Mycorrhizal Symbiosis*. Academic Press, New York.

Hayman, D. S. (1978). Endomycorrhizae. In *Interaction Between Non-Pathogenic Soil Microorganisms and Plants*, eds. Y. R. Dommergues and S. V. Krupa. Elsevier, Amsterdam, pp. 401-442.

Hayman, D. S. (1982). Practical aspects of vesicular-arbuscular mycorrhiza. In *Advances in Agricultural Microbiology*, ed. N. S. Subba Rao. Oxford and IBH Publishing Co. Pvt. Ltd., New Delhi, pp. 325-373.

Hayman, D. S. (1974). Plant growth responses to vesicular-arbuscular mycorrhiza. VI. The effect of light and temperature. *New Phytologist* 73: 71-80.

Hayman, D. S. and M. Tavares. (1985). Plant growth responses to vesicular arbuscular mycorrhiza. XV. Influence of soil pH on the symbiotic efficiency of different endophytes. *New Phytologist* 100: 367-377.

Hooker, J. E., M. Jaizme-Vega, and D. Atkinson. (1994). Biocontrol of plant pathogens using vesicular-mycorrhizal fungi. In *Impact of Arbuscular Mycorrhizas on Sustainable Agriculture and Natural Ecosystems*, eds. S. Gianinazzi and H. Schüepp. Birkhauser, Basel, Boston, Berlin, pp. 191-200.

Ho. I. and J. M. Trappe. (1973). Translocation of ^{14}C from *Festuca* plants to their endomycorrhizal fungi. *Nature, New Biologist* 244: 30-31.

Jakobsen, I. and L. Rosendahl. (1990). Carbon flow into soil and external hyphae from roots of mycorrhizal cucumber plants. *New Phytologist* 115: 77-83.

Jalali, B. L. and I. Jalali. (1991). Mycorrhizae in plant disease control. In *Handbook of Applied Mycology, Vol.1, Soil and Plants*. Marcel Dekker, New York, pp. 131-154.

Jastrow, J. D. and R. M. Miller. (1991). Methods for assessing the effects of biota in soil structure. *Agriculture, Ecosystem and Environment* 34: 279-303.

Johnson, N. C., F. L. Pfleger, R. K. Crookston, S. R. Simmons, and P. J. Copeland. (1991). Vesicular-arbuscular mycorrhizae respond to corn and soybean cropping history. *New Phytologist* 117: 657-663.

Khalil, S., T. F. Loynachan, and H. S. Mcnabb, Jr. (1992). Colonization of soybean by mycorrhizal fungi and spore populations in Iowa soils. *Agronomy Journal* 84: 832-836.

Kormanik, P. P., W. C. Bryan, and R. C. Schultz. (1977). The role of mycorrhizae in plant growth and development. In *Physiology of Root Microorganisms Association*, ed. H. Max Vines. Proceeding of a Symposium of the South Sector American Society of Plant Physiology, Atlanta, USA, pp. 1-10.

La Rue, J. H., W. D. McClellan, and W. L. Peacock. (1975). Mycorrhizal fungi and peach nursery nutrition. *California Agriculture* 29: 6-7.

Levy, Y. and J. Kirkun. (1980). Effect of vesicular-arbuscular mycorrhiza on *Citrus jambhiri* water relations. *New Phytologist* 85: 25-31.

Lewis, D. H. (1973). Concepts in fungal nutrition and the origin of biotrophy. *Biological Reviews* 48: 261-278.

Marx, D. H. (1972). Ectomycorrhizae as a biological deterrents to pathogenic root infections. *Annual Review of Phytopathology* 10: 429-454.

Marks, G. C. and R. C. Foster. (1973). Structure, morphogenesis and ultrastructure of ectomycorrhizae. In: *Ectomycorrhizae: Their Ecology and Physiology*, eds. G. C. Marks, and T. T. Kozlowski. Academic Press, New York, pp. 1-41.

Menge, J. A. (1983). Utilization of vesicular-arbuscular mycorrhizal fungi in agriculture. *Canadian Journal of Botany* 61: 1015-1024.

Menge, J. A. (1984). Inoculum production. In *VA Mycorrhiza*, eds. C. L. Powell and D. J. Bagyaraj. CRC Press, Inc., Boca Raton, Florida, pp. 187-203.

Mikola, P. (ed). 1981. *Tropical Mycorrhiza Research*, Oxford University Press, Oxford.

Miller, R. M. (1985). Mycorrhizae. *Restoration and Management Notes*, 3: 14-20.

Miller, R. M. and J. D. Jastrow. (1992). The application of VA mycorrhizae to ecosystem restoration and reclamation. In *Mycorrhizal Functioning*, ed. M. Allen. Chapman and Hall, Inc. New York, pp. 438-467.

Miller, R.M. and J. D. Jastrow. (1994). Vesicular-arbuscular mycorrhizae and biogeochemical cycling. In *Mycorrhizae and Plant Health*, eds. F.L. Pfleger and R. G. Linderman. APS Press, The American Phytopathological Society, St. Paul, Minnesota, pp. 189-212.

Morandi, D., J. A. Bailey, and V. Gianinazzi-Pearson. (1984). Isoflavonoid accumulation in soybean roots infected with vesicular-arbuscular mycorrhizal fungi. *Physiology and Plant Pathology* 24: 357-364.

Mosse, B. and P. B. Tinker. (eds.). (1975). *Endomycorrhizas*, Academic Press, London.

Nelson, C. E. and G. R. Safir. (1982). Increased drought tolerance of mycorrhizal onion plants caused by improved phosphorous nutrition. *Planta* 154: 407-413.

Newman, E. I. (1988). Mycorrhizal links between plants: their functioning and ecological significance. *Advances in Ecological Research* 18: 243-270.

Newman, E. I. and W. R. Eason. (1989). Cycling of nutrients from dying roots to living plants, including the role of mycorrhizae. In *Ecology of Arable Lands*, eds. M. Charholm and L. Bergström. Kluwer Academic Publishers, pp. 133-137.

Nicolson, T. H. (1967). Vesicular-arbuscular mycorrhiza-a universal plant symbiosis. *Science Progr. Oxford* 55: 561-568.

Paulitz, T. C. and R. G. Linderman. (1991). Mycorrhizal interactions with soil organisms. In *Handbook of Applied Mycology, Vol. 1, Soil and Plants*, eds. D. K. Arora, B. Rai, K. G. Mukerji and G. R. Knudsen. Marcel Decker Inc., New York, Basal, Honk Kong, pp. 77-129.

Pierce, F. J. and R. Lal. (1991). Soil management in the 21st century. In *Soil Management for Sustainability*, eds. R. Lal and F. J. Pierce. Soil and Water Conservation Society of America, Ankeny, pp. 175-179.

Reagnold, J. P., R. I Papendick, and J. F. Parr. (1990). Sustainable agriculture. *Science America* 262: 112-120.

Robson, A. D. and L. K. Abbott. (1989). The effect of soil acidity on microbial activity in soils. In *Soil Acidity and Plant Growth*, ed. A. D. Robson. Sydney, Academic Press, pp. 139-165.

Russel, E. W. (1973). *Soil Conditions and Plant Growth*. London: Longman.

Schenck, N. C. and G. S. Smith. (1982). Responses of six species of vesicular-arbuscular mycorrhizal fungi and their effects on soybean at four soil temperatures. *New Phytologist* 92: 193-201.

Sanchez-Dias, M. and M. Honrubia. (1994). Water relations and alleviation of drought stress in mycorrhizal plants. In *Impact of Arbuscular Mycorrhizas on Sustainable Agricultural and Natural Ecosystems*, S. Gianinazzi and H. Schüepp (eds.). Birkhauser, Basel, Boston, Berlin, pp. 167-178.

Schonbeck, F. (1979). Endomycorrhiza in relation to plant diseases. In *Soil-Borne Plant Pathogens*, B. Schippers and W. Gams (eds.). Academic Press, New York, pp. 271-280.

Simon, L. (1995). Systematique molecularie des champignons endomycorhizogènes. In *La Symbiose Mycorhizienne Etat des Lonnaissances*, J. A. Fortin, C. Charest and Y. Piche (eds.). Orbis, Frelighsburg, Canada, pp. 21-33.

Skinners, M. F. and G. D. Bowen. (1974). The uptake and translocation of phosphate by mycelial strands of pine mycorrhizae. *Soil Biology and Biochemistry* 6: 53-56.

Smith, S. E. (1974). *Mycorrhizal Fungi*. CRC Critical Reviews in Microbiology, June, pp. 273-313.

Smith, S. E. and V. Gianinazzi-Pearson. (1988). Physiological interactions between symbionts in vesicular mycorrhizal plants. *Annual Review of Plant Physiology and Molecular Biology* 39: 221-224.

Smith, S. E. (1982). Inflow of phosphate into mycorrhizal and non-mycorrhizal plants of *Trifolium subterraneum* at different levels of soil phosphate. *New Phytologist* 90: 293-303.

Son, C. L. and S. E. Smith. (1988). Mycorrhizal growth responses: Interaction between photon irradiance and phosphorous nutrition. *New Phytologist* 108: 305-314.

Sreenivasa, M. N. and D. J. Bagyaraj. (1989). Use of pesticides for mass production of vesicular-arbuscular mycorrhizal inoculum. *Plant and Soil* 119: 127-132.

Stewart, B. A., R. Lal, and S. A. El-Swaify. (1991). Sustaining the resource base on an expanding world agriculture. In *Soil Management for Sustainability*, eds. R. Lal and F. J. Pierce. Soil and Water Conservation Society of America, Ankeny, pp. 125-144.

Strullu, D. G., and C. Plenchette. (1991). The entrapment of *Glomus* sp. in alginate beads and their use as root inoculum. *Mycology Research* 95: 1194-1196.

Swaminathan, K. and B. C. Verma. (1979). Responses of three crop species to vesicular arbuscular mycorrhizal infection on zinc deficient Indian soils. *New Phytologist* 82: 481-487.

Sylvia, D. M. (1986). Spatial and temporal distribution of vesicular-arbuscular mycorrhizal fungi associated with *Uniola paniculata* in Florida foredunes. *Mycologia* 78: 728-734.

Sylvia, D. M. and L. H. Neal. (1990). Nitrogen affects the phosphorous response of VA mycorrhiza. *New Phytologist* 115: 303-310.

Sylvia, D. M. and S. E. Williams. (1992). Vesicular-arbuscular mycorrhizae and environmental stress. In *Mycorrhizae in Sustainable Agriculture*, eds. G. J. Bethlenfalvay and R. G. Linderman. American Society of Agronomy, Special Publication No. 54, Madison, WI, pp. 101-124.

Sylvia, D. M. (1994). Vesicular-arbuscular mycorrhizal fungi. In *Methods of Soil Analysis, Part 2 Microbiological and Biochemical Properties*, eds. R. W. Weaver et al. Soil Science Society of America, Madison, WI, pp. 351-378.

Sylvia, D. M. and A. G. Jarstfer. (1994). Production of inoculum and inoculation with arbuscular mycorrhizal fungi. In *Management of Mycorrhizas in Agriculture, Horticulture and Forestry*, eds. A.D. Robson, L.K. Abbott and N. Malajczuk. Kluwer Academic Publisher, Netherlands, pp. 231-238.

Timonen, S., R. D. Finlay, S. Olsson, and B. Söderström. (1996). Dynamics of phosphorous translocation in intact ectomycorrhizal systems: Nondestructive monitoring using *B*-scanner. *Federation of European Microbiological Society: Microbiological Ecology* 19: 171-180.

Trappe, J. M., R. Molina, and M. A. Castellano. (1984). Reactions of mycorrhizal fungi and mycorrhizal formation to pesticides. *Annual Review of Phytopathology* 22: 331-359.

Trappe, J. M. (1987). Phylogenic and ecological aspects of mycotrophy in angiosperms from an evolutionary standpoint. In *Ecophysiology of VA Mycorrhizal Plants*, ed. G. Safir. Boca Raton (USA): CRC Press, pp. 5-25.

Trewavas, A. J. (1985). Growth substances, calcium and regulation of cell division. In *The Cell Division Cycle in Plants*, eds. J. A. Bryant and D. Francis. Cambridge University Press, Cambridge, UK, pp. 133-156.

Van Veen, J. A. and P. J. Kuikman. (1990). Soil structural aspects of decomposition of organic matter by micro-organisms. *Biogeochemistry* 11: 213-233.

Verma, A. (1995). Arbuscular mycorrhizal fungi–The state of art. *Critical Review on Biotechnology* 15: 179-199.

Nature Farming with Vesicular-Arbuscular Mycorrhizae in Bangladesh

Md. Amin U. Mridha
Hui-lian Xu

SUMMARY. Systematic research on the effect of vesicular-arbuscular mycorrhizal fungi with different agricultural crops, sand dune plants and plantation crops has been carried out in Bangladesh for last few years. Large numbers of VA-mycorrhizal fungi have been detected in different soil types for identification and inoculum production and for use in nurseries and the field as an alternative to chemical fertilizers and pesticides and to progress towards nature farming. The VAM fungi not only absorb and translocate immobile nutrients like phosphorus, zinc and copper (especially phosphorus) through external hyphae, but also play important roles in inter- and intra-specific transfer of carbon, phosphorus, and nitrogen from plant to plant. VAM fungi may also protect plants from certain root-infecting pathogens, improve plant-water relations, enhance the establishment and growth of micro-propagated plants, and increase plant tolerance to salinity. VAM fungi can improve the ecological and environmental conditions of the country and its agriculture by reducing the farmer's dependence and expenditure on chemical fertilizers and pesticides. Effective mycorrhizal inoculum can be

Md. Amin U. Mridha is Professor, Department of Botany, University of Chittagong, Chittagong, Bangladesh.

Hui-lian Xu is Senior Crop Scientist, International Nature Farming Research Center, 5632 Hata, Nagano 390-1401, Japan.

Address correspondence to: Md. Amin U. Mridha at the above address (E-mail: Mridha@abnetbd.com).

[Haworth co-indexing entry note]: "Nature Farming with Vesicular-Arbuscular Mycorrhizae in Bangladesh." Mridha, Md. Amin U., and Hui-lian Xu. Co-published simultaneously in *Journal of Crop Production* (Food Products Press, an imprint of The Haworth Press, Inc.) Vol. 3, No. 1 (#5), 2000, pp. 303-312; and: *Nature Farming and Microbial Applications* (ed: Hui-lian Xu, James F. Parr, and Hiroshi Umemura) Food Products Press, an imprint of The Haworth Press, Inc., 2000, pp. 303-312. Single or multiple copies of this article are available for a fee from The Haworth Document Delivery Service [1-800-342-9678, 9:00 a.m. - 5:00 p.m. (EST). E-mail address: getinfo@haworthpressinc.com].

introduced directly to the field or indirectly through the production of VAM-infected seedlings from nurseries. The gap of knowledge on the occurrence of VAM fungi in association with different agricultural and forest crops is still immense in Bangladesh. This article has been prepared to draw attention of agricultural scientists, plant physiologist, forest managers, policy-makers and different government and non-government officials towards nature farming systems utilizing VAM and possibility of utilizing VAM in production system. *[Article copies available for a fee from The Haworth Document Delivery Service: 1-800-342-9678. E-mail address: getinfo@haworthpressinc.com <Website: http://www.HaworthPress.com>]*

KEYWORDS. Bangladesh agriculture, nature farming, VAM, vesicular-arbuscular mycorrhizae

INTRODUCTION

Bangladesh is predominantly an agrarian country, depending mainly on agricultural crops and forest products for its economic development. Although agro-ecological conditions are favorable for crop production, crop yields are low due to several factors including insufficient and improper use of fertilizers and pesticides. However, even recommended rates of chemical fertilizers and pesticides can adversely affect the soil microbial population, and particularly the VAM fungi populations, both qualitatively and quantitatively. Basic and applied research is urgently needed in this field to explore the possibility of utilizing both the indigenous and introduced VAM fungi more effectively and efficiently in crop production.

Bangladesh is one of the most densely populated countries in the world. It is essential to improve crop production for the burgeoning human population and to meet the increasing demands for food. Because of economic and environmental constraints, it is necessary to develop least-expensive and technologically simple methodologies for immediate benefit. Mycorrhizal technology as a nature farming technique can be one of the alternatives to improve crop production, farm profitability and environmental quality.

VAM RESEARCH IN BANGLADESH

Despite the extensive studies on beneficial effects of mycorrhizae worldwide, research on mycorrhizae in Bangladesh was not conducted until recently. Systematic research on VAM was initiated in 1988 at Chittagong University by Mridha and his colleagues (Mridha, 1988; Mridha and Mohammed,

1989; Mridha and Killham, 1992). Little work was done on VAM fungi in Bangladesh before 1990. Research by the Bangladesh Forest Research Institute on ectomycorrhiza reported improved growth of pine seedlings under nursery conditions (Mridha and Killham, 1992). Khan et al. (1988) studied the influence of VAM fungi on lentil and mungbean under field conditions and reported improved crop growth by VAM. A review article on the status of mycorrhizal research in Bangladesh was presented at the First Asian Conference on Mycorrhiza in 1988 (Mridha, 1988) with emphasis on the need for research. A large number of plant species including agricultural crops and sand dune plants have been studied for their mycorrhizal associations (Mridha and Mohammed, 1989). Many agricultural crops especially legumes and sand dune plants are highly mycotrophic. Studies on the interaction of VAM and *Rhizobium* on the growth of yard long bean (*Vigna unguiculata*) indicated that dual inoculation significantly improved plant growth (Mridha et al., 1991). Mridha collected numerous soil samples throughout the country in collaboration with Aberdeen University, Scotland, U.K. and isolated a vast number of mycorrhizal fungi (Mridha and Killham, 1992). The mycorrhizal association of some leguminous crops in Bangladesh has also been examined in details (Mridha et al., 1992). Research on the inter-plant nutrient transfer between lentil and wheat was conducted using ^{15}N and results were presented at a conference on Management of Mycorrhiza in Agriculture, Horticulture and Forestry held in Australia. In 1994, in collaboration with Hannover University, Germany, Mridha developed the inoculum production technology currently being used in the mycorrhiza laboratory and reported the early flowering of some leguminous pulse crops after inoculation with VAM fungi. A large number of root and soil samples were collected from rice seedbeds and analyzed for their mycorrhizal associations and spore populations of VAM fungi. Mridha et al. (1998a) reported that VAM fungi occurred extensively on rice seedlings under seedbed conditions and a large population was found during the study. Mycorrhizal fungi of tea soils and tea plants, leguminous crops, and weeds of tea plantations were also assessed (Begume et al., 1998a, b; Mridha et al., 1998b). The results indicated that most of the plants studied were mycotrophic. VAM fungal isolates from the tea soil were used to prepare an inoculum in the mycorrhiza laboratory for inoculation in the tea nursery. The close association of VAM fungi with leguminous plants of tea gardens indicated that inter-plant nutrient transfer may be possible between leguminous plants and tea plants grown in the same field. Results indicate that the available indigenous VAM fungi can be managed and manipulated for improving tea plantation performance, and suggest that this technology can be implemented to enhance the soil quality of tea plantations by using VAM as a biofertilizer.

An extensive assessment of VAM fungi on rubber plants, associated leguminous cover crops and weeds in several plantations of Chittagong and Sylhet, two major rubber growing areas of the country were carried out (Mahmud and Mridha, 1998; Mahmud et al., 1998).

A number of VAM fungi were isolated and preserved in the laboratory for future uses. Mass-scale inocula have been produced for use in the rubber nurseries. Rubber plants, leguminous cover crops and most of the weeds were found to be highly mycotrophic. The results of the present study indicated that the available indigenous VAM fungi can be managed and manipulated for better growth of rubber plantations and inter-plant nutrient transfer may be possible when leguminous cover crops are grown with rubber plants.

Biodiversity of VAM fungi of vegetable crops has been studied (Mridha et al., 1998b). Again, a large number of VAM fungi were isolated from a number of soil samples and the mycorrhizal infectivity to different vegetable crops was assessed. A wide range of diversity of infection and spore population were recorded with respect to soils, crops, locations and seasons. The interaction of VAM and EM (a mixed microbial inoculant including 80 species of beneficial microorganisms) was studied in the mycorrhiza laboratory at Chittagong University (Mridha et al., 1996). Improved crop yields were obtained with the dual inoculation of the test organisms.

BENEFITS OF MYCORRHIZAS

It is reported that mycorrhizal fungi improve crop yields, especially in infertile soils (Hayman, 1982, Howeler et al., 1987). Many crops are grown in acid soils, where their establishment is frequently limited by low availability of phosphorus. In this case, appropriate mycorrhizal fungi can greatly improve crop yields by increasing the phosphorus uptake by plants (Howeler et al., 1987). In addition to phosphorus uptake, VAM fungi can also enhance the uptake of relatively immobile micronutrients, particularly zinc and copper (Killham and Firestone, 1983; Lambert et al., 1979; Gnekow and Marschner, 1989; Swaminathan and Verma, 1979; Gildon and Tinker, 1983; Pacovsky, 1986).

VAM fungi can mediate inter-plant transfer of phosphorus (Francis et al., 1986; Newman and Ritz, 1986), carbon (Newman, 1988; Read et al., 1985) and nitrogen (Read et al., 1985; Kessel et al., 1985; Haystead et al., 1988; Barea et al., 1988, 1989; McNeil and Wood, 1990). The flow of nitrogen usually occurs from legume to non-legume when there are legumes and non-legumes growing in nitrogen deficient soils (Haystead et al., 1988).

Several studies have reported the potential for using VAM fungi to improve establishment and growth of different micro-propagated species after transplanting (Ravolanirina et al., 1988). Some researchers have reported that

VAM can decrease the severity of diseases caused by root pathogenic fungi, bacteria and nematodes (Jalali and Chand, 1988).

The effective role of VAM fungi in land rehabilitation has been well documented (Allen and Allen, 1988; Sylvia and Will, 1988; Habte et al., 1988; White et al., 1989). The VAM fungi, by maintaining the uptake of slowly diffusing nutrients under water stress conditions, can help plants resist drought stress (Azcon et al., 1988). VAM fungi can help plants become established in saline soils (Hirrel and Gerdeman, 1980; Pond et al., 1984) and in nutrient deficient soils or degraded (eroded) habitats, in coal wastes, eroded desert and disturbed soils (Hall and Armstrong, 1979; Khan, 1981).

USE OF VAM FOR ORGANIC FARMING IN BANGLADESH AGRICULTURE

It is widely acknowledged that VAM technology can improve soil and crop productivity by allowing farmers to reduce their inputs of chemical fertilizers and/or by enhancing plant survival, thus offsetting ecological and environmental concerns. In addition, the general role of VAM in enhancing the growth and nutrition of most crop plants in Bangladesh is now well recognized. Mycorrhizal fungi have particular value for legumes because of their need for an adequate phosphorus supply, not only for optimum growth but also for nodulation and nitrogen fixation (Azcon-Aguilar et al., 1979; Crush, 1974; Hayman, 1986; Jehne, 1984; Smith et al., 1979; Waidyanatha et al., 1979).

Both improved nitrogen fixation in legumes by *Rhizobium* and increased uptake of phosphorus from VAM fungal associations can indirectly reduce the chemical fertilizer requirement and the problems related to water and air pollution by chemicals as residuals to the soil-root zone. The reduction of fertilizer requirements by using efficient isolates of *Rhizobium* and VAM fungi with different leguminous agricultural crops grown in Bangladesh is of great value.

Intercropping as well as mixed cropping systems and crop rotations with inoculated (*Rhizobium* and VAM) legumes and non-legumes may benefit from the use of VAM fungi in Bangladesh. Appropriate VAM fungi may be incorporated in rice seedbed substrates and transfer of seedlings to the field is a simple inoculation technique currently suitable in Bangladesh.

Some of the agronomically important trees, which have VAM fungi association, include citrus, tea, coffee, rubber and oil palm. Where these plants are growing in nurseries, VAM inoculation may greatly facilitate establishment and early growth after transplanting to the field site. Nursery production of ornamental seedlings and cuttings by treating the rooting and growing media with appropriate inocula is another important area where VAM can be used.

Mycorrhizal inocula produced in the mycorrhiza laboratory at Chittagong University will be immediately useful in the nurseries of rubber and tea plantations as a means to increase root development, root volume and rooting percentage of plants propagated by cuttings, and to enhance the survival of transplanted seedlings by reducing water stress. Inoculation of leguminous plants in tea plantations and leguminous cover crops in rubber plantation with appropriate *Rhizobium* and VAM inocula can facilitate inter-plant nutrient transfer between tea/rubber and leguminous plants.

The practical application of VAM may be integrated into disease management in Bangladesh. Appropriate VAM fungi can be applied in both seedbeds and crop fields to ensure mycorrhiza infected plants, so as to prevent primary and secondary infection by pathogenic organisms and nematodes.

Mycorrhizal fungi appear to have beneficial effects on soil aggregation and may be an important means of controlling soil erosion. Extramatrical mycelia of VAM fungi have been reported to bind sand grains in sandy soils and dunes, and many sand dune plants are known to be mycorrhizal. This technology can be implemented in Bangladesh to enhance soil conditions in the coastal and other sandy areas of the country. There are saline areas in Bangladesh where appropriate osmotic-tolerant mycorrhizal fungi may be introduced to enable cultivation of suitable plant species and enhance crop production. In addition to the examples given, there are many other commercially- and economically-important crops grown in Bangladesh that may benefit from the application of mycorrhizal technology.

CONCLUSIONS

Vesicular-arbuscular mycorrhizal technology can be used profitably to facilitate the uptake and utilization of available nutrients by crops, thereby improving agricultural production in Bangladesh. This technology may be incorporated directly into cropping systems, e.g., crop rotations, intercropping, mixed cropping and integrated pest management programs. Mycorrhizal technology can be applied immediately for the production of VAM-infected seedlings in seedbeds and nurseries. Mycorrhizal research and its practical use as a low-input technology in nature farming and organic farming to improve crop production in Bangladesh are urgently needed to stop further deterioration of agricultural and forest lands due to subsistence agriculture, monocropping, excessive use of inorganic fertilizers, soil erosion, salinity and several other adverse factors. Such research will likely save money and energy now being expended for the production of chemical fertilizers and pesticides. It will also help the country to meet its ecological and environmental goals.

REFERENCES

Allen, E. B. and M. F. Allen. (1988). Facilitation of succession by the nonmycotrophic colonizer *Salsola kali* (Chenopodiaceae) on a harsh site: Effects of mycorrhizal fungi. *American Journal of Botany* 75: 257-267.

Azcon-Aguilar, C., R. Azcon and J. M. Barea. (1979). Endomycorrhizal fungi and *Rhizobium* as biological fertilizers for *Medicago sativa* in normal cultivation. *Nature* (London): 279:325-327.

Azcon, R., F. El-Atrach and J. M. Barea. (1988). Influence of mycorrhiza vs. soluble phosphate on growth, nodulation and N_2 fixation (^{15}N) in alfalfa under different levels of water potential. *Biological Fertilizer and Soils*. 7: 28-31.

Barea, J. M., C. Azcon-Aguilar and R. Azcon. (1988). The role of mycorrhiza in improving the establishment and function of the *Rhizobium* under field conditions. In *Nitrogen Fixation by Legumes in Mediterranean Agriculture*, eds. D. P. Beck and L. A. Materon. Berlin: ICARDA and Martinus Nijhoff Dordrecht, pp. 153-162.

Barea, J. M., F. EL-Atrach and R. Azcon. (1989). Mycorrhiza and phosphate interactions as affecting plant development, N_2-fixation, N-transfer and N-uptake from soil in legumes-grass mixtures by using a ^{15}N dilution technique. *Soil Biology and Biochemistry* 21: 581-589.

Begume, F., M. A. U. Mridha and N. Begume. (1998a). Association of vesicular-arbuscular mycorrhizal fungi in rice seedlings. *Bangladesh Journal of Plant Pathology* (In press).

Begume, F., M. A. U. Mridha and K. T. Osman. (1998b). Status of Vesicular-arbuscular mycorrhizal fungi in the soils of different tea gardens of Bangladesh. *Tea Research Journal of Sri Lanka* (Submitted).

Crush, J. R. (1974). Plant growth response to vesicular-arbuscular mycorrhiza. VII. Growth and nodulation of some herbage legumes. *New Phytologists* 73: 743-749.

Francis, R., R. D. Finlay and D. J. Read. (1986). Vesicular-arbuscular mycorrhiza in natural vegetation systems. VI. Transfer of nutrients in inter- and intra-specific combinations of host plants. *New Phytologists* 120:103-111.

Gildon, A. and P. B. Tinker. (1983). Interactions of vesicular-arbuscular mycorrhizal infections and heavy metals in plants. II. The effects of infection on uptake of copper. *New phytologists* 95: 263-268.

Gnekow, M. A. and H. Marschner. (1989). Role of VA-mycorrhiza in growth and mineral nutrition of apple (*Malus pumila* var. Domestica) rootstock cuttings. *Plant and Soil* 119: 285-293.

Habte, M., R. L. Fox, R. Aziz and S. A. EL-Swaify. (1988). Interaction of vesicular-arbuscular mycorrhizal fungi with erosion in an oxisol. *Applied Environmental Microbiology* 45: 945-950.

Hall, I. R. and P. Armstrong. (1979). The effect of vesicular-arbuscular mycorrhizas on growth of clover, lotus and ryegrass in some eroded soils. *New Zealand Journal of Agriculture Research* 22: 479-484.

Hayman, D. S. (1982). Practical aspects of vesicular-arbuscular mycorrhiza. In *Advances in Agricultural Microbiology*, ed. N. S. Subbra Rao. New Delhi: Oxford and IBM publishing Company, pp. 325-373.

Hayman, D. S. (1986). Mycorrhizae of nitrogen fixing legumes. *MIRCEN Journal* 2: 121-145.

Haystead, A., N. Malajczuk and T. S. Grove. (1988). Underground transfer of nitrogen between pasture plants infected with vesicular-arbuscular mycorrhizal fungi. *New Phytologists* 108: 417-423.

Howeler, R. H., E. Sieverding and F. Saif. (1987). Practical aspects of mycorrhizal technology in some tropical crops and pastures. *Plant and Soil* 100: 249-283.

Hirrel, M. C. and J. W. Gerdeman. (1980). Improved growth of onion and bell pepper in saline soils by two vesicular-arbuscular mycorrhizal fungi. *Proceedings of the Soil Science Society of America* 44: 654-655.

Jalali, B. L. and Chand, H. (1988). Role of VAM in biological control of plant diseases. In *Mycorrhizae for Green Asia-Proceedings of the First Asian Conference on Mycorrhizae*, eds. A. Mohadevan, N. Raman and K. Natarajan. Madras (India): The First Asian Conference on Mycorrhizae, pp. 209-215.

Jehne, W. (1984). Mycorrhizas and Stylosanthes. In *The Biology and Agronomy of Stylosanthes*, eds. H. M. Stace and L. A. Edge. Sydney: Academic Press, pp. 227-241.

Kessel, C. V., P. W. Singleton and H. J. Hoben. (1985). Enhanced N-transfer from soybean to maize by vesicular-arbuscular mycorrhizal (VAM) fungi. *Plant Physiology* 79: 562-563.

Khan, A. H., A. Islam, R. Islam, S. Begume and S. M. I. Huq. (1988). Effect of indigenous VA mycorrhizal fungi on nodulation, growth and nutrient of lentil (*Lens culinaris*) and black gram (*Vigna mungo*). *Journal of Plant Physiology* 133:84-88.

Khan, A. G. (1981). Growth response of endomycorrhizal onions in non-sterilized coal waste. *New Phytologists* 87:363-370.

Killham, K. S. and M. K. Firestone. (1983). Vesicular-arbuscular mycorrhizal mediation of grass response to acidic and heavy metal deposition. *Plant and Soil* 72: 39-48.

Lambert, D. H., D. F. Baker and H. Colo. (1979). The role of mycorrhizae in the interactions of phosphorus with zinc, copper and other elements. *Soil Science Society of America Journal* 43: 976-980.

Mahmud, R. and M. A. U. Mridha. (1998). Status of infection caused by vesicular-arbuscular mycorrhizal fungi in the roots of *Hevea brasiliensis* in Raozan, Dabua and Balisera rubber gardens. *Life Science Journal of Bangladesh* (In press).

Mahmud, R., M. A. U. Mridha and A. R. Rashid. (1998). VA-mycorrhizal fungi associated with leguminous cover crops grown in rubber gardens. *Journal of Rubber Research Institute* (Malaysia) (In press).

McNeil, A. M. and M. Wood. (1990). Fixation and transfer of nitrogen from white clover to ryegrass. *Soil Use Management* 6: 84-86.

Mridha, M. A. U. (1988). Status of mycorrhizal research in Bangladesh. In *Mycorrhizae for Green Asia–Proceedings of the First Asian Conference on Mycorrhizae*, eds. A. Mahadeva, N. Raman and K. Natarajan. Madras (India): The First Asian Conference on Mycorrhizae, pp. 1-2.

Mridha, M. A. U. and A. Mohammed. (1989). Mycorrhizal association of some crop plants of Bangladesh. Paper presented in the *Indian National Conference of Mycorrhizae*, Harayana, India.

Mridha, M. A. U., M. Mitra and A. Mohammed. (1992). A preliminary report on the occurrence of VA-mycorrhizal association in some plants of Bangladesh. Paper presented in the 3rd legume conference, Kew, U.K.

Mridha, M. A. U., M. Mitra and Z. Parveen. (1991). VAM-*Rhizobium* interaction on productivity and nutrient content of yard long bean (*Vigna unguiculata* L. walp sub species *sesquipedalis*). *Mycorrhiza, Proceedings of the 3rd European Symposium on Mycorrhiza*, Sheffield, England, UK.

Mridha, M. A. U. and K. S. Killham. (1992). Application of VA-mycorrhizal research to crop production in Bangladesh In *Advances in Crop Science–Proceedings Crop Science Society of Bangladesh*, eds. L. Rahman and M. A. Hashem. pp. 367-383.

Mridha, M. A. U., R. Mahmud, and A. R. Rashid. (1996). Interaction of EM and VAM on growth of *Sesbania rostrata*. *Proceedings of the Fifth EM Conference.* Thailand.

Mridha, M. A. U., M. A. Barshed, H. L. Xu, and H. Umemura. (1998a). Ecological study on vesicular-arbuscular mycorrhizal fungi in soils of tea plantations. *Japanese Society for Horticultural Science Autumn Meeting*, Oct. 6-8, 1998, Niigata. *Journal of Japanese Society for Horticultural Science* 67 (Extra 2):346.

Mridha, M. A. U., A. Sultana, S. Sultana, H. L. Xu and H. Umemura. (1998b). Biodiversity of VA-mycorrhizal fungi with some vegetable crops. In *World Security and Production Technologies for Tomorrow*, eds. T. Horie, S. Geng, T. Amano, T. Inamura and T. Shiraiwa. Kyoto University, Kyoto (Japan), pp. 330-324.

Newman, E. I. (1988). Mycorrhizal links between plants: Their functioning and ecological significance. *Advances in Ecology Research* 18: 243-270.

Newman, E. I. and K. Ritz. (1986). Evidence on the pathways of phosphorus transfer between vesicular-arbuscular mycorrhizal plants. *New Phytologist* 104: 77-78.

Pacovsky, R. S. (1986). Micronutrient uptake and distribution in mycorrhizal and phosphorus fertilized soybeans. *Plant and Soil* 95: 379-388.

Pond, E. C., J. A. Menze and W. M. Jarrel. (1984). Improved growth of tomato in salinized soil by vesicular-arbuscular mycorrhizal fungi collected from saline soils. *Mycologia* 79: 74-84.

Ravolanirina, F., V. Gianinazzi-Pearson and S. Gianinazzi. (1988). Preliminary studies on in-vitro endomycorrhizal inoculation of micro-propagated tree species of nutritional value. In *Proceedings of the Asian Seminar on Trees and Mycorrhiza*, ed. F. S. P. Ng. Kuala Lumpur (Malaysia): The Asian Seminar on Trees and Mycorrhiza, pp. 91-102.

Read, D. J., R. Francis and R. D. Finlay. (1985). Mycorrhizal mycelia and nutrient cycling in plant communities. In *Ecological Interactions in Soil*. London:Oxford Blackwell Scientific, pp. 193-217.

Smith, S. E., D. J. D. Nicholas and F. A. Smith. (1979). The effect of early mycorrhizal infections on nodulation and nitrogen fixation in *Tifolium subterraneum* L. *Australian Journal of Plant Physiology* 6: 305-316.

Swaminathan, K. and B. C. Verma. (1979). Response of three crop species to vesicular-arbuscular mycorrhizal infection on zinc deficient Indian soils. *New Phytologists* 82: 481-487.

Sylvia, D. M. and M. E. Will. (1988). Establishment of vesicular-arbuscular mycorrhizal fungi and other microorganisms on beach replenishment site in Florida. *Applied Environmental Microbiology* 54: 348-352.

Waidyanatha, V. P. De, S., N. Yogaratnam and W. A. A. Ariyaratne. (1979). Mycorrhizal infection on growth and nitrogen fixation of *Pueraria* and *Stylsanthes* and uptake of phosphorus from two rock phosphates. *New Phytologists* 82: 147-152.

White, J. A., L. C. Munn and S. E. W. William. (1989). Edaphic and reclamation aspects of vesicular-arbuscular mycorrhizae in Wyoming red desert soils. *Soil Science Society of America Journal* 53: 86-90.

Effects of Organic and Chemical Fertilizers and Arbuscular Mycorrhizal Inoculation on Growth and Quality of Cucumber and Lettuce

Hui-lian Xu
Ran Wang
Md. Amin U. Mridha

SUMMARY. Arbuscular mycorrhizae were inoculated into phosphorus-deficient soil fertilized with either organic or chemical fertilizer with cucumber (*Cucumis sativus* L.) as the first crop and lettuce (*Lactuca sativa* L.) as the second crop but without additional fertilization and AM inoculation. AM increased dry matter and fruit yield of cucumber significantly in the unfertilized, organic-fertilized and P-deficient plants compared with the fully chemical-fertilized plants. AM inoculation increased the available phosphorus in plant and soil by around 30% for all treatments except for those chemically-fertilized. The rate of AM infection did not differ significantly among the fertilization treatments,

---

Hui-lian Xu is Senior Crop Scientist, International Nature Farming Research Center, 5632 Hata, Nagano 390-1401, Japan.

Ran Wang is Professor, Laiyang Agricultural University, Shandong, China.

Md. Amin U. Mridha is Professor, Department of Botany, Chittagong University, Chittagong, Bangladesh.

Address correspondence to: Hui-lian Xu at the above address (E-mail: huilian@janis.or.jp).

The authors thank U. K. Aryal, S. Kato, K. Yamada, M. Fujita, K. Katase and H. Umemura for their technical assistance and advice.

[Haworth co-indexing entry note]: "Effects of Organic and Chemical Fertilizers and Arbuscular Mycorrhizal Inoculation on Growth and Quality of Cucumber and Lettuce." Xu, Hui-lian, Ran Wang, and Md. Amin U. Mridha. Co-published simultaneously in *Journal of Crop Production* (Food Products Press, an imprint of The Haworth Press, Inc.) Vol. 3, No. 1 (#5), 2000, pp. 313-324; and: *Nature Farming and Microbial Applications* (ed: Hui-lian Xu, James F. Parr, and Hiroshi Umemura) Food Products Press, an imprint of The Haworth Press, Inc., 2000, pp. 313-324. Single or multiple copies of this article are available for a fee from The Haworth Document Delivery Service [1-800-342-9678, 9:00 a.m. - 5:00 p.m. (EST). E-mail address: getinfo@haworthpressinc.com].

but the infection intensity was higher in unfertilized, organic-fertilized and phosphorus-deficient treatments than chemical-fertilized treatment. The residual effects of AM-inoculated to cucumber were evident for lettuce in all pre-treatments that were unfertilized and un-inoculated for the second cropping. Without P-fertilization, neither crop could grow optimally even when the soil was inoculated with AM, suggesting that AM could not serve as a substitute for phosphorus fertilizer. However, the other beneficial effects of AM on crop growth and yield could not be fulfilled with phosphorus fertilizer. *[Article copies available for a fee from The Haworth Document Delivery Service: 1-800-342-9678. E-mail address: getinfo@haworthpressinc.com <Website: http://www.HaworthPress.com>]*

KEYWORDS. Cucumber, lettuce, organic fertilizer, nature farming, organic farming, phosphorus, arbuscular mycorrhizae

INTRODUCTION

Policies for agriculture and environment in many countries are tending towards more sustainable systems because of concerns over environmental degradation and food pollution by excessive inputs of agrichemicals. The biological balance in soil microorganisms is also impaired by chemical farming practices. Beneficial microorganisms have lost their diversity while new and stronger pathogenic microorganisms appear frequently (Ayres et al., 1996; Ing, 1996; Mridha et al., 1999). Therefore, research on beneficial microorganisms has drawn attentions from agricultural scientists. One important group of such beneficial microorganisms is arbuscular mycorrhizae (AM). AM fungi are ubiquitous, zygomycetous soil fungi that colonize the roots of most species of vascular plants (Morton and Benny, 1990; Trappe, 1987). The benefits of AM result from fungus-root interactions with many complex and dynamic processes involved. The benefits of AM include mainly the improved uptakes of macronutrients, especially phosphorus, and micronutrients such as zinc, increased tolerance of stress, and beneficial alterations of phytohormones (Jarstfer and Sylvia, 1993; Smith and Read, 1997). With their hyphae extending away from the root, the AM fungi absorb inorganic nutrients, phosphorus or zinc and copper, which are translocated to roots of the host plant in exchange for photosynthetically-fixed carbon, which is absorbed by the arbuscules or vesicles in cells of the host root. AM fungi are often important in plant growth and influence plant competition, especially in nutrient-limiting soils (Allen, 1991). Usually immediate availability of nutrients is limited at early stages of crop growth in organic farming systems (Xu et al., 1999). The potential of AM application in sustainable agriculture has been demonstrated recently (Sieverding, 1991; Gianinazzi and Schuepp,

1994; Pfleger and Linderman, 1994; Smith and Read, 1997). The beneficial effect of AM on nutrient availability would be helpful in organic or nature farming systems. The ability of AM to enhance plant growth is influenced by soil nutrient conditions (Allen, 1991) and their specific interaction with the host-plant (Azcon et al., 1981; Ianson and Linderman, 1993). However, it is not clear whether the effect of AM inoculation can last for two or more growth seasons. Therefore, in the present study, the effect and the sustainability of AM were examined with the inoculated first crop of cucumber with different fertilizers, which was followed by the second crop of lettuce without fertilization and inoculation.

MATERIALS AND METHODS

Plant Materials and Treatments

For the first cropping, seeds of cucumber (Cucumis sativus L. cv. Nankyo-ku-1) were directly sown in polyethylene pots each filled with 2 kg of sterilized (48 h at 85°C) Andosol soil. For the second cropping, seedlings of lettuce (*Lactuca sativa* L.) were transplanted into the pots one month after the cucumber crop finished. The soil treatments were designed as follows: (1) No-Fertilizer: without fertilization and without AM inoculation; (2) No-Fertilizer + AM: without fertilization but with AM-inoculated; (3) Organic: 80 g of organic fertilizer which was fermented anaerobically using rice bran, oil cake meal and fish processing by-product; (4) Organic + AM: the same amount of organic fertilizer as in treatment (3) with AM inoculated; (5) N-K: with chemical fertilization of nitrogen [3.7 g ammonium sulfate, 2.0 g long-period coated urea] and potassium (3.5 g potassium sulfate); (6) N-K + AM: the same fertilization as in (5) with AM inoculated; (7) N-P-K: with the same amount of nitrogen (N) and potassium (K) as in (5) plus phosphorus (P) (9.1 g superphosphate); and (8) N-P-K + AM: the same chemicals as in (7) with AM inoculated. The amounts of N-P-K in treatments (7) and (8) were equivalent to the usable amounts (70%) of the total N-P-K in organic materials of treatments (3) and (4).

Sugar, Organic Acid and Vitamin C Determination

Fresh leaf tissue was homogenized with distilled water in a ratio of 1:4. The homogenate was centrifuged at 13000 × g for 15 min and the supernatant passed through a 0.45-μm filter. The extract was used to determine sugars, organic acids and ascorbic acid (vitamin C). Sugars were measured by a system of high performance liquid chromatography (HPLC) with a Refractive Index Detector (RI-930, Jasco, Tokyo, Japan) and a Shodex SC1011 column (Shoko, Tokyo, Japan) at a column temperature of 80°C and flow rate of 1 ml min^{-1} with deionized water as the effluent. Vitamin C was determined by a reflectometer (RQflex, Merck).

Chlorophyll and Carotenoid Determinations

Leaf disc with a fresh mass of 1 g was excised from the 5th leaf from the top. The 1 g leaf disc was ground in 4 ml of 100% acetone and diluted to a final volume of 20 ml in a mesflask. The leaf extract was centrifuged under 0°C at 3000 × g for 15 min. The supernatant was read at 663 nm, 647 nm and 470 nm with 80% acetone as blank using a spectrophotometer (U2000A, Hitachi, Tokyo, Japan). Contents of chlorophyll *a*, chlorophyll *b*, chlorophyll *a* + *b* and carotenoid were calculated from the light absorption readings (A) as

$$\text{Chl } (a + b) = 8.02A_{663} + 20.20A_{645};$$

$$\text{Chl } a = 12.7A_{663} - 2.69A_{645};$$

$$\text{Chl } b = 22.9A_{645} - 4.68A_{663}; \text{ and}$$

$$\text{Carotenoid} = [(1000 \times A_{470}) - (1.82 \times \text{Chl } a) - (85.02 \times \text{Chl } b)]/198$$

Salt and Nutrient Concentrations Determination

The whole plants were sampled and dried at 85°C for 48 h and then stored in a desiccator for use of nutrient analysis. Mineral cations in plants and soil were determined by the method of atomic absorption spectrophotometer (180-30, HITACHI, Tokyo, Japan). The concentrations of total nitrogen and carbon were measured by MT-700 Carbon-Nitrogen Recorder (Yanaco, Tokyo, Japan). Concentrations of nitrate-N and phosphorus were determined by a colorimetric method.

Determination of AM Infection

Roots of cucumber and lettuce were sampled at harvesting time. The soil samples with roots were collected from several plants of each treatment with a soil corer about 5 cm away from the stem of each plant at a depth of 10 to 15 cm. Roots were sieved out with a 2-mm sieve and washed with tap water. The presence of hyphae, arbuscules and vesicles were observed under a microscope (Olympus, Tokyo, Japan) after clarifying with 2.5% KOH, acidified in 1% HCl and stained in 0.05% aniline blue in lactoglycerol (Koske and

Gemma, 1989). The intensity of infection of AM fungi was estimated as follows: Poor × only mycelium was present; Moderate × mycelia and vesicles were present; and Abundant × mycelia, vesicles and arbuscules were present.

RESULTS AND DISCUSSION

Effect on the Growth of Cucumber at the First Cropping

Plants in unfertilized and phosphorus-deficient treatments could not grow well enough to produce fruit (Table 1). Compared with plants fertilized with full N-P-K chemical fertilizers, plants in organic-fertilized treatments grew worse in the early stage but better in the late stage with higher final dry matter production and higher fruit yield. This might be attributed to low nutrient availability at the beginning and high nutrient sustainability of the organic fertilizer. The organic fertilizer is an anaerobically fermented mixture of oil seed sludge and rice bran. Most of the organic nitrogen was not mineralized at the beginning. This leads to a problem for the early stage growth of crops as shown with tomato (Wang et al., 1999) and sweet corn (Xu et al., 1999).

AM inoculation increased fruit yield for all treatments of unfertilized, phosphorus-deficient organic and full N-P-K chemical plants. The effect on plant growth was lower in the treatment of full chemical fertilization. AM fungi are of particular importance for immobile nutrients such as phosphorus, which can be changed to available status by the activity of AM (Hetrick, 1989; Bolan, 1991). As shown in Table 2, available phosphorus concentration was higher in both plant and soil of the AM inoculation treatments than

TABLE 1. Effect of AM inoculation, organic fertilizer and chemical fertilizer on dry matter production and fruit yield of cucumber plants

Treatment	Plant dry matter (g plant^{-1})	Fruit yield (g plant^{-1})
No Fertilizer	11.8 ± 0.2	0.0 ± 0.0
No Fertilizer + AM	13.3 ± 0.3	0.0 ± 0.0
Organic	33.5 ± 0.9	103.1 ± 6.3
Organic + AM	40.0 ± 0.7	158.0 ± 14.2
NK	14.2 ± 0.3	0.0 ± 0.0
NK + AM	16.9 ± 0.3	0.0 ± 0.0
NPK	31.2 ± 0.7	92.9 ± 8.1
NPK + AM	32.8 ± 0.7	112.9 ± 8.5

TABLE 2. Effect of AM inoculation, organic fertilizer and chemical fertilizer on available phosphorus (P) in cucumber plants and pot soil

Treatment	P content in plant		P content in pot soil		Total P in plant & soil	Increase by AM
	(mg g^{-1} DM)	(mg $plant^{-1}$)	(mg kg^{-1} DM)	(mg pot^{-1})	(mg pot^{-1})	(%)
No Fertilizer	43.0 ± 5.7	60.2	6.5 ± 1.0	76.7	138.9	
No Fertilizer + AM	58.9 ± 4.4	82.5	7.3 ± 1.5	97.1	179.6	29.3
Organic	121.1 ± 21.8	169.5	8.0 ± 0.9	350.4	519.9	
Organic + AM	114.7 ± 19.3	160.6	9.0 ± 0.8	502.2	662.8	27.5
NK	56.7 ± 2.4	79.4	4.9 ± 0.8	82.8	162.2	
NK + AM	62.0 ± 2.4	86.8	7.9 ± 1.0	112.2	199.0	22.7
NPK	127.2 ± 26.5	178.1	12.0 ± 2.0	486.0	664.1	
NPK + AM	130.4 ± 6.1	182.6	10.4 ± 1.8	458.6	641.2	−3.4

the treatments without AM inoculation. The effect of AM was significant for the striking in organic-fertilized, phosphorus-deficient and unfertilized plants but less so for the full N-P-K chemical fertilizer. The main benefit from AM to plants is increased P-availability. The role of AM in increasing phosphorus efficiency is reviewed in details by Bolan (1991). The limitations for phosphorus uptake are the concentration and diffusion of the ions in the soil and the release of the phosphorus from the soil particles. The inflow of the phosphorus through AM hyphae is six times faster than through root hairs (Sanders and Tinker, 1973). However, the main reason for the increased phosphorus uptake is associated with the increased exploration of the soil by the hyphae, which decrease the distance in phosphate diffusion and increase the surface for nutrient absorption (Moawad and Vlek, 1997). In the treatment of full N-P-K fertilizer, the phosphorus applied (superphosphate) was highly available for plants. Thus, the effect of AM on available phosphorus could not be observed in this treatment. However, AM might improve uptake of other nutrients of plants and consequently increase plant growth and fruit yield. In this study, the concentrations per unit of dry mass of K, Ca and Mg were higher in AM-inoculated plants under all fertilizer regimes with the exception of the full N-P-K treatment (Table 3). If the cation concentration is shown per plant, larger data are also shown for the chemical-fertilized plants (% data in parenthesis). Moreover, the AM hyphae grow out of the root into the surrounding soil serving as root extensions. The hyphae make the nutrient uptake more effective than that by the root itself. Therefore, AM might increase growth and yield through increasing uptakes of all nutrients by root extension effect (Moawad and Vlek, 1997). In some cases, AM presence is more important than the soil phosphorus level. Kormanik (1985) has found that mycorrhizal walnut seedlings at 25 ppm phosphorus grow as well as non-mycorrhizal seedlings with 75 or 150 ppm of available soil phosphorus. The available nitrogen in the soil showed no clear difference between AM-inoculated and un-inoculated plants (Table 4). The potassium concentration in soil was lower in the AM-inoculated treatments (Table 4). This might be due to the enhanced absorption of K by the plant as shown in Table 3.

AM infection was not found in the sterilized soil even though pots were randomly placed together with those of inoculated soils. The rate of AM infection was slightly higher in the unfertilized and phosphorus-deficient-treatments (Table 5). The intensity of AM infection was weaker in N-P-K fertilized treatment than in other treatments. This suggests that a high phos-

TABLE 3. Effects of AM inoculation, organic fertilizer and chemical fertilizer on cation concentrations (g kg^{-1}) in cucumber plants

Treatment	K	Increase (%)	Ca	Increase (%)	Mg	Increase (%)
No Fertilizer	36.2 ± 9.3		30.1 ± 6.8		6.9 ± 1.8	
No Fertilizer + AM	42.2 ± 4.9	16.5 (31.3)	33.6 ± 1.8	11.6 (25.8)	7.4 ± 0.5	7.2 (20.8)
Organic	28.2 ± 1.9		25.7 ± 2.1		9.4 ± 0.8	
Organic + AM	29.0 ± 2.3	2.8 (29.8)	28.8 ± 2.8	12.1 (44.1)	8.9 ± 1.6	−5.6 (23.2)
NK	47.8 ± 7.8		27.1 ± 2.0		5.8 ± 0.5	
NK + AM	48.8 ± 5.0	2.1 (19.4)	38.3 ± 1.4	41.3 (68.1)	8.2 ± 1.5	41.3 (68.1)
NPK	44.9 ± 6.9		30.5 ± 2.9		5.9 ± 0.4	
NPK + AM	44.5 ± 5.2	−0.8 (8.8)	29.7 ± 1.3	−2.3 (6.9)	5.1 ± 0.6	−15.6 (10.2)

Data in ( ) show the increments per plant including fruit.

TABLE 4. The content of total carbon, total nitrogen, available nitrogen and K (mg kg^{-1}) in the soil after cucumber harvested

	Total C (mg kg^{-1})	Total N (mg kg^{-1})	C:N	NH_4^+	NO_3^-	T. Av-N	K
			----------	----------	(mg kg^{-1})	----------	----------
No Fertilizer	32.7 ± 0.49	2.9 ± 0.05	11.51	9.1 ± 1.52	3.3 ± 1.84	12.41	220.5 ± 7.58
No Fertilizer + AM	32.3 ± 0.28	2.8 ± 0.05	11.70	10.0 ± 1.85	3.6 ± 0.18	13.53	182.3 ± 5.50
Organic	35.2 ± 0.13	3.3 ± 0.05	10.56	17.7 ± 3.05	9.1 ± 0.18	26.76	143.7 ± 3.82
Organic + AM	37.3 ± 0.64	3.7 ± 0.09	10.15	17.5 ± 1.15	8.1 ± 0.00	25.62	133.9 ± 28.41
NK	32.4 ± 0.05	2.8 ± 0.01	11.45	9.6 ± 0.00	8.3 ± 1.87	17.88	283.8 ± 31.39
NK + AM	32.6 ± 0.26	2.9 ± 0.05	11.33	10.1 ± 0.35	10.3 ± 0.00	20.38	273.6 ± 28.79
NPK	32.9 ± 0.72	2.8 ± 0.09	11.60	8.5 ± 1.10	3.7 ± 0.54	12.20	230.8 ± 64.47
NPK + AM	32.6 ± 0.06	2.8 ± 0.01	11.58	9.1 ± 1.25	3.7 ± 0.00	12.79	201.4 ± 12.76

phorus level does not favor the infection and development of AM. The plants with good phosphorus nutrition develop a hard root surface that prevents infection by microorganisms.

Effect on the Growth of the Second Cropping Lettuce

The plant dry mass and concentrations of sugar and vitamin C for lettuce the second crop are presented in Table 6. Without any fertilization after the first crop of cucumber, the highest yield was obtained with organic fertilizer treatments, suggesting that the organic fertilizer had good sustainability. Un-

TABLE 5. Effect of AM inoculuation, organic fertilizer and chemical fertilizer on proportion and intensity of AM in cucumber roots (mean ± SE; n = 25)

Treatment	Percent infection	Intensity of infection (%)		
		Poor	Moderate	Abundant
No Fertilizer	0.0	0.0	0.0	0.0
No Fertilizer + AM	84.5 ± 6.1	34.2 ± 8.8	28.6 ± 6.2	21.7 ± 8.5
Organic	0.0	0.0	0.0	0.0
Organic + AM	64.0 ± 11.0	29.9 ± 2.4	19.3 ± 6.5	14.8 ± 6.4
NK	0.0	0.0	0.0	0.0
NK + AM	89.2 ± 4.9	24.2 ± 11.2	22.5 ± 4.2	42.5 ± 11.9
NPK	0.0	0.0	0.0	0.0
NPK + AM	70.0 ± 12.8	56.7 ± 6.5	11.3 ± 4.8	4.0 ± 1.5

TABLE 6. Effect of AM inoculation, organic fertilizer and chemical fertilizer on fresh mass, dry matter and the concentration of sugar and vitamin C of lettuce (g kg^{-1} DM)

Treatment	Dry mass (g $plant^{-1}$)	Sucrose	Glucose	Fructose	T. sugars	Vitamin C
		(g kg^{-1} DM)				
No Fertilizer	5.7 ± 0.41	119.0 ± 34.81	49.3 ± 5.00	102.1 ± 9.65	270.4	4.22 ± 0.365
No Fertilizer + AM	6.5 ± 0.38	118.8 ± 24.94	49.7 ± 15.46	110.5 ± 21.58	279.1	4.80 ± 0.272
Organic	33.8 ± 2.39	82.9 ± 5.92	87.0 ± 8.16	144.7 ± 8.45	314.7	2.49 ± 0.464
Organic + AM	40.3 ± 2.88	80.3 ± 21.73	101.7 ± 10.26	172.9 ± 7.05	354.9	3.05 ± 0.156
NK	7.1 ± 0.72	75.8 ± 52.75	41.9 ± 1.83	93.5 ± 9.07	211.1	3.65 ± 0.104
NK + AM	11.3 ± 2.25	75.9 ± 33.62	50.3 ± 11.92	106.0 ± 13.64	232.2	4.31 ± 0.384
NPK	11.4 ± 2.88	61.3 ± 18.50	38.3 ± 11.22	97.7 ± 4.79	197.2	2.35 ± 0.600
NPK + AM	15.1 ± 4.52	67.7 ± 16.76	40.6 ± 3.73	99.6 ± 7.46	207.9	2.95 ± 0.762

der the same condition, the plants grown in the treatment with AM-inoculated in the first crop showed higher fresh and dry mass as well as higher concentrations of total sugar and vitamin C than those in the treatments without AM inoculation. Though the phosphorus in the plant and soil without phosphorus fertilization was deficient, the available phosphorus in soil and plants was a little higher in the AM-inoculated treatment than in the treatment without AM (Table 7). This suggested that the effect of AM on phosphorus efficiency sustained to the second cropping. However, in both the first and the second croppings, plants without phosphorus fertilization could not grow to the normal extent even when inoculated with AM. This suggested that AM effect was limited. AM inoculation was just a supplementary measure and could not be used as an alteration of phosphorus fertilization. However, the importance of AM in other aspects can not be fulfilled by phosphorus fertilization.

Interestingly, it was found that AM inoculation increased the total concentration of soluble sugars (sucrose, fructose and glucose) and the concentration of vitamin C. The effects of increasing vitamin C were consistent with the results of available phosphorus (Table 6). It is well known that good phosphorus nutrition is favorable for sugar metabolism and translocation. The sugar concentration was higher in the organic-fertilized plants than in chemical-fertilized plants, in agreement with the result of tomato (Wang, 1999). Effect of AM on chlorophyll and carotenoids showed no clear trends (Table 8). In some fertilization treatments, chlorophyll was lower in AM-inoculated lettuce plants. This might be related to the dilution of substances by increased plant mass.

The results of AM infection are shown in Table 9. There was little or no

TABLE 7. Effect of AM inoculation, organic fertilizer and chemical fertilizer on available phosphorus (P) in lettuce plants and pot soil

Treatment	P content in plant		P content in pot soil		Total P in plant & soil	Increase by AM
	($mg\ g^{-1}$ DM)	($mg\ plant^{-1}$)	($mg\ kg^{-1}$ DM)	($mg\ pot^{-1}$)	($mg\ pot^{-1}$)	(%)
No Fertilizer	1.03 ± 0.085	5.9	13.7 ± 0.00	19.2	25.1	
No Fertilizer + AM	1.32 ± 0.151	8.6	13.8 ± 0.02	19.3	27.8	10.7
Organic	1.58 ± 0.302	53.3	29.1 ± 1.54	40.7	94.1	
Organic + AM	1.87 ± 0.040	75.1	25.3 ± 2.29	35.4	110.5	17.5
NK	1.18 ± 0.102	8.4	11.5 ± 2.29	16.1	24.5	
NK + AM	1.36 ± 0.050	12.7	13.8 ± 0.02	19.3	31.9	30.6
NPK	1.61 ± 0.127	18.2	29.7 ± 1.66	41.5	59.8	
NPK + AM	1.84 ± 0.227	27.8	33.7 ± 1.55	47.2	74.9	25.4

TABLE 8. Effect of AM inoculation, organic fertilizer and chemical fertilizer on concentration of chlorophylls and carotenoids (g kg^{-1} DM)

Treatment	Chl.*a*	Chl.*b*	Chl.*a* + Chl.*b*	Carotenoids
No Fertilizer	1.19 ± 0.121	0.40 ± 0.016	1.58	0.31 ± 0.12
No Fertilizer + AM	1.04 ± 0.056	0.46 ± 0.021	1.52	0.38 ± 0.11
Organic	1.64 ± 0.365	0.53 ± 0.184	2.17	0.60 ± 0.08
Organic + AM	1.40 ± 0.112	0.57 ± 0.131	1.87	0.56 ± 0.04
NK	2.30 ± 0.615	0.80 ± 0.301	3.10	0.30 ± 0.07
NK + AM	1.92 ± 0.230	0.78 ± 0.093	2.70	0.35 ± 0.03
NPK	0.94 ± 0.065	0.36 ± 0.043	1.30	0.26 ± 0.04
NPK + AM	1.08 ± 0.182	0.38 ± 0.073	1.46	0.31 ± 0.02

TABLE 9. Effect of AM inoculation, organic fertilzier and chemical fertilizer on proportion and intensity of AM in lettuce roots (Mean ± SE; n = 25)

Treatment	Percent infection	Intensity of infection (%)		
		Poor	Moderate	Abundant
Control	0.0	0.0	0.0	0.0
AM	46.0 ± 3.0	20.0 ± 4.6	22.7 ± 5.3	3.3 ± 3.3
Organic	0.0	0.0	0.0	0.0
Organic + AM	52.0 ± 13.8	12.7 ± 5.2	24.7 ± 4.0	18.0 ± 7.02
NK	0.0	0.0	0.0	0.0
NK + AM	63.3 ± 11.7	16.7 ± 3.7	33.3 ± 6.6	13.3 ± 5.33
NPK	0.0	0.0	0.0	0.0
NPK + AM	57.3 ± 14.4	23.3 ± 9.4	29.3 ± 5.8	4.7 ± 4.67

AM infection found in the lettuce crop with sterilized soil. The rate of AM infection was lower in the treatment without any fertilization. Poor soil nutrition might inhibit the development of AM. Differences in infection intensity were found between organic and full N-P-K fertilizer. Poor infection was found more frequently but the abundant infection was less frequent in roots of full N-P-K fertilizer and no-fertilizer than in root of other treatments.

CONCLUSION

In the first cropping of cucumber, AM inoculation increased dry matter production and fruit yield and the effects were larger in unfertilized, organic-fertilized and P-deficient plants than in N-P-K fertilized plants. AM increased total available phosphorus in plant and soil by around 30% in all treatments except the full N-P-K fertilization. This suggests that increased available phosphorus may be the main account for increased growth. The high rates of phosphorus do not favor AM infection and development. The effects of AM inoculation sustained the second crop of lettuce. The effects were shown for all fertilizer treatments made in the first crop. This might be due to the limitation of phosphorus in all treatments of the second crop. In both the first and the second seasons, crops without phosphorus fertilization could not grow normally even inoculated with AM. AM effect was not an alternative to phosphorus fertilization. However, the importance of AM in other aspects can not be replaced by phosphorus fertilization.

REFERENCES

Allen, M. F. (1991). *The Ecology of Mycorrhizae.* Cambridge, New York: Cambridge University Press.

Ayres, P. G., T. S. Gunasekera, M. S. Rasanayagam and N. D. Paul. (1996). Effects of UV-B radiation (280-320) on loliar saparoreophs and pathogens. In *Fungi and Environmental Change*, eds. J. C. Frankland, N. Magan and G. M. Gadd. Cambridge, New York: Cambridge University Press, pp. 32-50.

Azcon, R., J. M. Barea and A. Ocampo. (1981). Factors affecting the vesicular-arbuscular infection and mycorrhizal dependency of thirteen wheat cultivars. *New Phytologist* 87: 677-685.

Bolan, N. S. (1991). A critical review on the role of mycorrhizal fungi in the uptake of phosphorus by plants. *Plant and Soil* 134: 189-207.

Gianinazzi, S. and H. Schuepp. (1994). *Impact of Arbuscular Mycorrhizas on Sustainable Agriculture and Natural Ecosystems.* Basel, Switzerland: Birkhauser Verlag.

Hetrick, B. A. D. (1989). Acquisition of phosphorus by VA mycorrhizal fungi and the growth responses of their host plant. In N*itrogen, Phosphorus and Sulphur Utilization by Fungi*, eds. I. Boddy, R. Marchant and D. J. Reid. New York: Cambridge University Press, pp. 205-226.

Ianson, D. C., and R. G. Linderman. (1993). Variation in the response of nodulating pigeonpea (*Cajanus cajan*) to different isolates of mycorrhizal fungi. Symbiosis 15: 105-119.

Ing, B. (1996). Red data lists and decline in fruiting of macromycetes in relation to pollution and loss of habitat. In *Fungi and Environmental Change*, eds. J. C. Frankland, N. Magan and G. M. Gadd. Cambridge, New York: Cambridge University Press, p 61-69.

Jarstfer, A. G. and D. M. Sylvia. (1993). Inoculum production and inoculation strategies from vesicular-arbuscular mycorrhizal fungi. In *Soil Microbial Ecology*, ed. F. B. Metting Jr. New York: Marcel Dekker, pp. 349-378.

Kormanik, P. P. (1985). Effects of phosphorus and vesicular-arbuscular mycorrhizae on growth and leaf retention of black walnut seedlings. *Canadian Journal of Forestry Research* 15: 688-693.

Koske, R. E. and J. N. Gemma. (1989). A modified procedure for staining roots to delete VA-mycorrhizae. *Mycological Research* 92: 486-505.

Moawad, A. M. and P. L. G. Vlek. (1997). Potential contribution of (vesicular-) arbuscular mycorrhiza to nutrient efficient crops. *Proceedings of the BTIG Workshop on Oil Palm Improvement through Biotechnology*. Bogor (Indonesia): Indonesian Ministry of Agriculture, pp. 48-58.

Morton, J. B. and G. L. Benny. (1990). Revised classification of arbuscular mycorrhizal fungi (Zygomycetes); a new order, Glomales, two new suborders, Glominae and Gigasporinae, and two families, Acaulosporaceae and Gigasporaceae, with an emendation of Glomaceae. *Mycotaxon* 37: 471-491.

Mridha, M. A. U., H. L. Xu, R. Wang, K. Yamada, M. Fujita, K. Katase and H. Umemura. (1999). Arbuscular mycorrhiza with crops grown in nature farming soils. *Japanese Journal of Crop Science* 68 (Extra issue): 26-27.

Pfleger, F. L. and R. Linderman. (1994). *Mycorrhizae and Plant Health*. Minnesota: APS Press.

Sanders, F. E. and B. P. Tinker. (1973). Phosphate flow into mycorrhizal roots. *Pesticide Science* 4: 385-395.

Sieverding, E. (1991). *Vesicular-Arbuscular Mycorrhizal Management in Tropical Agrosystem*. Berlin: GTZ Publishers.

Smith, S. E. and D. J. Read. (1997). *Mycorrhizal Symbiosis*, 2nd edn. London: Academic Press.

Trappe, J. M. (1987). Phylogenetic and ecological aspects of mycotrophy in the angiosperms from an evolutionary standpoint. In *Ecophysiology of VA Mycorrhizal Plants*, ed. G. Safir. Boca Raton (USA): CRC Press, pp. 5-25.

Wang, R., H. L. Xu, M. A. U. Mridha and H. Umemura (1999). Effects of organic fertilization and microbial inoculation on leaf photosynthesis and fruit yield of tomato plants. *The 207th Annual Meeting of Japanese Society of Crop Science*, April 2-4, 1999, Tokyo. Japanese Journal of Crop Science 68 (Extra 1): 28-29.

Xu, H. L., R. Wang, S. Kato, K. Yamada, M. Fujita. K. Katase, T. Higa. and H. Umemura. (1999). Sweet corn growth and physiological responses to organic fertilization. In *World Food Security and Crop Production Technologies for Tomorrow*, eds. T. Horie, S. Geng, T. Amano, T. Inamura and T. Hirasawa. Tokyo: Crop Science Society of Japan, pp. 305-306.

Nodulation Status and Nitrogenase Activity of Some Legume Tree Species in Bangladesh

Uma K. Aryal
M. K. Hossain
Md. Amin U. Mridha
Hui-lian Xu

SUMMARY. A study was conducted to observe the nodulation status and measure nitrogenase activity of some important legume tree species in nursery. Of the thirteen species surveyed eight belong to Mimoceae, two Papilionaceae and three Caesalpiniaceae. All the species in Mimoceae and Papilionaceae were found nodulated whereas all the members of Caesalpiniaceae were non-nodulated. Among nodulated seedlings, highest number of nodules per seedling was recorded in *Leucaena leucocephala* (82) followed by *Acacia auriculiformis* (55), *Acacia mangium* (52), *Albizia lebbeck* (46), *A. procera* (41), *Dalbergia sissoo* (32) and *Acacia catechu* (29). Nitrogenase activity was highest in *L. leucocephla* (4913.59 nmole C_2H_4 h^{-1}) followed by *Albizia procera*

Uma K. Aryal is Visiting Microbiologist, International Nature Farming Research Center, 5632 Hata, Nagano 390-1401, Japan.

M. K. Hossain is Associate Professor of the Institute of Forestry and Environmental Sciences, University of Chittagong, Bangladesh.

Md. Amin U. Mridha is Professor, Department of Botany, University of Chittagong, Bangladesh.

Hui-lian Xu is Senior Crop Scientist, International Nature Farming Research Center, Nagano 390-1401, Japan.

Address correspondence to: Uma K. Aryal at the above address.

[Haworth co-indexing entry note]: "Nodulation Status and Nitrogenase Activity of Some Legume Tree Species in Bangladesh." Aryal, Uma K. et al. Co-published simultaneously in *Journal of Crop Production* (Food Products Press, an imprint of The Haworth Press, Inc.) Vol. 3, No. 1 (#5), 2000, pp. 325-335; and: *Nature Farming and Microbial Applications* (ed: Hui-lian Xu, James F. Parr, and Hiroshi Umemura) Food Products Press, an imprint of The Haworth Press, Inc., 2000, pp. 325-335. Single or multiple copies of this article are available for a fee from The Haworth Document Delivery Service [1-800-342-9678, 9:00 a.m. - 5:00 p.m. (EST). E-mail address: getinfo@haworthpressinc.com].

(2080 nmole C_2H_4 h^{-1}). Seedling height, nodule fresh weight, root fresh weight and nitrogenase activity (per nodule per h, per gram nodule fresh weight per h, per gram root fresh weight per h and per gram root dry weight per h) were also highest in *L. leucocephala*. *[Article copies available for a fee from The Haworth Document Delivery Service: 1-800-342-9678. E-mail address: getinfo@haworthpressinc.com <Website: http://www.HaworthPress.com>]*

KEYWORDS. Leguminous tree, nitrogenase activity, nitrogen fixation, nodulation, *Rhizobium*, soil degradation

INTRODUCTION

Leguminosae is the third largest family of flowering plants with approximately 750 genera and 20,000 species distributed worldwide (Dixon and Wheeler, 1986). At present more than 640 tree species are known as nitrogen-fixing legumes (Halliday and Nakao, 1982). The ability of many tree species to form nodules and fix atmospheric nitrogen symbiotically in response to infection by *Rhizobium* imparts considerable ecological and agronomic importance to the family (Mahmood, 1994). Nitrogen-fixing tree species (NFTS) are an ideal class of trees for afforestating degraded sites (Mac Dickens, 1994) because they are able to establish and thrive in nitrogen deficient soils. In addition to their nitrogen-fixing capacity, NFTS grow quickly and tolerate a variety of adverse soil conditions. It is widely believed that 75% of nitrogen is contributed by the root nodules of leguminous plants (Lawrie, 1981). The economic utilization of NFTS is ranked with grasses in their importance (Allen and Allen, 1981; Mahmood, 1994) and the introduction of legumes into forest ecosystems holds some promise for maintaining soil nitrogen without the use of inorganic nitrogenous fertilizer (Mishra and Prasad, 1980; Prichett, 1979). They may be useful for revitalization of impoverished soils by incorporation of organic matter with wide carbon to nitrogen ratios and for transfer of minerals from deep layers to the surface layers of the soil (Felker and Bandurski, 1979).

To increase the efficacy of biological nitrogen fixation (BNF), it is needed to study the nodulation status and nitrogen-fixing ability of the existing legume flora in different parts of the world (Mahmood, 1994). The global study on nodulation compiled by Allen and Allen (1981) show that at the species level only 15% of the legumes have been examined and of them 30% of the species in Caesalpiniaceae, 90% in Mimoceae and 98% in Papilionaceae were found to be nodulated. Although tree legumes are used for a wide range of purposes and even planted extensively in the field, the nodulation potential of many tree legumes has not been studied Legumes have formed a large part of the tree flora of Bangladesh and planted extensively for multipurpose uses. Information about tree legumes in Bangladesh were scattered in the literature; but now 68 tree species of Leguminosae in 34 genera have been listed consulting different literature (Khatun, 1987). However, the gap of knowledge about the nodulation potential and nitrogen-fixing ability of legume tree species occurring in Bangladesh is immense and extensive study is yet to be started. Therefore, the purpose of this study was to assess the nodulation potential and nitrogen-fixing ability of some promising legume tree species used extensively for afforestation programs in Bangladesh.

MATERIALS AND METHODS

Nodule Morphology

Seedlings were collected from the nursery of the Institute of Forestry and Environmental Sciences, Chittagong University (IFESCU) and Bangladesh Forest Research Institute (BFRI), Chittagong. Seedlings were randomly selected with the intact root systems, removed from the polyethylene bags, and washed in running water to remove soil particles from the roots. Then a morphological study of nodules was conducted on five randomly-selected seedlings of each species. Special care was taken to distinguish root nodules from malformations such as those caused by nematodes, insects or other pathogenic organisms. The presence or absence of nodules and degree of nodulation on the root systems were recorded. The color, size, shape, structure and distribution of the nodules were also noted. The degree of nodulation was classified as sparse (1-30 nodules/seedling), moderate (31-50 nodules/seedling) and abundant (more than 50 nodules/seedling).

Measurement of Nitrogen Fixing Ability

Nitrogenase activity of the nodules of seven tree species, i.e., *Acacia auriculiformis, Acacia mangium, Dalbergia sissoo, Acacia catechu, Leucaena leucocephala, Albizia procera* and *Albizia lebbeck* was determined by assaying the reduction of acetylene as described by Hardy et al. (1968). Seedlings were grown in the nursery of the Institute of Forestry and Environmental Sciences, Chittagong University. Soil samples were collected from the bottom of the hills of the Chittagong University Campus at different locations, well-sieved (< 3 mm) and mixed with cow manure at a ratio of 3:1. These hills consist of moderate to strongly acidic soils of loam to sandy clay loam in texture (Osman et al., 1992) with an average pH of 5.5 (Badruddin et al., 1989). Seeds were obtained from BFRI. Five seedlings of each species (six-month old) were randomly selected and removed from the culture bags after measuring plant height. Seedlings were washed in tap water and then in distilled water several times to remove soil particles. The root nodules were counted with their fresh weight recorded, kept in Erlenmeyer flask with 10 ml of acetylene (C_2H_2) added and incubated for one hour. The ethylene formed by the reduction of acetylene was measured by removing 1 ml of the gas mixture from the flasks and analyzed by gas chromatography. The ethylene was quantified from the peak height after reference to a standard ethylene gas. Fresh weight and dry weight of roots were also recorded (oven-dried at 80°C for 48 hours). The nitrogenase activity was expressed in terms of nmole C_2H_4 per plant per h, per nodule per h, per gram root fresh weight per h and per gram root dry weight per h.

Seedlings of *Cassia fistula, Cassia siamea* and *Delonix regia* were also raised in the nursery with the soil inoculated with *Rhizobium* broth at the rate of 10 ml per pot to study nodulation. *Rhizobium* strains were isolated from the nodules of *A. procera*, *A. lebbeck*, *L. leucocephala*, *A. mangium*, and *A. auriculiformis* in Yeast Extract Mannitol Agar (YEMA) media as described by Vincent (1970). Then, these *Rhizobium* isolates were mixed together in YEM broth to produce inoculum and inoculated into the soil before seeds were sown. Fifteen-day old seedlings were again inoculated with the same *Rhizobium* broth at a rate of 10 ml per seedling. Seedlings were examined 45, 60 and 90 days after inoculation, but no nodulation was observed.

RESULTS

The nodulation status, color, shape, size, structure and distribution of nodules of the species surveyed are reported in Table 1 and measurements of height, nodule number, nodule fresh weight, root fresh and dry weight and nitrogenase activity are presented Table 2. All species of Mimoceae and Papilionacae were nodulated whereas all members of Caesalpiniaceae were non-nodulated. *Albizia procera, Albizia lebbeck* and *Dalbergia sissoo* were moderately-nodulated and sparse nodulation was only recorded in *Acacia catechu*. In the other species, abundant nodules were found. Most of the nodules were brown to pink in color and the shape varied from globose to semi-globose and elongate to elongate with branching. The size varied from 2.1 mm × 3.2 mm in *L. leucocephala* to 5.1 mm × 7.6 mm in *A. mangium*. Variations in nodule structure and distribution were also observed among the species studied (Table 1).

The highest number of nodules were recorded in *L. leucocephala* (82) followed by *A. auriculiformis* (55), *A. mangium* (52), *A. lebbeck* (46), *A. procera* (41), *D. sissoo* (32) and *A. catechu* (29). Nitrogenase activity per

TABLE 1. Nodulation status, color, shape, size (mm), structure and distribution of nodules of some legume tree species.

No.	Botanical name	Nodulation status	Color	Shape	Size (mm)	Structure	Distribution
	FAMILY MIMOCEAE						
1.	*Albizia lebbeck* (L.) Benth	Moderate	Brown to white	Globose	3.1 × 3.6	Advance determinate	Secondary roots
2.	*Albizia procera* (Roxb.) Benth	Moderate	Brown	Semi globose	2.9 × 4.1	Advance indeterminate	Secondary roots
3.	*Samania saman* (Jacq.) Merr.	Abundant	Brown to dark brown	Elongate to elongate with clusters	3.0 × 4.2	Advance indeterminate	Secondary roots
4.	*Albizia falcataria* (L.) Forsberg.	Abundant	Brown	Elongate to elongate. with branching; some are semi globose type	2.1 × 4.3	Advanced indeterminate	Primary roots
5.	*Acacia mangium* Willd.	Abundant	Pink	Elongate to elongate with branching	5.1 × 7.6	Advance indeterminate	Secondary roots
6.	*Acacia auriculiformis* A. Cunn. Ex. Benth.	Abundant	Pink	Elongate to elongate with branching	5.0 × 7.2	Primitive indeterminate to advanced indeterminate	Secondary roots
7.	*Acacia catechu* Willd.	Sparse	Brown	Globose to semi-globose	2.4 × 2.9	Primitive to advanced indeterminate	Secondary roots
8.	*Leucaena leucocephala* (Lam.) de Wit	Abundant	Pink	Elongate to elongate with cluster	2.1 × 3.2	Primitive to advanced indeterminate	Primary and secondary root
	PAPILIONACEAE						
9.	*Sesbania sesban* (Linn.) Merr	Abundant	Pink	Globose with reticulate surface	2.2 × 2.6	Advanced determinate	Secondary roots
10.	*Daibergia sissoo* (Roxb.)	Moderate	Brown	Globose, some are circular	1.9 × 2.3	Aeschynomenoid	Secondary roots
	CAESALPINIACEAE						
11.	*Cassia fistula* Linn.	Not applicable					
12.	*Cassia siamea* (Lamk.)	Not applicable					
13.	*Delonix regia* Bolf Raf	Not applicable					

TABLE 2. Plant height, nodule number, nodule fresh weight, root fresh weight, root dry weight and nitrogenase activity (nmole C_2H_4 produced) of seven legume tree species.

No.	Name of species	Sub family	Plant and root parameters					Nitrogenase activity (nmole C_2H_4 produced)				
			Ht. (cm)	Nod. num.	Nod. Fr. wt. (g)	Root fr. wt. (g)	Root dry wt. (g)	$Plant^{-1}$ h^{-1}	g^{-1} h^{-1} (Fr. nod.)	$Nod.^{-1}$ h^{-1}	g^{-1} h^{-1} (Fr. root)	g^{-1} h^{-1} (Dry root)
1.	*A. auriculiformis*	Mimoceae	79.6	55	1.34	16.4	6.1	1291	963	23.4	78.7	211.6
2.	*A. mangium*	Mimoceae	75.3	52	1.22	12.7	4.9	1125	922	21.6	88.4	229.1
3.	*D. sissoo*	Papilionaceae	66.8	32	0.82	15.4	8.7	366	447	11.5	23.8	42.0
4.	*L. leucocephala*	Mimoceae	105.5	82	1.98	18.4	8.8	4914	2482	59.9	266.9	561.5
5.	*Albizia lebbeck*	Mimoceae	89.6	46	1.12	14.7	6.7	2080	1857	45.2	141.3	309.5
6.	*Albizia procera*	Mimoceae	81.6	41	1.05	16.1	7.8	1888	1797	46.0	117.6	243.6
7.	*Acacia catechu*	Mimoceae	49.6	29	0.345	10.2	5.4	235	680	8.09	23.0	43.5

plant was highest in *L. leucocephala* (4913.59 nmole h^{-1}) and lowest in *A. catechu* (234.7 nmole h^{-1}). Nitrogenase activity per gram nodule fresh weight per h, per nodule per h, per gram root fresh and dry weight per h were also highest in *L. leucocephala*.

DISCUSSION

Compared with the findings of other workers, the results of our study were similar to those of Allen and Allen (1981), Halliday and Nakao (1982) and Faria et al. (1989). The results of Singh and Pokhriyal (1998) and Tewari (1998) are also in agreement with our observations. The reports of Athar and Mahmood (1990), Lim and Burton (1982), Mahmood and Athar (1985), Mahmood and Iqbal (1994) are very encouraging and support our results.

Failure to find nodules on a given plant at any time does not necessarily mean that the plant is always non-nodulated (Mahmood, 1994). The non-nodulating habit of *Cassia fistula, Cassia siamea* and *Delonix regia* confirms the results of Allen and Allen (1981), Faria et al. (1989), and Mahmood and Iqbal (1994). Ty (1996) has also reported the non-nodulating nature of *Cassia* and *Delonix*.

Nodule biomass plays an important role in the nitrogen-fixation activity of the plants. Nitrogenase activity per plant in *Leucaena* was 280.6%, 336.8%, 1242.3%, 136.2%, 160.3% and 1993.5% higher than *A. auriculiformis, A. mangium, D. sissoo, A. lebbeck, A. procera* and *A. catechu*. Pokhriyal et al. (1987) has also reported 221.7%, 262.5%, 133.3% higher nitrogenase activity per plant in *L. leucocephala* than *A. lebbeck, A. catechu* and *D. sissoo*. They also found 64%, 62% and 785.7% higher nodule weight per plant in *Leucaena* compared with *Albizia, Acacia* and *Dalbergia*. In the present study, nodule number per plant and nitrogenase activity per nodule were also markedly higher in *Leucaena* than other species. Tewari (1998) also found the highest nitrogenase activity (nmole C_2H_4 per plant per h) in *L. leucocephala* among several species from a nodulation survey. *L. leucocephala* has been found to develop more effective and active nodules, which is in agreement with the findings of Tewari (1998).

Plant heights also showed a similar trend to that of nitrogenase activity per plant in all seven species. The highest value for plant height was observed in *Leucaena*, which was 32.6%, 40.0%, 17.8%, 29.3%, 57.8% and 112.8% higher than *A. auriculiformis, A. mangium, A. lebbeck, A. procera, D. sissoo* and *A. catechu*, respectively. These findings also support those of Pokhriyal et al. (1987) and Gibson (1976). Nitrogen is the major nutrient required for growth of tree crops. Bino (1998) has reported an increase in mean nitrogen from 0.48% to 0.53%, phosphorous from 6.65 mg per kg to 8.82 mg per kg and mean organic carbon from 6.79% to 6.81% in the surface soil after NFTS were planted. Nitrogen-fixing efficiency of the legumes provides a substantial amount of fermentable organic matter for satisfactory microbial activity (Perera et al., 1992). Solubility of nitrogen is also an essential factor that determines the rate of microbial activity (Hoover, 1986). Growing legume plants in eroded and degraded lands increases the solubility rate of nitrogen and hence microbial activity. Greater nitrogen accretion due to legumes has been reported recently by Singh and Pokhriyal (1998). Higher accretion of nitrogen ensures higher photosynthetic rates and availability of more photosynthates in roots results in more energy available for nitrogen fixation and nitrate assimilation in the root nodules (Tewari, 1998) and as a consequence greater growth and nodulation.

Gordon and Wheeler (1978) have reported a significantly positive correlation of net rate of photosynthesis with both nodule fresh weight per plant and nitrogenase activity in *Alnus glutinosa*. The amount of photosynthates available is considered to be one of the major factors controlling rates of nitrogen-fixation. Increases in height, nodulation and nitrogenase activity in *Leucaena* may also be attributed to the availability of more effective strain of *Rhizobium* in the soil. The metabolites produced by these microorganisms increase root cell permeability resulting in enhanced root exudation (Barber and Martin, 1976). Increased root exudation would also mean that compounds, such as flavonoids that are involved in triggering nodulation activity in *Rhizobium* (Phillips et al., 1995), would be more readily available, leading to higher densities of *Rhizobium* nodules per gram of root. In addition, increased leaching of root exudates implies that plants divert a greater proportion of the available photosynthates to the root systems (Andrade et al., 1998). This would mean that more photosynthates would be available for the *Rhizobium* population infecting these roots. A higher nodulation could result in a higher turnover of nitrogen in *Leucaena*, a greater rhizobial activity in soil and greater amounts of nitrogen and phosphorus to increase soil fertility. This case may be reverses in other species showing poor performance.

CONCLUSION

The environmental benefits from using biological nitrogen-fixation are seen to be associated with the replacement of chemical based technologies with a biological system. There is an increasing awareness in many areas that the development of ecologically-sustainable production systems are essential for maintenance of long-term production at sufficient levels to meet increasing demands from increasing populations. On the other hand, deforestation, soil erosion, monocropping, unplanned management of forest land and several other factors have caused serious degradation of our forest lands. Researches have shown that declines in soil fertility due to land degradation can be checked and soil sustainability can be maintained by planting nitrogen-fixing tree species. Knowledge of the symbiotic association of different microorganisms with tree legumes is still very limited and more extensive research is needed to find ways and means of exploiting legume trees in afforestation programs For achieving successful afforestation, we should screen and grade all the nitrogen-fixing tree species depending upon their nitrogen-fixing capacity at different sites. As the species included in the present study are extensively used in different forestry programs in Bangladesh, these findings would be useful for successful afforestation.

REFERENCES

Allen, O. N. and E. K. Allen. (1981). *The Leguminosae: A Source Book of Characteristics, Uses and Nodulation*. Madison: The University of Wisconsin Press.

Andrade, G., F. A. A. M. De Leij, and J. M. Lynch. (1998). Plant mediated interactions between *Pseudomonas fluorescens*, *Rhizobium leguminosarum* and arbuscular mycorrhizae on pea. *Letters in Applied Microbiology* 26: 311-316.

Athar, M. and A. Mahmood. (1990). A qualitative study of the nodulating ability of legumes of Pakistan, List 4. *Tropical Agriculture* (Trinidad) 67: 53-56.

Badruddin, A. Z. M., M. K. Bhuiyan and M. Mustafa. (1989). Comparative growth study of three fuel-wood species grown in the agricultural marginal lands of Chittagong University Campus. *Chittagong University Studies, Part II: Science* 13: 137-142.

Barber, D. E. and J. K. Martin. (1976). The release of organic substances by cereal roots in soil. *New Phytologist* 76: 69-80.

Bino, B. K. 1998. Biomass yield and nitrogen-fixing trees and shrubs in Papua New Guinea. In *Nitrogen Fixing Tree for Fodder Production*, eds. T. N. Daniel and J. M. Roshetko. Arkansas: FACT Net, Winrock International, pp. 86-99.

Dixon, R. O. D. and C. T. Wheeler. (1986). *Nitrogen Fixation in Plant*. New York: Blackwell Chapman and Hall.

Faria, S. M. De, G. P Lewis, J. I. Sprent and J. M. Sutherland. (1989). Occurrence of nodulation in the Leguminosae. *New Phytologist* 11: 609-619.

Felker, P. and R. S. Bandursk. (1979). Potential uses of leguminous trees for minimal energy input in agriculture. *Economic Botany* 33: 172-184.

Gibson, A. H. (1976). Recovery and compensation by nodulated legumes to environmental stresses. In *Symbiotic Nitrogen Fixation in Plants*, ed. P. S. Nutman. London: Cambridge University Press, 385 p.

Gordon, J. C. and C. T. Wheeler. (1978). Whole plant studies on photosynthesis and acetylene reduction in *Alnus glutinosa*. *New Phytologist* 80: 179-186.

Halliday, J. and P. L. Nakao. (1982). *The Symbiotic Affinities of Woody Plants under Consideration as Nitrogen Fixing Trees*. Honolulu: University of Hawaii, 85 p.

Hardy, R. W. F., E. R. Holsten, E. R. Jackson and R. C. Burns. (1968). The acetylene-ethylene assay for nitrogen fixation: laboratory and field evaluation. *Plant Physiology* 43: 1185-1207.

Hoover, W. H. (1986). Chemical factors involved in nylon fiber digestion. *Journal of Dairy Science* 69: 55-66.

Khatun, B. M. R. (1987). *Tree Legumes of Bangladesh* (Bulletin 4). Forest Botany Division, Forest Research Institute (FRI), Chittagong, Bangladesh, 86 p.

Lawrie, A. C. (1981). Nitrogen fixation by native Australian legumes. *Australian Journal of Botany* 29: 143-157.

Lim, G. and T. C. Burton. (1982). Nodulation status of the Leguminosae. In *Nitrogen Fixation*, Volume II, ed. W. J. Broughton. Oxford: Clarendon Press, pp. 1-34.

MacDickens, K. G. (1994). *Selection and Management of Nitrogen Fixing Trees*. Morrilton (Arkansas): FAO Bangkok and Winrock International.

Mahmood, A. and P. Iqbal. (1994). Nodulation status of leguminous plants in Sind. *Pakistan Journal of Botany* 26: 7-20.

Mahmood, A. and M. Athar. (1985). Nodulation studies on legumes of Pakistan. In *Nitrogen and Environment*, eds. K. A. Malik, M. Naqvi and M. I. H. Aleam. Faisalabad (Pakistan): NIAB, pp. 225-239.

Mahmood, A. (1994). Nodulation and nitrogen fixation in some tree legumes of Pakistan. In *Plants for Human Welfare*, eds. S. Hadiuzzaman, A. A. A. Mushi, N. Sadaque, M. S. Khan and A. Aziz. *Proceeding of the Eight Biennial Botanical Conference*, 12-13 December, 1994, Dhaka, Bangladesh, pp. 40-45.

Mishra, J. and U. N. Prasad. (1980). Agri-silvicultural studies on raising of oil seeds like *Sesamum indicum* Linn. (Til), *Arachis hypogeae* Linn. (Groundnut) and *Glycine max* (soybean) as cash crops in conjunction with *Dalbergia sissoo* and *Tectona grandis* Linn. at Mondar (Rachi). *Indian Forester* 106: 675-695.

Osman, K. T., M. S. Islam and S. M. S. Haque. (1992). Performance of some fast growing trees in the University of Chittagong, Campus. *Indian Forester* 118: 858-859.

Perera, A. N. F., V. M. K.Yaparatne and J. V. Bruchem. (1992). Characterization of protein in some Sri Lankan tree fodder and agro-industrial by products by nylon bag degradation studies. In *Livestock and Feed Development in the Tropics*, eds. M. N. M. Ibrahim, R. D. Jong, J. Van Vruchem and H. Purnomoi. *Proceedings of the International Seminar in Malang*, Indonesia, October, 1991.

Phillips, D. A., W. R. Streit, C. M. Joseph. (1995). Plant signals to soil microbes: regulation of rhizosphere colonization. In *Nitrogen Fixation: Fundamentals and Applications*, eds. I. A. Tikhonovich, N. A. Provorov, V. I. Romanov and W. E. Newton. Dordrecht: Kluwer Academic Publishers, pp. 293-297.

Pokhriyal, T. C., A. S. Ratusi, S. P. Pant, S. P. Pande and S. K. Bhatnaghar. (1987). Nitrogen fixation in *Albizia*, *Acacia*, *Dalbergia* and *Leucaena leucocephala*. *Indian Forester* 5: 366-369.

Prichett, L. W. (1979). *Properties and Management of Forest Soils*. New York: John Wiley and Sons.

Singh, K. C. and T. C. Pokhriyal. (1998). Effects of some leguminous and non-leguminous nitrogen fixing herb, shrub, climber and tree species on soil accretion. *Annals of Forestry* 6: 119-122.

Tewari, D. N. 1998. Nitrogen fixing tree species research. In *Nitrogen Fixing Trees for Fodder Production* (Forest Farm and Community Tree Research Reports–special issue), eds. J. N. Daniel and J. M Roshetko. Arkansas: FACT Net, Winrock International, pp. 244-256.

Ty, H. X. (1996). Symbiont screening for nitrogen fixing trees on strong acid and acid sulfate soils in Vietnam. In *Use of Mycorrhizae and Nitrogen Fixing Symbionts in Revegetating Degraded Sites* (FORSPA publication No. 14), ed. FORSPA. Bangkok (Thailand): FORSPA, pp. 51-56.

Vincent, J. M. (1970). *A Manual for the Practical Study of Root Nodule Bacteria*. Oxford: Blackwell Scientific Publication.

Application of Microbial Fertilizers in Sustainable Agriculture

Zhengao Li
Huayong Zhang

SUMMARY. This paper summarizes the conception and development of microbial fertilizers and inoculants in China that are used to improve soil quality and the growth, yield and protection of crops. The results presented show the relative effects of these microbial products applied singly or with chemical fertilizers on soil fertility, crop yield and protection. The paper discusses current trends in the development and use of microbial fertilizers in China and suggests areas of high priority research. *[Article copies available for a fee from The Haworth Document Delivery Service: 1-800-342-9678. E-mail address: getinfo@haworthpressinc.com <Website: http://www.HaworthPress.com>]*

KEYWORDS. Nodule bacteria, phosphobacteria fertilizer, silicate bacteria fertilizer, compound microbial fertilizer, azotobacter fertilizer

INTRODUCTION

Soil microorganisms play important roles in controlling plant diseases, eliminating plant pests and converting some nutrients into more available form for plants. They also have vital functions in promoting soil fertility,

Zhengao Li and Huayong Zhang are Soil Scientists, Institute of Soil Science, Academia Sinica, P.O. Box 821, Nanjing 210008, China.

[Haworth co-indexing entry note]: "Application of Microbial Fertilizers in Sustainable Agriculture." Li, Zhengao, and Huayong Zhang. Co-published simultaneously in *Journal of Crop Production* (Food Products Press, an imprint of The Haworth Press, Inc.) Vol. 3, No. 1 (#5), 2000, pp. 337-347; and: *Nature Farming and Microbial Applications* (ed: Hui-lian Xu, James F. Parr, and Hiroshi Umemura) Food Products Press, an imprint of The Haworth Press, Inc., 2000, pp. 337-347. Single or multiple copies of this article are available for a fee from The Haworth Document Delivery Service [1-800-342-9678, 9:00 a.m. - 5:00 p.m. (EST). E-mail address: getinfo@haworthpressinc.com].

decomposing wastes and detoxifying polluted soils (Chen and Hu, 1990). Today, the application of beneficial microorganisms as fertilizers and inoculants is receiving considerable attention worldwide. Many kind of microbial fertilizers have been developed and applied. They are often referred to as bio-fertilizer or microbial inoculant, and are comprised of living microorganisms that can function as chemical fertilizer adjuvant, biocontrol agents and plant growth factors. All of these functions require the action of living microorganisms (Chen, 1996).

The study of manufacture and application of microbial fertilizers in China has been underway since the 1950s. At that time, the effects of microbial fertilizer could not be fully comprehended and properly evaluated for lack of basic and applied research. There were two opposing opinions on microbial fertilizers: (1) that microbial fertilizers would take the place of chemical fertilizers due to their significant effects in agriculture; and (2) that microbial fertilizers would not show positive effects because these beneficial microorganisms are already present in the soil. This opposing viewpoint resulted in a sporadic development and production of microbial fertilizer for a long time. Since 1980s interest in the development of microbial fertilizers has increased. For example, the National Specialized Conference in 1995 encouraged the development of microbial fertilizers and now these products have wide applications. It is estimated that there are more than 100 corporations that manufacture microbial fertilizers and their combined annual output is 100 thousand tons (Chen and Xiong, 1997). It has been reported that microbial fertilizers can increase crop yields, improve biodiversity and soil fertility, reduce the need for chemical fertilizers, recycle organic wastes and, consequently, abate environmental pollution. Microbial fertilizers are ideal fertilizers for "green food" and using microbial fertilizers is a simple and economical way to improve ecological agriculture.

The objectives of this chapter are to summarize the development and types of microbial fertilizers, review the application and effects of microbial fertilizers in China, and discuss the prospects for application of microbial fertilizers in sustainable agriculture.

TYPES OF MICROBIAL FERTILIZERS

The followings are the main types of microbial fertilizers and inoculants used in China today.

Nodule bacteria fertilizer. Nodule bacteria fertilizer is the microbial fertilizer with the longest history of use and beneficial effects (Olson, 1989; Song, 1994). Nodule bacteria fertilizer is derived from classical research on symbiotic nitrogen fixation by leguminous plants. Functional bacterial species of the genera *Rhizobium* and *Bradyrhizobium* infect the roots of leguminous plants forming nodules within which the bacteria fix atmospheric nitrogen (N) into plant-available form. The main species are *Bradyrhizobium japonicum* (*Sinorhizobium fredii/Rhizobium fredii*), *Bradyrhizobium* sp. (*Arachis hypogaea*), *Rhizobium huakuii*, *Rhizobium leguminosarum* bv. *viceae*, *Rhizobium* sp. (*astraglas*), *Rhizobium meliloti*, *Rhizobium trifolii*, *Rhizobium leguminosarum* bv. *phaseoli*, and *Bradyrhizobium* sp. (*vigna*) (Chen et al., 1981; Lou, 1962).

Azotobacter fertilizer. Azotobacter–the functional bacteria of azotobacter fertilizer, performs nitrogen fixation in a free living condition. The azotogen has lower efficiency of nitrogen-fixation than *Rhizobium*. However, it has no seasonal limit to its N-fixing activities and applications. There are various naturally-occurring species of azotobacters in nature. The most frequently used in China are *Azotobacter chroococcum*, *Klebsiella pneunoniae*, *Alcaligenes faecalis*, and *Enterobacter cloacae*. There are still some diazotrophic biocoenosis bacteria such as *Azospirillum* spp. and some photoautotrophs as *Rhodospirillum* and *Cyanobacteria*.

Phosphobacteria fertilizer. Some microorganisms such as *Bacillus megaterium*, *Bacillus cereus*, *Bacillus firmus*, *Bacillus brevis*, *Thiobacillus thiooxidans*, species of *Pseudomonas* and *Arthrobacter* can decompose phosphate (Ge and Wu., 1996). In recent years, vesicular-arbuscular mycorrhizae have been used as effective phosphobacterin sources to improve phosphorus status in the rhizosphere (Zhang et al., 1994; Li et al., 1994; Liang, 1994; Lin et al., 1994).

Silicate bacteria fertilizer. It is also known as bio-potash fertilizer. The functional bacteria are *Bacillus mucilaginosus*, *Bacillus circulans* and *Bacillus macerans*. They could decompose potassium minerals into soluble potassium ions. When applied to potassium-deficient soils, crop yields would be expected to increase. But the exact mechanism of this process needs further research (Xiong et al., 1993; Peng and Ye, 1995).

Antibiotic bacteria fertilizer. Besides improving soil fertility, this type of microbial fertilizer protects crops from diseases and promotes plant growth. A strain of antibiotic-producing bacteria–*Actinomyces microflavus* designated as "5406" has been widely used in China (Wu et al., 1994). The Plant Growth Promoting Rhizobacteria (PGPR) which is still under study belongs to this type of microbial fertilizer. The functional bacteria are wide-ranging and distributed over many Gram-negative genus including *Bacillus* and *Pseudomonas* (Weller, 1988). While some strains in this genus are PGPR, others may be Deleterious Rhizosphere Microorganisms (DRMO) (Schippers et al., 1987; Weaver and frederick, 1974). PGPR has given good results of practical application and experiments (Wu and Li, 1994) and has shown promise as a microbial fertilizer (Hu, 1995; Zhang et al., 1996).

Compound microbial fertilizer. In the recent years, many "new" compound microbial fertilizers have been developed. According to their components, they can be subdivided as follows:

- *Inoculants with mixed strains*. This was formulated by mixing different strains of the same functional bacterium. For example, Xu et al. (1991) selected three strains of *Bacillus cereus* to produce a compound inoculant.
- *Inoculants with different bacteria*. Examples involve mixing *Azotobacter* species with phosphobacteria or mixing potash bacteria with other bacteria. A new inoculant of this type involves a mixture of species of photosynthetic bacteria, lactic acid bacteria, yeasts and actinomycetes, which is called Effective Microorganisms or EM (Higa, 1996).
- *Functional bacteria combined with organic fertilizer or chemical fertilizer*. Microbial organic blended fertilizer, also called bio-active compound fertilizer, consists of microorganism, macro- and micro-elements, rare-earth elements and organic wastes (Deng et al., 1993; Yu et al., 1998; Chen et al., 1998). However, additional scientific research and development are needed to ensure proper composting of this type of organic fertilizer. The all-important principle here is to avoid antagonism among the selected microorganisms.

EFFECTS OF MICROBIAL FERTILIZER

The various types of microbial fertilizers can be produced as liquid, powder or granular products. Some of them can be used as basal fertilizer dressings, while others can be seed-coated or sprayed onto plant leaves. In the 1960s, *Rhizobium huakuii* was used as a fertilizer by Huazhong Agricultural University and the results showed yield increases of 38% and 10%, respectively, on new and old farmlands (Chen, 1994). In a study by Institute of Soil Science, Academia Sinica, *Rhizobium leguminosarum* bv. *viceae* was inoculated on vetch seeds and the yield increased by 20 to 50%, and in one case by 100% (Chen, 1994). In 1997, Cao (1997) inoculated *Rhizobium* to 4 kinds of different green manure crops and found that their yields increased by 18.1 to 44.2% and the total amount of N increased by about 30 kg ha^{-1} (Table 1). In recent years, the use of green manure crops has decreased and, therefore, application of *Rhizobium* inoculants to green manure crops is less than that in the 1950s. However, the application of inoculants to legume crops such as peanut and soybean has increased steadily. It was reported that the acreage of *Bradyrhzobium* sp. (*Arachis hypogaea*) application was about 670 thousand hectares for the period of 1949 to 1955. During the next 30 years, the acreage increased to 2.28 million hectares and the average yield increase exceeded 5%. In 1983 to 1985, the inoculated area in Jiangsu Province was 33.3

TABLE 1. Effects of *Rhizobium* inoculation on green manure crops.

Plant	Treatment	kg ha^{-1}			
		Fresh Weight	Dry Weight	N Content	N Amount
Trifolium incarnatum	Control	13597.5	3024.0	18.27	55.20
	Inoculated	19612.5	4548.0	17.94	81.60
Trifolium repens	Control	12973.5	2958.0	22.60	66.90
	Inoculated	16294.5	3213.0	26.03	83.70
Astragalus sinicus	Control	11262.0	2014.5	15.47	31.20
	Inoculated	13909.5	2304.0	26.83	61.80
Vicia villosa	Control	12387.0	3069.0	22.47	69.00
	Inoculated	14634.0	3481.5	30.13	104.85
Vicia sativa	Control	5695.5	1461.0	20.13	29.40
	Inoculated	7542.0	1789.5	23.45	42.00

Adapted from Cao, 1997.

thousand hectares and the yield increase was 11%. Obviously, nitrogen-fixing bacteria have great potential to increase the yield of crops (Zhang et al., 1996).

In the recent years, the application of silicate bacteria has been extended from cotton and tobacco to rice, wheat, corn, peanut, soybean, sweet potato, potato, apple, peach, grape, tomato, cabbage, chili and beanstalk (Liu et al., 1996). Now silicate bacteria is used as a fertilizer for ten different crops in more than twenty provinces with a total treated area of 2.67 million hectares. According to statistical data, crops yields increased by about 13%; the yield of economic plants by 10 to 25%, and the yield of melon and vegetables by 20 to 30%, from application of silicate bacteria. All those practices received great economic benefit. It is shown in Table 2 that the nutrient contents of plants were improved as bio-potash fertilizer was used (Liu et al., 1996).

Table 3 and Table 4 show that the application of a microbial fertilizer compounded with *Bradyrhzobium* sp. (*Arachis hypogaea*) and *Bacillus subtilis* can reduce the incidence of peanut nematodiasis by 34.5% and increase peanut yield by 23.4% (Li et al., 1998).

With the development of modern agriculture, chemical fertilizers rather than organic fertilizers become the most popular fertilizers. But chemical fertilizers have some negative effects leading to decreased soil organic matter, soil fertility and the quality of agricultural products. Microbial fertilizers mixed with organic-inorganic fertilizers are more efficient than chemical fertilizers alone and show fewer negative effects in agricultural production and the environment. As shown in Table 5, when beneficial microorganisms are added into chemical + organic fertilizer with 10% NPK, the yields of

TABLE 2. Effect of bio-potash fertilizer on the nutrient contents of crops.

Crop	Treatment	N Content	P Content	K Content
Cotton	bio-potash fertilizer	3.89	0.55	2.45
	CK	3.41	0.42	2.06
Tobacco	bio-potash fertilizer	3.03	0.30	3.87
	CK	2.35	0.37	2.36
Corn	bio-potash fertilizer	3.57	0.69	4.53
	CK	3.33	0.57	3.85
Sweet potato	bio-potash fertilizer	2.65	0.74	3.35
	CK	2.59	0.70	2.00

Adapted from Liu et al., 1996

TABLE 3. Effect of a compound microbial fertilizer (*Bradyhrizobium* sp. and *Bacillus subtilis*) on the incidence and control of peanut nematodiasis in four villages.

Site	Index			Effects (%)
	Experiment	Control	Difference	
Village 1	50.95	64.90	13.95	21.50
Village 2	48.00	75.00	27.00	36.00
Village 3	37.50	67.60	30.10	44.44
Village 4	36.65	50.70	14.05	28.00
Total	173.10	258.10	85.00	
Average	43.28	64.53	21.25	34.48

TABLE 4. Effect of a compound microbial fertilizer (*Bradyrhizobium* sp. and *Bacillus subtilis*) on peanut yields (kg ha^{-1}) in four villages.

Treatment	Village 1	Village 2	Village 3	Village 4	Average	Increase (%)
Experiment	2783	1087	3368	2173	2352.0	23.43
Control	2200	665	2745	2011	1905.5	
Difference	578	422	623	160	446.5	

Logarithmic statistic: $t = 8.109 > t_{0.01}5.841$, beyond significance level
Adapted from Li et al., 1998

most crops increased significantly compared with chemical fertilizer alone: rice by 22.2%, peanut by 21.6%, watermelon by 34.3%, tea by 13.9%, rapeseed by 18% and leaf tobacco by 11.7% (Lin et al., 1997).

Compared with chemical + organic fertilizer only which contains the same nutrient level, the addition of microbial fertilizer increased yield by 4.7 to

TABLE 5. Effects on various crop of inorganic, organic and microbial fertilizers.

Crop	Inorganic	Inorganic + Organic ($N + P_2O_5 + K_2O = 10\%$)	Inorganic + Organic + Microbial	Inorganic ($N + P_2O_5 + K_2O = 10\%$)
Rice	4444.5	5191.5	5433.0	5265.0
Wheat	5034.0	5164.0	5608.5	6490.5
Peanut	3645.0	4207.5	4432.5	4431.0
Watermelon	24123.0	26506.5	32419.5	29718.0
Tea	4635.0	5280.0	5625.0	5617.0
Rape	1479.0	1600.5	1744.5	1890.0
Tobacco	2119.5	2160.0	2367.0	2475.0
Cucumber		42004.0	50502.0	
Pimiento		25975.5	28317.0	
Water spinach		13854.0	15328.5	

22.3%. Large yield increase for watermelon and some vegetables were noted (Lin et al., 1997; Lin and Song, 1997). Results showed that the chemical + organic fertilizer inoculated with beneficial microorganisms give significant yield increases. Effective Microorganisms (EM) inoculants are applied in Japan, USA, France, China, Brazil, Thailand and about 40 other countries. The results of experiments on crop growth, poultry and livestock feeding and environmental protection also show good prospects (Li, 1991; Li, 1996).

Microbial fertilizers and inoculants may not always produce beneficial effects or significant results in every experiment. These products must be formulated with stable, efficient and compatible strains of microorganisms. It is vital to maintain proper substrate selection and optimum operating conditions to ensure product quality control with respect to inoculum density (i.e., population) and their activities.

TREND OF DEVELOPMENT

The development of the Chinese microbial fertilizer industry has evolved step-by-step during the past 40 years. The trends of evolvement of name (from nitrogen to bacterial fertilizer, then to microbial fertilizer) and the tendency of development can be summarized as followings (Chen and Ge, 1995):

- *The type of microbial fertilizer from single strain to multi-strains.* It was imperative to compound some kinds of strains because the desired microbial fertilizer efficiency was lacking in some single strains of bacteria. However, future research is needed to verify how many and what type of bacteria would be compatible in mixed cultures and compound fertilizers.

- *The genus of microbial fertilizer from bacteria without gemma to endospore-forming bacteria.* Most microbial fertilizers are nitrogen-fixing bacteria (including nodule bacteria) which belong to the bacteria without gemma. They cannot tolerate high temperature and dry conditions and other adversities. They cannot be stored for a long time and only used in liquid form. Endospore-forming bacteria are easily stored, transported and applied in dust or granule form.
- *The function of microbial fertilizer from simple to multi-function.* Experiences from Plant Growth Promoting Rhizobacteria (PGPR) show that, the microbial fertilizers can enhance fertilizer efficiency, but can also improve plant nutrition, stimulate growth in plants and suppress plant diseases. The multiple functions of microbial fertilizers have become the main focus of microbial fertilizer development.
- *The form of microbial fertilizer from inoculant to a compounded microbial fertilizer.* The inoculation approach often needs a long time to work and may not meet the plants' demands in time. The compound microbial fertilizer and mixed beneficial bacteria with organic or inorganic fertilizer are quick-acting and long-term-acting fertilizers.
- *Extending the scope of microbial fertilizers from leguminous plants to other plants.* The requirement of microbial fertilizer on legume crops is limited because legume crops can be infected by natural *Rhizobium* and the legume planting area is relatively small.

CONCLUSION

Microbial fertilizers have comprehensive and multiple functions that will foster a mode sustainable agriculture, especially in abating environmental pollution and long-term adverse effects on soil fertility. But microbial fertilizers do not work in every situation. The factors that limit its function are: type of soil, pH, soil fertility; plant species, quality and quality control of microbial fertilizers.

In the future, the objective of research on microbial fertilizers should concentrate on increasing crop yield and decreasing adverse social effects of pollution. Several important areas of future research are:

- *The study of basic theory.* For instance, bacteria of the genus *Rhizobium* are very selective in choosing roots of a particular legume species to infect. It is not clear how bacteria recognize host roots. Some research shows that a plant protein belonging to lectins may be involved in the recognition process. Research on lectins may lead to ways and means of "bridging" the cross inoculation barriers in legumes (Subba Rao, 1982).

- *The search for new microorganisms.* In fact, besides Leguminosae, there are species of plants under the following genera which also bear nodules formed by species of actinomycetes: *Coriaria*, *Alnus*, *Myrica*, *Casuarina*, *Hippophae*, *Elaeagus*, *Shepherdis*, *Ceanothus*, *Dryas*, *Purshia*, *Cerocarpus*, *Discaria* and *Arctostaphylos* (Quispel, 1974).
- *The construction of genetic engineering bacteria.* Genetic linkage maps of *Rhizobium* spp. were established before 1979 (Sharnmugam and Hennecke, 1979). This provides concept for scientists to transfer genes of nitrogen-fixation from *Rhizobium* to other bacteria.

REFERENCES

Cao, J. Q. (1997). Effect of *Rhizobium* inoculation on several leguminous green manure crops in a newly cultivated low hilly land of red soil. In *Research on Red Soil Ecosystem* (IV) (in Chinese), Ecological Experimental Station of Red Soil, Chinese Academy of Sciences. Nanchang (China): Jiangxi Science and Technology Press, pp. 223-226.

Chen, C. Y. and S. G. Xiong. (1997). The present and development of microbial fertilizer. *Journal of China Agricultural University* (in Chinese) 2:12-15.

Chen, B. S., J. Z. Hong, J. P. Xu and Y. Zhou. (1998). Application of active microbial fertilizer Biol. I.F.E.™ to cotton fields. *Jiangsu Agricultural Sciences* (in Chinese) 3: 47-49.

Chen, H. K. (1996). On the connotation of the term "microbial fertilizer"–some suggestions for the drawing up of quality standards and management of commodity production. In *Production, Application and Development of Microbial Fertilizers* (in Chinese), ed. C. Ge. Beijing, China: China Agricultural Science and Technology Press, p. 6.

Chen, H. K., F. L. Li, W. X. Chen and Y. Z. Cao. (1981). *Soil Microbiology* (in Chinese). Shanghai, China: Shanghai Science and Technology Press, pp. 132-173.

Chen, T.W. and C. Ge. (1995). The tendency of microbial fertilizer development in China. *Soil Fertilizer* (in Chinese) 6:16-20.

Chen, W. X. (1994). The resource of biological nitrogen fixation in China and its prospect in sustainable agriculture. In *Soil Science and Sustainable Development of Agriculture* (in Chinese), Soil Science Society of China. Beijing, China: China Science and Technology Press, pp. 177-182.

Chen, W. X. and Z. J. Hu. (1990). *Soil and Environmental Microbiology* (in Chinese). Beijing, China: Beijing Agricultural University Press, 288 p.

Deng, B. X., J. C. Zhang and G. L. Zhou. (1993). Applied effects of biological complex fertilizers on orange. *Hubei Agricultural Sciences* (in Chinese) 8:16-19.

Ge, C. and W. Wu. (1994). The production, application and problem of Chinese microbial fertilizer. *Chinese Agricultural Science Bulletin* (in Chinese) 10: 24-28.

Higa, T. (1996). *An Earth Saving Revolution–A Means to Resolve Our World's*

Problems Through Effective Microorganisms (EM). Tokyo, Japan: Sunmark Publishing Inc., 336 p.

Hu, X. J. and X. J. Zhang. (1995). A study on rhizosphere bacterial promoting rape growth. *Hubei Agricultural Sciences* (in Chinese) 3: 24-26.

Liang, X. T. (1994). Investigation on ectotrophic mycorrhizal resource of main tree species in Guangxi and their application. *Acta Pedologica Sinica* (in Chinese) 31 (supplement): 134-140.

Li, H. X., Z. Y. Huang and F. M. Bei. (1998). The preliminary research on the increasing production effects of compound microbial fertilizer in peanut. *Soil Fertilizer* (in Chinese) 2: 35-37.

Li, W. J. and Y. Z. Ni. (1996). *Studies and Applications of Effective Microorganisms Bio-Technique* (in Chinese). Beijing, China: China Agricultural Science and Technology Press, 219 p.

Li, X. L., W. L. Zhou and Y. P. Cao. (1994). Acquisition of phosphorus by VA-mycorrhizal hyphae from soil with high compatibility. *Acta Pedologica Sinica* (in Chinese) 31 (supplement): 195-203.

Li, Z. G. (1996). Practical research on compound formulations of microbial inoculant. In *Production, Application and Development of Microbial Fertilizers* (in Chinese), ed. C. Ge. Beijing, China: China Agricultural Science and Technology Press, pp. 151-155.

Lin, D. Y. and Y. J. Song. (1997). Increase production and exploit prospect of compound bio-fertilizer. *Guangxi Agricultural Sciences* (in Chinese) 6: 283-285.

Lin, X. G., W. Y. Hao, T. H. Wu and Y. Q. Shi. (1994). Biological characters of the 11 different VAM fungal species. *Acta Pedologica Sinica* (in Chinese) 31 (supplement): 114-121.

Liu, R. C., F. T. Li, C. R. Hao, W. Y. Wang, Z. Y. Yang, A. M. Zhang, G. Q. Cao and L. F. Wang. (1996). Application of silicate bacterial inoculant in agricultural production. In *Production, Application and Development of Microbial Fertilizers* (in Chinese), ed. C.Ge. Beijing, China: China Agricultural Science and Technology Press, pp. 66-74.

Lou, L. H., W. A. Mei, S. F. Chen, C. C. Ye and X. Q. Zhou. (1962). *Application of Microorganism on Transform of Soil Nutrition* (in Chinese). Beijing, China: Science Press, pp. 186-194.

Olsen, P. E. and W. A. Rice. (1989). *Rhizobium* strain identification and quantification in commercial inoculants by immunoblot analysis. *Applied and Environmental Microbiology* 55: 520-522.

Peng, S. P. and F. J. Ye (1995). Applied effects of silicate bacterial agent on cotton. *Hubei Agricultural Sciences* (in Chinese) 2: 34-35.

Quispel, A. (1974). The endophytes of root nodules in non-leguminous plants. In *The Biology of Nitrogen Fixation*, ed. A. Quispel. Amsterdam, the Netherlands: A North Holland Publishing Co., pp. 499-520.

Schippers, B., A. W. Bakker and P. A. Bakker. (1987). Interactions of deleterious and beneficial rhizosphere microorganisms and the effect of cropping practices. *Annual Review of Phytopathology* 15: 339-358.

Shanmugam, K. T. and H. Hennecke. (1979). Microbial genetics and nitrogen fixation. In *Recent Advances in Biological Nitrogen Fixation,* ed. N. S. Subba Rao. New Delhi, India: Oxford and IBH Publishing Co.

Song, Y. Q. (1994). The application of bio-fertilizer in Liaoning Province. *Journal of Microbiology* (in Chinese) 14: 45-51.

Subba Rao, N. S. (1982). Outlook for the future. In *Biofertilizers in Agriculture*. New Delhi, India: Oxford and IBH Publishing Co. pp. 153-167.

Weaver, R. W. and L. R. Frederick. (1974). Effect of inoculation rate on competitive nodulation of glycine max L. Merrill: II. Field studies. *Agronomy Journal* 66: 233-235.

Weller, D. M. (1988). Biological control of soil-borne plant pathogens in the rhizosphere with bacteria. *Annual Review of Phytopathology* 26: 379-407.

Wu, G. Y. and H. Q. Li. (1994). The application and function of antibiotic bacterial fertilizer. *Soil Fertilizer* (in Chinese) 3: 45-46.

Xiong, C. C., X. Li, H. Q. Wang and J.C. Wu. (1993). Applied effects of biological potassium fertilizer on paddy soils of red soil in the south of Hubei. *Hubei Agricultural Sciences* (in Chinese) 6:11-13.

Xu, Z. Y., J. O. Zhang, F. H. He, C. S. Zhang and A. Q. Yang. (1991). The effect of liquid preparation of *Bacillus cereus* on rape (*Brassica napus* L.). *Oil Crops of China* (in Chinese) 4:55-58.

Yu, L. P., L. F. Cai, W. C. Wang, D. Y. Lin and Y. C. Liu. (1998). Application research on the compounded and blended biological fertilizer to groundnut. *Guangxi Agricultural Sciences* (in Chinese) 1: 23-24.

Zhang, L. H., Y. Q. Zou and Z. P. Lei (1994). The application of ectomycorrhizal preparation on cultivated seedlings of *Larix olgensis* and *Pinus koraiensis*. *Acta Pedologica Sinica* (in Chinese) 31 (supplement): 164-171.

Zhang, X. P., Q. Chen and D. Y. Li. (1996). Application of microbial fertilizers on "three high" agriculture. *Journal of Soil Agrochemistry* (in Chinese) 11: 43-47.

Zhang, Y. X., D. B. Wang and Y. F. Peng. (1996). Studies on the colonization of the biocontrol agent D_{93} strain on wheat roots. *Journal of Huazhong Agricultural University* (in Chinese) 15:18-23.

Beneficial Microorganisms and Metabolites Derived from Agriculture Wastes in Improving Plant Health and Protection

Dezhong Shen

SUMMARY. There is growing interest in the presence of certain naturally-occurring, beneficial microorganisms in agricultural wastes (e.g., processing wastes, composts and anaerobic slurries) that have considerable potential to enhance the growth, health and protection of crops. The numbers of these organisms can be increased through incubation procedures and they can be applied as mixed or pure culture inoculants to soils and plants in waste materials or as sprays and suspensions. The various mechanisms of disease control and plant protection imparted by these beneficial microorganisms may be related to (a) microbe-microbe interactions, (b) plant-microbe interactions, (c) metabolites produced and /or, (d) induced systemic-acquired resistance. Research is needed to elucidate their exact mechanisms or modes-of-action to determine the optimum time, rate and frequency of application for improving plant health and protection. *[Article copies available for a fee from The Haworth Document Delivery Service: 1-800-342-9678. E-mail address: getinfo@haworthpressinc.com <Website: http://www.HaworthPress.com>]*

KEYWORDS. Beneficial microorganisms, agricultural wastes, compost, anaerobic slurries, plant health, disease control mechanisms, sustainable agriculture

---

Dezhong Shen is Professor, College of Natural Resources and Environment, China Agricultural University, Beijing, China 100094.

[Haworth co-indexing entry note]: "Beneficial Microorganisms and Metabolites Derived from Agriculture Wastes in Improving Plant Health and Protection." Shen, Dezhong. Co-published simultaneously in *Journal of Crop Production* (Food Products Press, an imprint of The Haworth Press, Inc.) Vol. 3, No. 1 (#5), 2000, pp. 349-366; and: *Nature Farming and Microbial Applications* (ed: Hui-lian Xu, James F. Parr, and Hiroshi Umemura) Food Products Press, an imprint of The Haworth Press, Inc., 2000, pp. 349-366. Single or multiple copies of this article are available for a fee from The Haworth Document Delivery Service [1-800-342-9678, 9:00 a.m. - 5:00 p.m. (EST). E-mail address: getinfo@haworthpressinc.com].

INTRODUCTION

Crop health is very important for modern agriculture. It can assure adequate food supplies for meeting the needs of increasing populations. Large quantities of wastes are produced during food production and processing which contain an abundance of organic matter, nutrients and other components that can be recycled beneficially as soil conditioner and biofertilizers. Plants on earth have been nourished with wastes for millions of years since pre-historic times and Asian agriculture has a history of about 4000 years according to King (1911). However, during the last 50 years, modern agriculture has used increasing amounts of chemical fertilizers and pesticides while largely neglecting the recycling of organic wastes. Today, large amounts of wastes pollute the rivers and lakes, degrade the ecosystem and harm human and animal health. This type of agricultural production is not sustainable, neither agronomically nor environmentally. In this view, new approaches to recycling agricultural wastes have been considered, especially methods which are simple, practical, and economical. Microbial technologies, which can link agricultural production closely with waste management, are beginning to receive considerable attention.

Microbial inoculants and their metabolites that can enhance plant disease control and plant health have the following advantages over conventional chemicals: (1) they are considered to be safer than many of the chemicals now in use; (2) they do not accumulate in the food chain; (3) self-replication circumvents repeated application; (4) target organisms seldom develop resistance as often happens when conventional synthetic chemicals are used; and (5) naturally-occurring biocontrol agents are seldom considered harmful to agro-ecosystems (Gould, 1990). Consequently, there is increasing commercial and environmental interest in the use of microbial inoculants or their metabolites as alternatives to synthetic chemicals for controlling the spread and severity of a range of crop diseases (Dowling and O'Gara, 1994). Options include applying these naturally-occurring agents singly or in combination with synthetic chemicals.

This paper reviews some of the beneficial microorganisms and their metabolites that are present in certain agricultural wastes, and their potential roles and mechanisms for controlling plant diseases and improving plant health.

MANUFACTURE AND APPLICATION OF MICROBIAL PRODUCTS FROM AGRICULTURAL WASTES FOR IMPROVING PLANT HEALTH

Microbial Products from Agricultural Wastes

Because of the wide diversity of microbial species, their metabolism and interspecific relationships, certain microorganisms and their metabolites play important roles in providing available nutrients for plant growth ("biofertilizers"), antagonism, competition, and microparasitism against plant pathogens ("biocontrol"), and degradation of pollutants in the environment ("biodegradation") (Shen, 1997). The numbers and activities of such beneficial microorganisms can be maximized in pure cultures, or through natural fermentation of organic wastes including straw, stalk, leaves, bark, sawdust, animal waste, and food processing by-products in either solid or liquid form.

Pure Culture Products. Many types of microorganisms can act as beneficial crop inoculants for plant growth promotion and biological disease control, both topics of which have been extensively reviewed (Baker and Dickman, 1993; Bashan and Levanony, 1990; Kloepper, 1993, Lugtenberg et al., 1991; Lam and Gaffney, 1993; Okon and Hadar, 1987; Schippers et al., 1987; Schroth and Beck, 1990; Walter and Paau, 1993). Research on the application of microbial inoculants in crop production has been extensive. Often the research has involved the utilization of bacteria, e.g., *Pseudomonas* spp., *Bacillus* spp. and fungi such as *Trichoderma* spp. and *Gliocladium virens* (Baker and Dickman, 1993; Chung and Hoitink, 1990; Dowling and O'Gara, 1994; Kwok et al., 1987; Kloepper, 1993; Lam and Gaffney, 1993; Mei, 1991; Walter and Paau, 1993).

Fermentation wastes and some of liquid wastes from food processing, such as starch production and bean curd processing, are excellent media for production of some pure cultures. These wastes are suitable for the growth of many bacteria after certain carbon and nitrogen sources are added and pH adjusted. For example, the liquid waste from bean curd processing contains a total sugar content of 0.4-0.5%, reducing sugar of 0.1%, crude protein of 0.13-0.15%, total nitrogen of 0.2% and total phosphorus of 0.5% (Chen, 1987). Pure cultures of *Bacillus* spp. can occur after adding a carbon source (starch or glucose) adjusting the pH, and incubating the inoculum for 24-36 hr; in this case, the total count of *B. cereus* reached 2-3 $\times$ 10^8 ml^{-1} (Liu and Shen, 1994).

Natural Fermentation Products. The common products of natural fermentation include compost and slurry from anaerobic digestion. Both are effective methods for treating of many organic wastes, especially animal wastes. Diverse microorganisms participate in the degradation and transformation process. The products contain organic matter, nutrients, metabolites and mixed cultures of microorganisms. The composting process is divided into three phases. The first phase occurs during the first 24-48 hours as temperature gradually rises to 40-50°C and when sugars and other easily biodegradable substances are utilized for microbial growth and metabolism. During the second phase, when the temperature range is 40-65°C, cellulose and other more resistant substances are degraded by thermophilic microorganisms. Lignin breaks down even more slowly. Many plant pathogens, weed seeds, and some biocontrol organisms–with the exception of *Bacillus* spp.–are either killed or rendered inviable by the heat generated during this high-temperature phase of the process. The third, or curing, phase begins as the supply of readily biodegradable compounds is exhausted. At this time, the mesophilic microorganisms (i.e., those that function in the range of 20-40° C) recolonize the compost. By this time, humic substances are accumulating in increasing quantities. Biocontrol organisms that recolonize composts after peak heating include bacteria such as *Bacillus* spp., *Enterobacter* spp., *Flavobacterium balustinum*, *Pseudomonas*, actinomycetes such as *Streptomyces* spp., and fungi such as *Gliocladium virens* (Hoitink et al., 1991).

In general, anaerobic digestion occurs in the following stages: liquefaction or polymer breakdown, acid formation, and methane formation. There are four main groups of bacteria involved in the process, namely, (1) acid-forming bacteria, (2) acetogenic bacteria, (3) acetoclastic bacteria, and (4) hydrogen-utilizing methane bacteria (Brown and Tata, 1985). In the first stage, complex organic polymers such as proteins, fats and carbohydrates are broken down and solubilized by extracellular enzymes. In the second stage, the monomeric components are further converted to acetic acid (acetates), hydrogen (H_2) and carbon dioxide (CO_2). Other volatile fatty acids, such as propionic and lactic acid, are also produced. In the third stage, the products of stage 2 are finally converted to methane (CH_4) and other end-products.

Application of the Microbial Products

Pure Culture Products–Yield-Increasing Bacteria (YIB). In China, many microbial products are widely used for plant nutrient sources, plant growth promotion, and plant disease and pest control. The products are often used as substitutes for some pesticides, hormones, antibiotics and chemical fertilizers. One of the best examples is Yield-Increasing Bacteria (YIB) (Mei, 1991; Chen et al., 1996), which is the commercial name for a mixed culture microbial inoculant comprised of different strains of *Bacillus* spp., including *B. cereus*, *B. subtilis*, and *B. firmus*, *B. licheniformis*, *B. coagulans*, *B. brevis*, *B. sphaericus*. The microbial product is applied by either dressing seeds, spraying leaves of crops or soaking roots of crops in a spore suspension. Beneficial effects on the crop include earlier seed germination, larger root systems, and greener leaves. The application of YIB reduces severity of some diseases, including seedling diseases, stem blight, downy mildews, anthracnose rot, blackleg, and dry rot (Chen et al., 1996). The percentage decrease in disease incidence (PDIDI) from YIB treatment, compared to controls, ranged from about 36% in sheath and culm blight of rice (*Rhizoctonia solani*) to 85% in rice seedling blight (*Xanthomonas campestris* pv *oryzae*) (Table 1).

Compost. Composting of organic waste for use as soil conditioners and biofertilizers has been conducted in countries of the Far East, as described in

TABLE 1. Percentage decreases in disease incidence (PDIDI) of crops in China after treatment with yield-increasing bacteria (YIB).

Disease	Pathogen	PDIDI
Anthracnose of apple	*Glomerella cingulata*	80.3
Downy mildew of cucumber	*Pseudoperonospora cubensis*	36.2
Dry rot of Chinese chestnut	*Diplodina* sp.	61.6
Fusarium wilt of cotton	*Fusarium oxysporum* f. sp. *vasinfectum*	50.5
Head smut of corn	*Ustilago maydis*	77.0
Millet downy mildew	*Sclerophthora macrospora*	65.0
Root rot of sweet potato	*Ceratocystis fimbriata*	56.5
Scab of wheat	*Gibberella zeal*	60.2
Seedling blight of rice	*Xanthomonas campestrid* pv *oryzae*	85.5
Seedling disease of cotton	*Rhizoctonia solani*	70.2
Sheath & culm blight of wheat	*Rhizoctonia cerealis*	71.4
Sheath & culm blight of rice	*Rhizoctonia solani*	35.7
Take-all of wheat	*Gaeumannomyces graminis* var. *tritici*	53.7

Adopted from Chen et al., 1996
* PDIPI = percent decrease in disease incidence and calculated by determining the percent reduction in disease on YIB-treated plants compared with non-treated controls in field trials.

Frankin H. King's "Farmers of Forty Centuries" (King, 1911). With the advent of modern farming methods (especially during the last three decades), composting has largely disappeared from farms. Now, with increasing concerns about environmental degradation, commercial compost production from food industry wastes and agricultural wastes and municipal sewage sludge have expanded dramatically. Since the early 1970s, composts have been considered as peat substitutes and used effectively for control of plant diseases caused by soil-borne plant pathogens, such as *Fusarium* spp., *Phytophthora* spp., *Pythium* spp., and *Rhizoctonia solani* (Hoitink, 1991). In the nursery industry, disease-suppressive, compost-amended container media have been effective enough to replace methyl bromide (Quarles and Grossman, 1995). Scientists at the university of Bonn in Germany reported that compost extracts were effective in controlling many plant diseases, such as *Pseudopeziza tracheiphila*, *Plasmopara viticola*, *Uncinula necator*, *Botrytis cinerea* in grape, *Venturia inaequalis* in apple, *Erysiphe polygoni* in bean, *Phytophthora infestans* in potato and tomato, *Erysiphe graminis* in barley, *Botrytis cinerea* in strawberry, *Erysiphe beta* in sugar beet and *Pseudoperonospora cubensis* in cucumber (Budde and Weltzien, 1990; Traenkner, 1993; Weltzien et al., 1987; Winterscheidt et al., 1990). Moreover, some

Chinese researchers have used compost extracts to control plant diseases of cucumber, wheat, tomato and apple (Qiao et al., 1998).

Anaerobically Digested Slurry. Traditionally, products from anaerobic digestion have been used as renewable energy sources and nutrient sources. The slurries have also been found to decrease the incidence of many plant diseases (Table 2). For example, Shi et al. (1991) studied the effect of applications of anaerobically-digested effluents on wheat scab (*Fusarium graminearum*). When effluent is sprayed during full-bloom stage, the disease can be controlled effectively. Their results showed that the disease incidence decreased by 20.7% compared with the control (equal volume of water applied). The extent of biocontrol by the effluent was similar to the effect of the fungicide Benomyl. Li et al. (1992) investigated the use of anaerobic slurry supernatant by soaking sweet potato in storage. After six months, the disease incidence of treated sweet potato decreased by 47-61%, compared with untreated sweet potato. Pot experiments showed that a mixture of digested manure and soil decreased the incidence of watermelon Fusarium wilt. A high rate of digested manure applied to soil was 86% as effective as the non-pathogen control (Table 3). Similar results were obtained in the field test (Figure 1).

Germination power and germination percentage of seeds as well as seed-

TABLE 2. Disease suppression of crops in China after treatment with anaerobic digested slurry.

Disease	Pathogen	Reference
Anthracnose of cotton	*Glomerella gossypii*	Shen (1990)
Blast of rice	*Piricularia oryzae*	Huang et al. (1993)
Fusarium wilt of cotton	*Fusarium oxysporum* f. *vasinfectum*	Wu et al. (1993)
Fusarium wilt of cucumber	*Fusarium solani* f. *cucurbitae*	Zhou (1993)
Fusarium wilt of watermelon	*Fusarium oxysporum* f. *niveum*	Zhao and Zhou (1991) Shen et al. (1995)
Powdery mildew of wheat	*Erysiphe graminis*	Zhou (1993)
Scab of wheat	*Fusarium graminearum* (*Gibberella zeal*)	Shi et al. (1991)
Seedling rot or rice		Shen (1990)
Sheath blight of rice	*Rhizoctonia solani*	Shen (1990)
Soft rot of sweet potato	*Rhizopus nigricans*	Li et al. (1991)
Wilt of potato	*Pseudomonas solanacearum*	Shen (1993)
Yellow mosaic of barley	*Polymyxa gracmimanis*	Shen (1993)

TABLE 3. Incidence and control of watermelon wilt after treating soil with digested manure (Adapted from Shen et al., 1995).

Treatment	Mortality	Incidence	Disease index	Control effect
Non-inoculating pathogen	0	0	0	100
Inoculating pathogen	84	89	0.86	0
Composted manure	56	69	0.62	27
Digested manure (low)	58	89	0.70	18
Digested manure (medium)	31	44	0.36	58
Digested manure (high)	12	12	0.12	86

Note: low, medium and high level digested manure are respectively 1:7, 1:5, 1:3 (digested manure and soil volume ratio), digested manure (water content 29%) was obtained from anaerobic digestion processing at 55°C for 20 days.

FIGURE 1. Effect of anaerobically digested slurry applied to soil on disease index of watermelon wilt (Adapted from Shen et al., 1997)

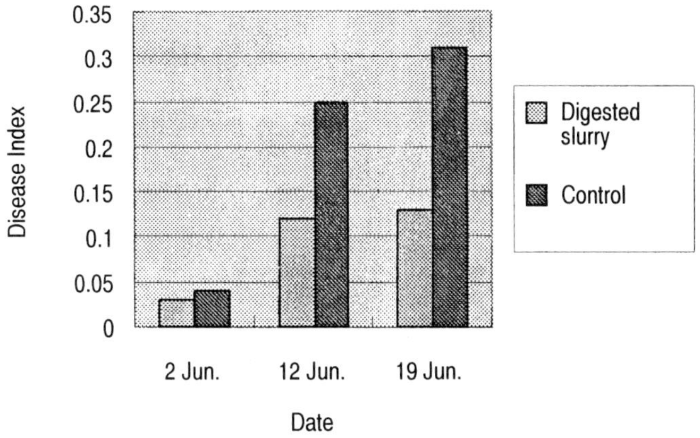

ling growth are increased by soaking seed with anaerobically-digested slurry (Huang et al., 1993). In regard to crop health, seed soaking also increased plant resistance to stress and disease. Spraying the supernatant phase of digested slurry on the leaves of orange trees can prevent freezing injury (Fang et al., 1993). Digested slurry increased the cold-resistance and decay-preventing properties of early season rice, and increased the seedling survival rate by 8-15% (Wu and Wei, 1993). The application of the supernatant of digested slurry can control sucking pests, such as aphids, spider mites and plant hoppers (Yang, 1993; Huang et al., 1993).

MECHANISM FOR CONTROLLING CROP DISEASES

Microbe-Plant and Microbe-Microbe Interactions

The complex relationships between the plant and microorganisms and among the microorganisms themselves constitute the key to plant health and growth. During seed germination and plant growth, a number of soil bacteria and fungi become intimately associated with the developing root-rhizosphere ecosystem. Many of these are soil saprophytes, doing little harm to healthy tissue. However, others cause serious diseases, being pathogenic on a variety of commercially-important plants. Another group of microorganisms living in close association with the plant root-rhizosphere are bacteria that obtain carbon, energy and nutrients from root exudates and in turn provide growth factors and nutrients to the plant (Davison, 1988).

There is a diversity of relationships among the microorganisms. The most obvious way is to exert antagonism toward of phytopathogens, which are common in the soil under certain conditions. Phytopathogenic bacteria occupy a particular ecological niche and beneficial bacteria compete with them for sites and nutrients. Pseudomonads and *Bacillus* spp. catabolize diverse nutrients and have fast generation times in the root zone or on leaves and hence, they are logical candidates as biological control agents through competition for nutrients, especially against the slower-growing pathogenic fungi. Some fungi can parasitize other living microorganisms for nutrients. Perhaps the best known microparasite is the fungus *Trichoderma*, hypha of which may penetrate resting structures or parasitize growing hypha (Campbell, 1989).

Yield-Increasing Bacteria (YIB). Studies indicate that YIB strains could colonize, migrate and multiply both inside and outside plant tissues. Soon after germination of rape seeds (24-48 hours) inoculated with an antibiotic-resistant mutant of YIB strain (No. 83-10), with 1×10^4 to 1×10^5 colony forming units (CFU), were recovered from individual roots (Ji and Chen, 1992). The bacteria from inoculation with strains, No.83-10 and A-47, on the seeds of cabbage, maize and wheat, were able to migrate to and colonize the stems and leaves, reaching a population of 1×10^2 to 1×10^8 CFU per leaf (Ji and Chen, 1992). Yu et al. (1992) have sprayed YIB strain, A-47, on the leaves of wheat and observed a rapid population increase after spraying especially around the stomata, wounds and sunken sites according to scan electron microscopic observation, antibiotic double marker test (rifampicin, streptomycin resistance), and leaf surface replica test. The spore-forming *Bacillus* spp. have metabolic predominance for long periods and occupy ecological niches. They compete with pathogens for space and nutrients and have the ability to suppress pathogen activity. The population of DRB (deleterious rhizobacteria) strains *Bacillus* I-12 and I-14 were greatly reduced in the rhizosphere of rapeseed after seed-soaking with YIB strain 83-10 suspension (Ji and Chen, 1992). Chen and Liu (1985) reported that the application of a YIB strain could eliminate the effect of some DRB pseudomonads in the rhizosphere of turnip. By spraying the suspension onto wheat leaves in the field and then isolating leaf samples weekly, the average numbers of *Cladosporium* spp. and *Alternaria* spp. on the leaves were as reduced by 53.9% and 66.4%, respectively, compared with non-treated check samples (Tang et al., 1992b).

Compost. Numerous reports reveal that microbiostasis (i.e., suppression of growth) and microparasitism (i.e., direct microbial attack) can explain the disease control activity of composts (Chen et al., 1988; Hadar et al., 1992; Kuter et al., 1983; Kwok et al., 1987; Lumsden et al., 1983). Microbiostasis is involved in the suppression induced by compost against *Pythium* spp. and *Phytophthora* spp., but competition for carbon can only partially explain the suppressive effect. *Pseudomonas* spp. provide biological control and predominate in both rhizosphere and edaphic niches of compost-amended substrates suppressive to *Pythium* root rot. In contrast, gram-positive pleomorphic genera and putative oligotrophs that are incapable of inducing biological control are most abundant in consistently conducive, highly decomposed sphagnum peat substrates. Populations of pseudomonads and other biocontrol agents decline as organic matter decomposes, and suppressiveness to *Pythium* root rot is lost (Boehm and Hoitink, 1992; Boehm et al., 1993). In mature compost, sclerotia of *Rhizoctonia solani* are killed by the hyperparasites of *Trichoderma*, and biological control prevails (Chung, and Hoitink, 1990).

Anaerobic Digested Slurry. Guo and Sun (1991) indicated that antagonism against certain plant pathogens by *Pseudomonas* in anaerobic slurry could inhibit infection and disease development.

Metabolites

Metabolites, i.e., anti-microbial compounds, play important roles in disease control. They are closely related to antagonism and competition among microorganisms. Many microorganisms produce anti-microbial compounds, which may act on pathogens by inducing fungistasis (i.e., suppression of growth), inhibition of germination, lysis of fungal mycelia, or by exerting a fungicidal effect (Gould, 1990). Fluorescent pseudomonads produce a variety of metabolites (see Figure 2), many of which are inhibitory to other microorganisms and some which are implicated in the biological control of plant pathogens (Dowling and O'Gara, 1994).

Table 4 lists important metabolites that induce disease suppression. Some of the metabolites implicated in biocontrol appear to be broad-ranging in their inhibitory action. For example, phloroglucinols and phenazines have been shown to inhibit a wide range of fungal pathogens in the laboratory.

FIGURE 2. The range of secondary metabolites and other compounds produced by fluorescent *Pseudomonas* spp. The compounds indicated by an asterisk have been implicated in the biocontrol ability of the producer strain (Adapted from Dowling and O'Gara, 1994).

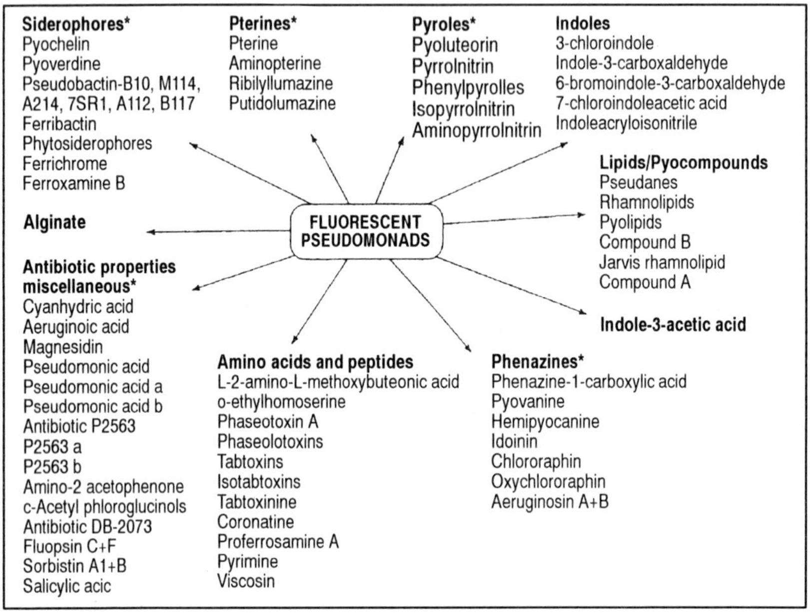

Siderophores exhibit both fungistatic and bacteriostatic effects in the laboratory under conditions of low iron. In the field, these iron-chelating compounds are thought to deprive the pathogen of iron, a limiting essential nutrient. Others have very specific effects and target particular pathogen. For example, agrocin 84, produced by *Agrobacterium radiobacter*, is specific for virulent strains of *Agrobacterium tumefaciens* (Dowling and O'Gara, 1994).

Other mechanisms can inhibit pathogen growth. For example, chitinase, an enzyme, produced by some bacteria can lyse the chitin-containing hyphal walls of a *Fusarium* sp. (Mitchell and Alexander, 1961). Volatile compounds, such as ammonia (NH_3) and hydrogen cyanide (HCN) produced by rhizosphere microorganisms are harmful to certain pathogens (Dowling and O'Gara, 1994).

Many microorganisms produce plant growth regulators (PGRs), including auxins, gibberellins, cytokinins and ethylene. Arshad and Frankenberger (1993) summarized PGR-producing microorganisms, their soil ecology and plant responses to the PGRs.

TABLE 4. Example of specific microbial metabolites implicated in the control of crop diseases.

Disease	Pathogen	Effective metabolite
Take-all of wheat	*Gaeumannomyces graminis* var. *tritici*	Phenazines c-Acetyl Phloroglucinols
Tan spot of wheat	*Pyrenophora triticirepentis*	Pyrrolnitrin
Pre-emergent damping-off of:		
cotton	*Pythium* spp.	Oomycin A
sugarbeet	*Pythium ultimum*	Pyoluteorin, 2,4-Diacetylphloroglucinol
Black root-rot of tobacco	*Thielaviopsis basicola*	Hydrogen cyanide, 2,4-Diacetylphloroglucinol
Crown gall of fruit trees	*Agrobacterium tumefaciens*	Agrocin 84
Flax wilt	*Fusarium oxysporum*	Pseudobactiin B 10
Damping-off	*Pythium* spp.	Ammonia

(Adapted from Dowling and O'Gara, 1994)

Yield-Increasing Bacteria. Some strains of Yield-Increasing Bacteria (YIB) can produce antibiotics and inhibitory substances (Tang et al., 1992c). A liquid culture of YIB strain A-47 inhibited spore germination in *Puccinia striiformis* f. sp. *tritici*, and *Cladosporium* spp. by 71.9% and 87.3%, respectively. The YIB culture also inhibited germ tube growth (Tang et al., 1992a). But inhibitory substances have not been identified.

Some of strains can secrete high concentrations of superoxide dismutase (SOD), an enzyme associated with resistance to stress, disease and aging. Enzymatic analysis showed that YIB strains also contained other enzymes, such as, amylase, proteinase and alcohol dehydrogenase (Mei, 1991).

The culture filtrate of YIB contains phytohormones, such as, indole acetic acid (IAA), zeatin and gibberellin (G3) (Yang et al., 1992). Peng and Xiong (1992) showed that a YIB strain grown on ^{14}C-glucose medium could secrete IAA, G3 and zeatin.

Compost. Inhibition of pathogens by compost extracts has been extensively studied. Qiao and Gao (1998) demonstrated the inhibition of pathogens by compost extract on Petri dishes and slides. Compost extract can directly inhibit growth and spore germination of some pathogens such as *Alternaria mali*, *Botrytis cinerea*, *Colletotrichum gossypii*, *Fusarium graminearum*, *Fusarium oxysporum* f. sp. *cucumerinum*, *Fusarium oxysporum* f. sp. *lycopersici*, *Pseudoperonospora cubensis*, *Puccinia recondita* f. sp *tritici*. However, the exact mechanisms of inhibition have not been identified.

Anaerobic Digested Slurry. Inhibition tests in Petri dishes showed that microbially digested slurry strongly inhibited the growth of plant pathogens *Helminthosporium sativum* and *Rhizoctonia solani*, and moderately inhibited the growth of *H. turcicum*, *H. maydis*, *H. sigmoideum* and *Colletotrichum gossypii* (Mo, 1988). Anaerobically digested slurry was shown to inhibit hyphal growth of *Fusarium graminearum*, as Benomyl does (Zhang, 1994). Shen (1993) found that ethylene and methane produced from digested slurry inhibited the activity of *Polymyza gracimimanis* responsible for the spread of yellow mosaic virus. Li et al. (1992) suggested that the control of sweet potato soft rot in storage by digested slurry depended on the ammonia ion concentration of the slurry.

Digested slurry contains plant hormones such as indole acetic acid (IAA), cytokinin, and gibberellin (Shen, 1990). It also contains fulvic acid, which is a very important active ingredient. Research has shown that fulvic acid has the following functions: (1) anti-transpiration–reducing the stomatal opening of leaves thereby decreasing the transpiration loss of water; (2) growth promotion–stimulating root system development, increasing cereal crop tillering, enhancing photosynthesis, and increasing resistance against stress; (3) chelation–chelating the micronutrients and enhancing the uptake and transportation in the plant; and (4) pesticide activity promotion–increasing the pesticide effect if mixed in pesticide, thereby reducing the dose and toxicity of the pesticide (Wang, 1992; Ma and Guo, 1993). It can also control some diseases, such as apple-tree canker (*Valsa mali*), and cotton wilt disease (*Fusarium oxysporum* f. sp. *vasinfectum* and *Verticillium dahliae*). Fulvic sodium is used to cure wounds after scraping canker of bark (Fan and Wang, 1987). Fulvic acids has been applied with irrigation in the field, sprayed on leaves, or mixed with seeds to control cotton wilt (Wang, 1987). Fulvic acid is a common component of soil organic matter and compared with humic acid, it has unique chemical characteristics, such as smaller molecular weight, more functional groups and is more water soluble, all of which contribute to its unusual physiological interaction with plants. In China, considerable research has been done with fulvic acid derived from weathering coals, and microbial metabolism.

Induce Systemic Acquired Resistance (SAR)

SAR refers to the changes that occur after a plant has been challenged by a potential pathogen. These changes can be either chemical or physical, and it is believed that most plants have the inherent ability to respond defensively by producing SAR components pathogen attack (Kuc, 1985). Chemical and biological factors can induce this effect in plants. Many phytotoxic chemicals, including salicylic acid, can induce SAR (Kessmann et al., 1994). Some plant enzymes, such as peroxidase, phenylalanine ammonia lyase (PAL), and chitinase, are indicators of SAR in plants.

Peroxidase activity increased after wheat plants were treated with YIB (Mei, 1991). The SOD activity of rape (*Brassica napus*) tissue also increased after rape was treated with a YIB filtrate (Xia and Ding, 1990).

The results of Zhang et al. (1996) suggest that compost-induced SAR helped to control *Pythium* root rot of cucumber. Plants grown in compost-amended mixes had increased peroxidase activity and enhanced peroxidase isozyme levels compared with plants grown in peat mixes. This is further evidence of the role of compost-induced SAR in the control of this disease.

Xie and Shen (1997) reported increases in phenylalanine ammonia lyase (PAL) activity on watermelon seedling treated with different sources of anaerobically digested slurry compared with untreated controls. The increased PAL level was related to the fulvic acid content in the anaerobically digested slurry. Fulvic acid can induce PAL activity. Wu and Wei (1993) studied the mechanism of enhancing cold resistance of rice seedlings, and showed that digested slurry can decrease ion exosmosis of the cell, increase free proline content of seedlings, moderate the decrease of ascorbic acid and peroxidase, and increase the unsaturated fatty acid percentage in total fatty acid. All of these were related to increased cold resistance.

Detoxification of Contaminants

The activity and growth of microorganisms in soil provides an opportunity to enhance the biodegradation of organic contaminants present in soil, such as pesticides residues, industrial solvents, gasoline and fuel oil (Skladany and Metting, Jr., 1993). Moreover, plants or plant-associated microflora convert pollutants into non-toxic materials (Cunningham et al., 1995; Schnoor et al., 1995). Shen (1992) showed that some YIB strains also have the ability to degrade phenols in soil.

CONCLUSION

Wastes produced from agricultural production and processing must be recycled effectively and beneficially if we are to achieve a more sustainable agriculture, environment and society. Microorganisms play very important roles in the transformation of wastes as they do in the transformation of soil organic matter. This paper points out that we have made considerable progress through research to (a) select and identify effective microbial species and strains that can transform wastes, and directly or indirectly improve plant heath and protection, (b) identify beneficial microorganisms in regard to microbe-microbe and plant-microbe interactions and (c) isolate and identify microbial metabolites as biocontrol agents against phytopathogens. Nevertheless, there is much more that we need to know if we are to expand our knowledge and develop practical technological applications for processing agricultural wastes while enhancing the growth and activities of the beneficial microorganisms they contain. It is needed to conduct research on (a) the exact mechanisms or mode-of-action on how beneficial microorganisms and /or their metabolites and enzyme systems affect microbe-microbe and plant-microbe interactions to improve plant health and plant protection; (b) the waste treatment methods and technologies and operating conditions that enhance the growth and activity of "key" beneficial microorganisms and the level of their metabolites; and (c) the conditions under which these beneficial "biocontrol" microorganisms can survive, grow, reproduce and remain effective after application to crops and soils. Thus, these three areas of research, i.e., mechanisms, optimum waste treatment methods to enhance microbial growth and activity, and maximizing their survival and growth in the environment, can help to achieve a more sustainable agriculture, environment and society in the future.

REFERENCES

Arshad, M. and W. T. Frankenberger. (1993). Microbial production of plant growth regulators. In *Soil Microbial Ecology: Applications in Agricultural and Environmental Management*, ed. F. B. Metting, Jr. New York: Marcel Dekker, pp. 307-348.

Baker, R and M. B. Dickman. (1993). Biological control with fungi. In: *Soil Microbial Ecology: Applications in Agricultural and Environmental Management*, ed. F. B. Metting, Jr. New York: Marcel Dekker, pp. 275-306.

Bashan, Y. and H. Levanony.(1990). Current status of *Azospirillum* inoculation technology: *Azospirillum* as a challenge for agriculture. *Canadian Journal of Microbiology* 36:591-608.

Boehm, M. J. and H. A. J. Hoitink. (1992). Sustenance of microbial activity in the potting mixes and its impact on severity of *Pythium* root rot of poinsettia. *Phytopathology* 82:259-264.

Boehm, M. J., L. V. Madden and H. A. J. Hoitink. (1993). Effect of organic matter decomposition level on bacterial species diversity and composition in relationship to *Pythium* damping-off severity. *Applied Environmental Microbiology* 59:4147-4179.

Brown, N. L. and P. B. S. Tata. (1985). *Biomethanation* (ENSIC Review no. 17/18). Bangkok (Thailand): Environmental Sanitation Information Center, Asian Institute of technology.

Budde, K. and H. C. Weltzien. (1990). Use of compost extracts and compost substrates to the control of *Erysiphe graminis*. Internationals Symposium-Schaderreger des Getreides. Halle/Saale, Germany. pp. 527-528.

Campbell, R. (1989). *Biological Control of Microbial Plant Pathogens*. Cambridge: Cambridge University Press.

Chen, T. 1987. *Introduction to Chemistry of Agroindustry Waste* (in Chinese). Beijing (China): Chemical Industry Press.

Chen, W., H. A. J. Hoitink and L. V. Madden. (1988). Microbial activity and biomass in container media for predicting suppressiveness to damping-off caused by *Pythium ultimum*. *Phytopathology* 77:755-760.

Chen, Y. and L. Liu. (1985). A preliminary research on the action of deleterious rhizobacteria of radish. *Acta Phytopathologia Sinica*. 15:235. (in Chinese).

Chen, Y., R. Mei, S. Lu, L. Liu and J. W. Kloepper (1996). The use of yield increasing bacteria (YIB) as plant growth-promoting rhizobacteria in Chinese agriculture. In *Management of Soil Borne Disease*, eds. R. S. Utkhede and V. K. Gupta. Ludhiana (India): Kalyani Publisher.

Chung, Y. R. and H. A. J. Hoitink. (1990). Interactions between thermophilic fungi and *Trichoderma hamatum* in suppression of *Rhizoctonia* damping-off in bark compost-amended container medium. *Phytopathology* 80:73-77.

Cunningham, S. D., W. R. Berti and J. W. Huang. (1995). Phytoremediation of contaminated soils. *Trend in Biotechnology* 13:393-397.

Davison, J. (1988). Plant beneficial bacteria. *Biotechnology* 6:282-286.

Dowling, D. N. and F. O'Gara. (1994). Metabolites of *Pseudomonas* involved in the biocontrol of plant disease. *Trend in Biotechnology* 12:133-141.

Fan, S. and G. Wang. (1987). Study of fulvic sodium for curing canker of apple-tree. *Humic Acid* 8: 18-24 (in Chinese).

Fang, R., J. Li, C. Xiong and B. Xie. (1993). Investigation and analysis of sprinkling digested slurry to the leaves surface of orange-tangerine trees in order to prevent frost impact(abstracts). In: *Biogas and Sustainable Agriculture*, Department of Environmental protection and Energy, Ministry of Agriculture, China. Bremen (Germany): Bremen Overseas Research and Development Association, pp. 290-291.

Gould, W. D. (1990). Biological control of plant root diseases by bacteria. In *Biotechnology of Plant-Microbe Interactions*, eds. J. P. Nakas and C. Hagedorn. New York: McGraw-Hill, pp. 287-372.

Guo, L. and S. Sun. (1991). Preliminary study on wheat take-all control with anaerobic digested slurry. *China Biogas* 9:32-33 (in Chinese).

Hadar, Y., R. Mandelbaum and B. Gorodecki. (1992). Biological control of soilborne plant pathogens by suppressive compost. In *Biological Control of Plant Diseases*, eds. E. S. Tjamos, G. C. Papavizas and R. J. Cook. New York: Plenum Press.

Hoitink, H. A. J., Y. Inbar and M. J. Boehm. (1991). Status of compost-amended potting mixes naturally suppressive to soilborne diseases of floricultural crops. *Plant Diseases* 75: 869-873.

Huang, D., Y. Chen and X. Chen. (1993). Studies on the techniques for comprehensive utilization of biogas waste in agriculture. In: *Biogas and Sustainable Agriculture*, Department of Environmental Protection and Energy, Ministry of Agriculture, China. Bremen (Germany): Bremen Overseas Research and Development Association, pp. 270-274.

Ji, P. and Y. Chen. (1992). Colonization of yield-increasing bacteria (YIB) on the surfaces of rapeseed and Chinese cabbage and the effect on deleterious rhizobacteria (DRB). *Chinese Journal of Microecology* 4:20-27 (in Chinese).

Kessmann, H., T. Staub, C. Hofmann, T. Maetzke, J. Herzog, E. Ward, S. Uknes and J. Ryals. (1994). Induction of systemic acquired disease resistance in plants by chemicals. *Annual Reviews of Phytopathology* 32: 439-456.

King, F. H. (1911). *Farmers of Forty Centuries*. Emmaus (Pennsylvania, USA): Rodale Press Inc.

Kloepper, J. W. (1993). Plant growth-promoting rhizobacteria as biological control agents. In: *Soil Microbial Ecology: Applications in Agricultural and Environmental Management*, ed. F. B. Metting, Jr. New York: Marcel Dekker, pp. 255-274.

Kuc, J. (1985). Expression of latent genetic information for diseases resistance in plants. In: *Cellular and Molecular Biology of Plant Stress*, eds. J. L. Key and T. Kosuge. New York: Alan R. Liss, pp. 303-318.

Kuter, G. A., E. B. Nelson, H. A. J. Hoitink and L. V. Madden. (1983). Fungal population in container media amended with composted hardwood bark suppressive and conducive to *Rhizoctonia* damping-off. *Phytopathology* 73:1450-1456.

Kwok, O. C. H., P. C. Fahy, H. A. J. Hoitink and G. A. Kuter. (1987). Interactions between bacteria and *Trichoderma hamatum* in suppression of *Rhizoctonia* damping off in bark compost media. *Phytopathology* 77:1206-1212.

Lam S. T. and T. D. Gaffney. (1993). Biological activities of bacteria used in plant pathogen control. In *Biotechnology in Plant Disease Control*, ed. I. Chet. New York: Wiley-Liss, pp. 291-320.

Li, S., B. Shen, X. Gu and Q. Fan. (1992). The effect of biogas fermentation liquid on control soft rot of sweet potato. *China Biogas* 10:18-21 (in Chinese).

Li, S., B. Shen and Q. Fan. (1991). Inhibition mechanism of solution of biogas fermentation to *Rhizopus nigricans*. *China Biogas* 9: 6-9 (in Chinese).

Liu, Z. and D. Shen. (1994). The growth of *Bacillus* spp. and liquid waste of bean-curd processing (unpublished).

Lugtenberg, B. B. J., L. A. de Weger and J. W. Bennett. (1991). Microbial stimulation of plant growth and protection from disease. *Current Opinion in Biotechnology* 2:457-464.

Lumsden, R. D., J. A. Lewis and P. D. Millner. (1983). Effect of composted sewage sludge on several soilborne plant pathogens and diseases. *Phytopathology* 73:1443-1448.

Ma, W. and Y. Guo. (1993). Some discussion on active ingredients in digested slurry. *China Biogas* 11: 50-51 (in Chinese).

Mei, R. 1991. *Yield-Increasing Bacteria*. Agriculture Press. Beijing, China (in Chinese).

Mitchell, R. and M. Alexander. (1961). The mycolytic phenomenon and biological control of *Fusarium* in soil. *Nature* 190:109-110.

Mo, Y. (1988). Antimicrobial test of biogas fluid on plant pathogenic microorganisms. *China Biogas* 6: 6-10 (in Chinese).

Okon, Y. and Y. Hadar. (1987). Microbial inoculants as crop-yield enhancers. *CRC Critical Review of Biotechnology* 6: 61-85.

Peng, G. and Y. Xiong. (1992). Study on the relation between hormone and YIB strain in plants with isotope-trace method. *Chinese Journal of Microecology* 4:114-116 (in Chinese).

Qiao, X., L. Ma and F. Gao. (1998). Mechanism of compost extract for controlling plant diseases. In *Plant Diseases Research and Control*, ed. Y. Liu. Beijing: China Agricultural Science and Technology Press, pp. 336-338 (in Chinese).

Quarles, W. and J. Grossman. (1995). Alternatives to methyl bromide in nurseries–disease suppressive media. *IPM Practitioner* 17:1-11.

Schippers, B., A. W. Bakker and P. A. H. M. Bakker. (1987). Interactions of deleterious and beneficial rhizosphere microorganisms and the effect of cropping practices. *Annual Reviews of Phytopathology* 25:339-358.

Schnoor, J. L., L. A. Licht, S. C. McCutcheon, N. L. Wolfe and L. H. Carreira. (1995). Phytoremediation of organic and nutrient contaminants. *Environmental Science and Technology*. 29: 318-323(A).

Schroth, M. N. and J. O. Beck. (1990). Concepts of ecological and physiological activities of rhizobacteria related to biological control and plant growth promotion. In: *Biological Control of Soil-Borne Plant Pathogens*, ed. D. Hornby. Wallingford (UK): CAB International, pp. 389-414.

Shen D. (1997). Microbial diversity and application of microbial products for agricultural purposes in China. *Agriculture, Ecosystems and Environment* 62: 237-245.

Shen D., Q. Zhou, Z. An, L. Li, H. Li and P. Xu. (1997). Effect of digested slurry on suppression to watermelon wilt in field. In: *Anaerobic Digested Slurry and Suppression of Fusarium Wilt of Watermelon* (Research report), ed. D. Shen. Beijing (China): China Agricultural University Press (in Chinese).

Shen, D. 1992. The ability of YIB to degradation of phenols. *Chinese Journal of Microecology*. 4:48 (in Chinese).

Shen, D., D. Yu, Z. Liu, B. Chen and J. Zhou. (1995). The effect of digestion residue on watermelon growth stimulation and disease resistance. *China Biogas* 13: 4-7 (in Chinese).

Shen, R. (1990). Comprehensive utilization of biogas fermentation residues. In: *China Biogas Years: 1980-1990*. ed. China Biogas Association. Beijing (China): China Science and Technology Press, pp. 45-49 (in Chinese).

Shen, R. (1993). Plant tolerance due to nitrogenous compounds in anaerobic fermentation of residues. In: *Biogas and Sustainable Agriculture*. Department of Environmental Protection and Energy, Ministry of Agriculture, People's Republic of China. Bremen Overseas Research and Development Association. Bremen, Germany, pp. 9-60.

Shi, Z., P. Zhang, P. Li and C. Huang. (1991). Studies on using biogas slurry to control wheat scab disease. *China Biogas* 9: 11-14 (in Chinese).

Skladany, G. J. and F. B. Metting, Jr. (1993). Bioremediation of contaminated soil. In: *Soil Microbial Ecology: Applications in Agricultural and Environmental Management*, ed. F. B. Metting, Jr. New York: Marcel Dekker, pp. 483-514.

Tang, W., S. Chang, Y. Wang and H. Chen, (1992a). Preliminary study on the mechanism reducing plant diseases by YIB. *Chinese Journal of Microecology* 4:67-72 (in Chinese).

Tang, W., Y. Chen and G. Yu. (1992b). The ecological effects of YIB on the phylloplane of winter wheat. *Chinese Journal of Microecology* 4:49-54 (in Chinese).

Tang, W., H. Pen and C. Cong. (1992c). Preliminary study on the fungistatic activity of volatile metabolites from *Bacillus cereus*. *Chinese Journal of Microecology* 4:86-91 (in Chinese).

Traenkner, A. (1993). Elements of organic farming compost extracts as a means of biological control of diseases. Bonn (Germany): Forschungsberichte-Landwirtschaftliche Fakultaet der Rheinischen Friedrich-Wilhelms-Universitaet, pp. 62-67.

Walter J. F. and A. S. Paau. (1993). Microbial inoculant production and formulation. In: *Soil Microbial Ecology: Applications in Agricultural and Environmental Management*, ed. F. B. Metting, Jr. New York: Marcel Dekker, pp. 579-594.

Wang, S. (1987). The effect of fulvic acids on controlling cotton wilt and increasing yield. *Humic Acid* 3:33-37 (in Chinese).

Wang, T. (1992). Application of fulvic acid in agriculture. *Humic Acid*:14-36.

Weltzien, H. C., N. Ketterer, C. Samerski, K. Budde and G. Medhin. (1987). Studies on the effects of compost extracts on plant health. *Nachrichtenbl. Deut. Pflanzenschutzd.* 39:25-28.

Winterscheidt, H., V. Minassian and H. C. Weltzien. (1990). Studies on biological control of cucumber downy mildew (*Pseudoperonospora cubensis*) with compost extracts. *Gesunde Pflanzen* 42: 235-238.

Wu, A., Y. Chen, N. Qian, J. Tang and H. Ruan. (1993). Studies on using bioslurry to cotton fusarium wilt. *China Biogas* 11: 39-41 (in Chinese).

Wu, B. and X. Wei. (1993). Studies on effect of soaking seeds with digested slurry to enhance cold-resistance of early season rice seedlings. In: *Biogas and Sustainable Agriculture*. Department of Environmental Protection and Energy, Ministry of Agriculture, China. Bremen (Germany): Bremen Overseas Research and Development Association, pp. 61-68.

Xia, L. and X. Ding. (1990). The effect of YIB on rice aging in physiological and biochemical aspects. *Human Agricultural Science and Technology* 4:15-17 (in Chinese).

Xie, Z. and D. Shen 1997. The mechanism of watermelon wilt control by anaerobic digested slurry. In: *Anaerobic Digested Slurry and Suppression of Fusarium Wilt of Watermelon* (Research report) ed. D. Shen. China Agricultural University, Beijing, China (in Chinese).

Yang, J., S. Lu and Y. Chen. (1992). Studies on the metabolites and mechanisms increasing crop yield by YIB (*Bacillus* spp.). *Chinese Journal of Microecology* 4: 28-34 (in Chinese).

Yang, Z. (1993). Field experiment of curing wheat aphids with digested slurry. In: *Biogas and Sustainable Agriculture*. Department of Environmental Protection and Energy, Ministry of Agriculture, China. Bremen (Germany): Bremen Overseas Research and Development Association, pp. 110-114.

Yu, G., W. Tang and Y. Chen. (1992). Colonization of YIB on the phylloplane on wheat. *Chinese Journal of Biological Control*. 8:83-86 (in Chinese).

Zhang, P. (1994). The inhibitive effect of anaerobic slurry supernatant on the growth of *Fusarium graminearum*. *China Biogas* 12:32-34 (in Chinese).

Zhang, W., W. A. Dick, and H. A. J. Hoitink. (1996). Compost-induced systemic acquired resistance in cucumber to *Pythium* root rot and anthracnose. *Phytopathology* 86:1066-1070.

Zhao, Z. and J. Zhou. (1991). The experiment of controlling watermelon wilt disease by anaerobic digested slurry. *China Biogas* 9: 36-38 (in Chinese).

Zhou, J. (1993). Studies on increasing crop yield and controlling diseases by digested slurry and sludge. In: *Biogas and Sustainable Agriculture*, Department of Environmental Protection and Energy, Ministry of Agriculture, China. Bremen (Germany): Bremen Overseas Research and Development Association, pp. 74-82.

Approaches to Biological Control of Nematode Pests by Natural Products and Enemies

Mohammad Akhtar

SUMMARY. In recent years, continuing environmental problems associated with the use of nematicides have resulted in a sense of urgency regarding the search for alternative methods of nematode control. Biological control of plant-parasitic nematodes with natural products from plants and animals, and soil organisms are alternative control tactics that are receiving increased interest among nematologists. Natural products include a number of plant parts, by-products, and residues when incorporating into soil, and plant-interculture with other crop plants, crop rotation with non-host or poor host of nematodes, green manuring and other organic manures. Nemato-toxic compounds of the different plants are released through volatilization, exudation, leaching and decomposition. Natural enemies as bio-control agents to nematodes include bacterial and fungal parasites, predatory nematodes and soil invertebrates. The beneficial effects of natural products have been generally considered to be due to direct or indirect stimulation of predators and parasites of nematodes. Very often, when there was a suppression in the nematode population, there was a consequent increase in crop production. *[Article copies available for a fee from The Haworth Document Delivery Service: 1-800-342-9678. E-mail address: getinfo@haworthpressinc.com <Website: http://www.HaworthPress.com>]*

Mohammad Akhtar is affiliated with the Department of Plant Protection, Institute of Agriculture, Aligarh Muslim University, Aligarh 202002, India.

[Haworth co-indexing entry note]: "Approaches to Biological Control of Nematode Pests by Natural Products and Enemies." Akhtar, Mohammad. Co-published simultaneously in *Journal of Crop Production* (Food Products Press, an imprint of The Haworth Press, Inc.) Vol. 3, No. 1 (#5), 2000, pp. 367-395; and: *Nature Farming and Microbial Applications* (ed: Hui-lian Xu, James F. Parr, and Hiroshi Umemura) Food Products Press, an imprint of The Haworth Press, Inc., 2000, pp. 367-395. Single or multiple copies of this article are available for a fee from The Haworth Document Delivery Service [1-800-342-9678, 9:00 a.m. - 5:00 p.m. (EST). E-mail address: getinfo@haworthpressinc.com].

KEYWORDS. Biological control, natural enemies, natural product, nematode control

INTRODUCTION

For decades, farmers have applied broadly adaptable technologies to farms and fields for improving crop production. However, pests, diseases and weeds are important biotic constraints in food production. Plant-parasitic nematodes are one of the major yield limiting factors around the world. The control of plant-parasitic nematodes is more difficult than other pests because they inhabit the soil and usually attack the underground plant parts. Moreover, they have world-wide distribution, extensive host ranges and may be associated with other pathogens or pests making the diagnosis of disease confusing. Frequently this result in an underestimate of the damage caused by plant-parasitic nematodes.

The excessive use of chemical fertilizers, pesticides, and weedicides in agriculture has caused environment imbalance and is causing problems to all living beings on the earth. The ever-increasing demand for food from the world's growing population necessitates greater sophistication and intensification. The importance of pest conbtrol in these process has been well recognized, but there is increasing concern about the over-use of pesticides, which has led to undesirable side-effects, such as deterioration of the environment, hazard to human health, the development of pest-resistance, resurgence of pests and determines effects on non-target organisms. There is now tremendous pressure on farmers to use methods of pest control which do not pollute or degrade the environment (Duncan 1991, Akhtar, 1995a). Biological resources such as natural products from plants and animals, organic amendments, natural biopesticides, natural enemies (biocontrol agents) and some cultural practices including mixed-cropping, crop rotation and green manuring offer alternative or supplemental control tactics to chemical control of nematode pathogen on agricultural crops (Singh and Sitaramaiah, 1970; Muller and Gooch, 1982; Akhtar, 1993, 1997; Akhtar and Alam, 1993b; Akhtar and Mahmood, 1993c). In this paper, recent approaches in the area of biological control of plant-parasitic nematodes with natural products and natural enemies (biocontrol agents) are reviewed.

NATURAL PRODUCTS

Organic matter amendments to soil have been used for centuries to improve soil fertility and crop yield. Other studies suggest that organic amendments can also be used as nematicidal agents (Akhtar, 1997). Linford et al. (1938) suggested that organic materials amendments to soil stimulated the activity of naturally occurring antagonists of nematode pests and argued that activity of these organisms provided control of plant-parasitic nematodes. Subsequent recognition of the potential of organic amendments for the control of plant-parasitic nematode required more complex studies.

Plant Products

Bioactive products of plant origin being less persistence in environment and safe to mammals and other non-target organisms. Such natural products include plant origin pesticides, biofertilizers, organic manures and green manures. A number of crops and weeds exhibit biochemical mechanisms to counteract the activity of nematodes. Numerous plant species, representing 57 families, possess nematicidal compounds (Sukul, 1992). Amendments with neem (*Azadirachta indica*) are the best known examples that act by releasing pre-formed nematicidal constituents into soil. Neem products, including leaf, seed kernel, seed powders, seed extracts, oil, sawdust and particularly oilcake, have been reported as effective for the control of several nematode species (Egunjobi and Afolami, 1976; Akhtar and Alam, 1991, 1993a; Akhtar and Mahmood, 1996a; Akhtar, 1998a, Akhtar, 1999a,b,c,). Neem possesses a large number of bitter compounds called azadirachtin, nimbin, salanin, etc. Neem seed-kernel and oilcake are richest source of meliacines and contain 0.2-0.3% azadirachtin and 30-40% oil, though neem leaves, seed coats and barks also contain these but in smaller quantities (Thakur et al. 1981).

Several countries today, are in pursuit of technologies based on neem products because of their mammalian safety and environment findings. Many neem-based formulations like Neemark, Neemaguard, Wellgro, Agricef, Neoconeem, Limonol, Achook, Nimin, Suneem, Jawan and Replin are available in the market, and are highly effective for controlling nematode pests of several agronomic and horticultural crops. The neem products affect pests by functioning as antifeedent, repellent and oviposition, deterrent and toxicant (Akhtar, 1997).

Neem products are highly biodegradable ensuring their non-resistance in the environment. These also do not induce pest resistance unlike chemical pesticides. Besides the nematicidal effects, triterpene compounds in neem oilcake inhibit the nitrification process and provide more available nitrogen to the plants for the same amount of fertilizer (Akhtar and Alam, 1993a). Nematode control effects of neem were observed when its by-products and commercial products were used as seed coatings and bare-root dip treatments on nematode pests (Akhtar and Alam, 1993a; Akhtar and Mahmood, 1993b; 1994a,b,c,d, 1995a,b, 1996a,b,c,d, 1997b; Akhtar, 1998a,c,d). Other available oilcakes such as castor (*Ricinus communis*), groundnut (*Arachis hypogea*) and mahua (*Madhuca indica*) have been reported effective in nematode control (Lear, 1959; Derrico and Maio, 1980; Bora and Phukan, 1983; Akhtar et al., 1990; Akhtar and Alam, 1991).

Marigolds (*Tagetes* spp.) that have been used for control of nematodes include African marigold, *T. erecta*, the French marigold, *T. patula*, and the South American marigold, *T. minuta*. These are effective in controlling nematode pests in certain circumstances (Bridge, 1996). Empirical observations indicated that many species of marigold, resistant to a number of nematode species, and can effectively control nematodes in agricultural crops when they are grown in rotation, interplanted with other crop, or used as soil amendments (Oostenbrink et al., 1957; Uhlenbroek and Bijloo, 1958; Yuhara, 1971; Hackney and Dickerson, 1975; Prot and Kornprobst, 1983; Siddiqui and Alam, 1987a,b, 1988; Akhtar and Alam, 1990a, 1992; Akhtar, 1998b). Cruciferous plants are receiving increased attention as novel tools for the control of plant-parasitic nematodes (McSorley and Frederick, 1995). Decomposition products from Compositae and Cruciferae can be suppressive to plant-parasitic nematodes (Akhtar and Alam, 1992). Many of these plants are recommended as intercrops grown in close vicinity to the cross plants, and the detrimental effects of these plants on crop yield often outweigh the beneficial effects of nematode control (Bridge, 1996). These plants have shown to have some nematicidal potential by decomposition of plant residues or release some chemicals from their roots (Akhtar and Mahmood, 1994a).

Mixed-Cropping/Intercropping

In general, most subsistence farmers in developing countries mix different crops in the same field and same time so as to control or reduce the risk of crop failure. Because pests and diseases on single crop (main crop) may result in its total failure, whereas mixed-cropping systems, even if main crop is affected their remain other (secondary) crops that withstand, yield and support the farmers. However, in the recent past, it has been reported that some antagonistic plants to nematodes, often reduce the population densities of plant-parasitic nematodes by exudation and leaching from roots that repell or kill the nematode pests.

Plants of the family Compositae and Cruciferae are receiving increased attention as mixed-cropping system for the control of plant-parasitic nematodes. Tyler (1938) and Steiner (1941) were the earliest workers to report the resistance of marigolds species to *Meloidogyne* spp. Interest in using nemato-toxic plants for the control of plant-parasitic nematodes was stimulated with the discovery of antagonistic effects of mustard (*Brassica hirta*) and black mustard (*B. nigra*) (Triffitt, 1929). Thereafter, several investigators successfully used marigold and mustard for reducing the other nematodes (Steiner, 1941; Ellenby, 1945; Oostenbrink, 1957). Unfortunately, the development of antagonistic plants as mixed-cropping in nematode control has been very slow or practically non-existent in developed countries. The development of nematicidal chemicals during 1940 to 1960s relegated research on these aspects of nematode control. Few attempts were made upto the 1980s to combat nematode disease with antagonistic plants.

Although various types of plant species supported mixed-cropping in the control of nematode pests. In Nigeria, mixed-cropping of maize and cowpea significantly reduced populations of *Pratylenchus sefaensis* relative to monocultures of maize, and populations of *M. javanica* and *Rotylenchulus reniformis* relative to monoculture of cowpea. Mixed-cropping of maize and cowpea could well prove an effective means of nematode pest control, with specific advantages for the maize crop in particular (Egunjobi et al., 1986). In Alabama, reduced densities of *Meloidogyne arenaria* have resulted following the use of a number of these crops, including velvet-bean (*Mucuna deeringiana*), jointvetch (*Aeschynomene americana*), hairy indigo (*Indigofera hirusuta*), crotalaria (*Cortalaria spectabilis*), horsebean (*Canavalia ensiformis*), sesame (*Sesamum indicum*) and castor (*Ricinus communis*) (Rodriguez-Kabana et al., 1987, 1988a,b). Several of these crops, including velvetbean, crotolaria, jointvetch and hairy indigo have reduced the numbers of *M. incognita* in the field in Florida (Reddy et al., 1986). This approach can be particularly effective with nematodes that have a limited host range, such as the sugar-beet cyst nematode (*H. schachtii*) and soybean cyst nematode (*H. glycines*). However, the wide host ranges of *Meloidogyne* spp. greatly restrict this option for alternate crop.

Marigolds have been used as a mixed-crop with other vegetable crops such as tomato, eggplant, okra, cabbage and cauliflower (Siddiqui and Alam, 1987a,b, 1988). In India, for example, marigold traditionally grown for its ornamental value and for religious ceremonies, and it is common practice to grow mustard as an intercrop in agricultural fields. Marigold plants characterized by the presence of α-terthienyl compounds, which acts as natural deterrents of nematode pests serve as bionematicide (Zechmeister and Sease, 1947; Uhlenbroek and Bijloo, 1958; Osman and Viglierchio, 1988).

Similarly, black mustard (*Brassica nigra*) and rocket-salad (*Eruca sativa*) as mixed-crop with potato significantly reduced the population level of plant-parasitic nematodes (Akhtar and Alam, 1991; Akhtar and Mahmood, 1996b). The nematicidal activities are considered to be due to the presence of allyl-isothiocynates in root-diffusates (Okopnyi et al. 1980). Besides these many other antagonistic plants are well known nematicidal in nature, therefore, further research is needed in mixed-cropping with antagonistic in nematode control.

Crop Rotation

Crop rotation is the best known and most widely used cultural practice in nematode control. The principle of crop rotation for nematode control is to reduce initial populations of damaging nematode species to levels that allow the following crop to become established and complete early growth before being heavily attacked (Nusbaum and Ferris, 1973; Bridge, 1996). Monoculture of a highly susceptible host plant for several years will increase populations of nematode pests and cause severe damage to crops, however, growing of a poor crop will significantly decline the nematode populations build-up. The overall object of any rotation is to allow sufficient intervals between susceptible host crop so that nematode population can suppress to a level that will allow the next susceptible crop to grow and yield at an acceptable rate. This may be achieved by alternating poor host, non-host, tolerant or, resistant or antagonistic crops with a susceptible crops for longest interval possible in the same field. Rotation will likely need to be developed specifically for different regions, soil types, nematode problems and crop cultivation systems. Farmers acceptance of crop rotation strategies in nematode control will involve practical consideration such as cost, planting requirement, availability of seeds, market value of secondary (antagonistic) crop, etc.

Green Manuring

Greenhouse studies have shown that green manure treatments can suppress population of plant-parasitic nematodes but there are few reports of successful application in commercial agriculture (Duddington and Duthoitt, 1960; Duddington et al., 1961; Mojtahedi et al. 1993; Jing, 1994). In addition to suppression of nematodes, a green manure crop can aid renovation of the field by reducing soil compaction, providing organic matter, and helping to control weeds. However, acceptance of green manure as a standard practice will depend upon efficacy and long-term benefits to crop production. Nematodes can occur deep in the soil profile, and second-stage juveniles have the capability of migrating upward to damage the plants roots. Thus depth and duration of control achieved by green manure play an important role in the success of green manure treatment. Mojtahedi et al. (1991) indicate that rapeseed as green manure significantly reduces potato damage caused by *Meloidogyne chitwoodi*. Field experiments in Pennsylvania have shown that rapeseed has potential as a preplant treatment for *Xiphinema americanum* in replant orchards when used as green manure (Jing, 1994). Rapeseed green manure treatment may provide an acceptable alternative for higher value crops such as tree fruit.

Ideally, a green manure crop should not serve as a host for the target nematode. Poor host status combined with the detrimental effects of green manure may reduce the population to manageable level. Economic considerations are important in commercial agriculture, and since green manure treatments do not provide a crop income also prevent the growing of alternative crops. In tree production, a green manure treatment may be cost efficient relative to fumigation or nematicidal application.

Organic Manure

Grass and hedge clippings, shrubbery trimmings, tree parts from urban homesites, domestic garbage and agroindustrial wastes such as oilcakes, crop-residues, sawdust as compost comprise a large quantity of plant derived organic wastes requiring environmentally safe disposal as well as help in improving fertility of soil. Reduction in population densities of plant-parasitic nematodes have been attributed to application of these organic wastes in some cases, which could provide an additional benefit for plant growth (Muller and Gooch, 1982; Akhtar and Alam, 1993b; Akhtar and Mahmood, 1996b). Van der Laan (1956) reported that the addition of farmyard and composted manures to soil reduced the population densities of *Heterodera rostochiensis* in potato roots. Steer and chicken manures reduced the numbers of cyst and citrus nematodes and resulted in increased yields of potato and citrus (Gonzalez and Canto-Sannz, 1993). Many studies have been conducted on the efficacy of poultry manure amendment to soil for the controlling plant-parasitic nematodes on vegetable crops (Derrico and Maio, 1980; Kaplan and Noe, 1993; Akhtar and Mahmood, 1997a). Compost manure, a mixture of animal and plant wastes, is a potential cause of pollution. Conversely, it is considered a valuable organic fertilizer rich in all macro- and micro-nutrients necessary for plant growth. Utilization of compost manure might offer an inexpensive alternative for both fertilization and plant-parasitic nematode control. Toxic products that have been produced by compost decomposition have been shown to have no immediate effects on nematode pests. Beside these, other biological wastes, e.g., sewage-sludge, municipal refuse or livestock waste and farmyard manure have been used for controlling plant-parasitic nematodes (Akhtar and Alam, 1993b). However, if the presence of chemical products is maintained for a long time, these may stimulate increased activity of biological control agents towards nematodes (McSorley and Gallaher, 1995).

Chitin

Particular attention has been given to chitin and chitinous wastes, as this amendment has been shown to have a strong nematicidal action and by the stimulation of chitinolytic organisms such as bacteria and actinomycetes that attack nematode eggshells (Mankau and Das, 1969; Miller, 1976; Mian and Rodriguez-Kabana, 1982; Mian et al., 1982; Rodriguez-Kabana et al., 1983; Culbreath et al., 1985, 1986; Spiegel et al., 1986, 1987, 1989). Since chitin can be phytotoxic and has low C:N (3:2), ratio, some studies have been undertaken on the effect of mixing chitinous materials with a source of available carbon such as hemicellulosic wastes, in order to immobilize the excess nitrogen that stimulates microbial activity (Huebner et al., 1983; Spiegel et al., 1987). The quantities of chitin amendments to soil needed to bring about successful control of nematode pests (Rodriguez-Kabana et al., 1984) are usually in excess of 5 t/ha; this means that chitin is not appropriate for such a use and, when it has been produced commercially, has proved more expensive than methyl bromide and their use would only be economical on more valuable crops.

NATURAL ENEMIES (BIOCONTROL AGENTS)

A number of soil organisms have been recognized as biological agents for the control of plant-parasitic nematodes (Poinar and Jansson, 1988). It is promosing because bacterial and fungal parasites have been observed infecting nematodes in several agricultural crops. These organisms have been found on a variety of nematode hosts and in many different climates and environmental conditions (Sayre and Starr, 1988).

Interest in using biological agents to control plant-parasitic nematodes was stimulated with the discovery of predatory nematodes (Cobb, 1917), nematode-trapping fungi (Linford, 1937; Linford and Yap, 1939), thereafter and a nematode parasite first described as *Duboscqia penetrans* (Thorne, 1940) but later renamed *Pasteuria penetrans*. Other biocontrol agents of nematodes include arthropods (Walter et al., 1987), predatory nematodes (Small, 1987) and a variety of other invertebrate organisms (Sayre and Walter, 1991). Although various types of organic matter supported growth of some of these organisms. The addition of such materials into soil did not promote nematode predation. Indeed, the nematicidal activities of the fungal parasite, *Hirsutella rhossiliensis* was actually decreased by the addition of soil amendments (Jaffee et al., 1994).

Commercial development of biological control will probably depend on the selection of a single agent or its products. The commercial development of biocontrol agents is limited because mass production methods have not been developed, with some exceptions (Cayrol, 1983). Since a vast array of potential nematode biocontrol agents typically occur in most soils (Rodriguez-Kabana and Morgan-Jones, 1988); various aspects of biological control of plant-parasitic nematodes have been reviewed a number of times (Mankau 1980; Jatala, 1986; Duncan, 1991; Stirling, 1991; Dickson et al., 1994; McSorley and Duncan, 1995).

Nematicidal efficacy of a naturally occurring biocontrol agent is to be likely affected by environmental conditions. Interactions between plant-parasitic nematodes and fungal, bacterial and invertebrate soil organisms are influenced differently by several biotic and abiotic factors (Sayre and Walter, 1991). Biocontrol agents of nematodes are unlikely to be as fast acting as nematicides for maintaining efficient nematode control, it is likely that they will have to be integrated with other methods (Kerry, 1990; Tzortzakakis and Gowen, 1994).

Fungi

Many soil-borne fungi have been demonstrated to be antagonists of nematodes. Special attention is being given to species of endoparasitic fungi but interest in sedentary endo-parasitic nematodes has increased rapidly in recent years (Kerry, 1990). These include nematode-trapping or predacious fungi, endoparasitic fungi, parasites of nematode eggs and cysts, and fungi that produce metabolites toxic to nematodes (Stirling, 1991). Some species of fungi are capable invading young females and, as the females become more exposed in the soil, they become increasingly vulnerable to fungi (Gintis et al., 1982; Kerry, 1989). More than 150 fungal species have been isolated from the females or cysts of *Heterodera glycines*, including *Exophiala tusarium* and species of *Glicladium*, *Neocosmospora*, *Paraphoma*, *Phoma*, *Stagonospora*, *Verticillium*, *Dictyochaeta* and *Pyrenchaeta* (Chen et al., 1996a). Natural control by some of these fungi have been reported on cyst and root-knot nematodes (Kerry, 1975; Stirling and Mankau, 1978; Kerry et al., 1982; Stirling and White, 1982; Godoy et al., 1983; Boag and Lopez-Llorea, 1989; Carris et al., 1989; Roccuzzo et al., 1993; 1996a,b). *Hirsutella rhossiliensis* is an endoparasite of vermiform nematodes (Jaffee and Zehr 1982; Ciancio et al., 1986), including second-stage juveniles of *H. avenae* (Stirling and Kerry, 1983) and *H. schachtii* (Jaffee and Muldoon, 1989).

In the recent past, much work has been conducted on *Paecilomyces lilacinus*, an ubiquitous soil hyphomycete which parasitizes eggs of *M. incognita* in laboratory tests (Dunn et al., 1982; Morgan-Jones et al., 1984) and provided biocontrol of cyst and root-knot nematodes in field soil (Jatala et al., 1980; Davide and Zorilla, 1983; Rodriguez-Kabana et al., 1984; Jatala,

1985). Rodriguez-Kabana et al. (1987) have successfully explored the feasibility of adding *P. lilacinus* with chitin or castor to soil for controlling *M. arenaria*. In other experiments *P. lilacinus* provided encouraging levels of control, its efficacy was variable and potential health hazards associated with this fungus are likely to prevent its widespread use (Minogue et al., 1984).

The effects of organic amendments to soil can be enhanced by saprophytic fungi which although not directly parasitic to nematodes, produce enzymes that destroy the nematode body structure (Galper et al., 1990). Galper et al. (1991) found that addition of 0.1% w/w collagen to soil supplemented with the collagenolytic fungus *Cunninghmaella elegans* reduced 90% root galling of tomato caused by *M. javanica* and reduced motility of *Rotylenchulus reniformis* and *Xiphinema index*. These effects were due to enzymes on toxins produced by the fungus (Janson and Nordbring-Hertz, 1988). Some parasites of nematode eggs are thought to produce metabolites that affect embryonic development and hatching (Jatala, 1986).

Organic manure, widely used in some countries, may provide a simple and convenient method of enhancing the activities of nematophagous fungi in soil. Some nematophagous fungi have been grown on waste organic materials which are colonized by, or combined with, the agent before they are introduced into soil, in order to enhance the effect of organic materials and provide better levels of control (Villanueva and Davide, 1984; Culbreath et al., 1986).

Bacteria

During the last decade there has been an increasing interest in bacterial antagonists of nematodes (Poinar and Hansen, 1986; Sayre, 1988; Oosterdorp and Sikora, 1989; Dickson et al., 1994). Rhizobacteria have shown potential as biological control agents. These agents can be applied to seed and may significantly reduce nematode invasion of roots (Oosterdorp and Sikora, 1989). *Pasteuria* spp., obligate parasite of plant-parasitic nematodes, has potential to be economically and environmentally practical biological control agents. Recent field observations (Bird and Brisbane, 1988; Sayre and Starr, 1988; Minton and Sayre, 1989; Dickson et al., 1991; Hewlett et al., 1994; Weibelzahl-Fulton et al., 1996; Chen et al., 1996c), and many greenhouse observations (Dube and Smart, 1987; Brown et al., 1985; Sharma, 1992) and microplot experiments (Oosterdorp et al., 1990) have demonstrated that *Pasteuria* spp. can provide effective control of plant-parasitic nematodes on some vegetables and leguminous crops. *Pasteuria* spp. are known to survive for prolonged periods under dry conditions (Stirling and Wachtel, 1980; Oosterdorp et al., 1990), but the effect of alternating drying and wetting on survival and infectivity of endospores is not known. This information will be important for practical applications of the bacterium in nematode-infested soil (Dickson et al., 1994). The commercial development of *Pasteuria* spp. as nematode biological control agents has been limited because methods for mass production have not been developed. Additional studies concerning the ecology of these biological control agent in agricultural ecosystems are needed before they can be used effectively in nematode control.

The actinomycetes have also been one of the main groups of interest, as they are known to produce antibiotics. It is possible that such antibiotics and/or other microbial metabolites produced by the soil microflora have adverse effects on nematodes. The nematicidal properties of avermectins produced by actinomycetes (*Streptomyces* spp.) have stimulated interest in 'natural' nematicides. The production of avermectins by a species of *Streptomyces* (Burg et al., 1979) shows that soil-borne organisms can produce highly nematicidal compounds and suggests that further work should be done on the production of such antibiotics in soil and their effects on nematodes.

Culture filtrates of *Clostridium butyricum*, containing formic, acetic, propionic and fatty acids, and particularly butyric acid were found toxic to nematodes (Johnston, 1959; Hollis and Rodriguez-Kabana, 1966, 1967). Rodriguez-Kabana et al. (1965) reported that hydrogen sulphide, produced by *Desulfovibrio desulfuricans* acts as potent nematicide.

Predatory Nematodes

Plant-parasitic nematodes generally occur with other nematode communities, including predacious nematodes. Predatory nematodes comprise four main taxonomic groups namely *mononchids*, *dorylaimids*, *aphelenchids* and *diplogasterids*. Predacious nematodes generally feed on all nematodes and also on other members of the soil microfauna. Interest in using predatory nematodes for suppressing populations of plant-parasitic nematodes in the soil is receiving less attention. However, few studies have investigated predatory nematodes as biocontrol agents in the soil (Akhtar, 1989, 1995). Esser (1987) found dorylaimid predators to be the most efficient and highly potent as biological control agents. The most advantageous and encouraging aspect of the *dorylaimid*, *nygloamid* and *diplogasterid* predators is that its easy to maintain their populations simply by adding organic matter to agricultural fields; as they are so polyphagous in nature, and they will remain abundant in soil without prey nematodes (Webster, 1972). The diplogasterids are the most readily cultured of the predacious nematodes, being easily maintained on simple nutrient media containing bacteria (Yeates, 1969). Lal et al. (1983) noted a two-fold increase in the populations of several predatory nematodes, such as species of *Dorylaimus*, *Discolaimus* and *Mononchus* when green manure was added to the soil. They concluded that the associated decrease in incidence of root-knot nematodes was due to the increased populations of predatory nematodes. Further suppression of the development of root-knot symptoms in tomato and chilli was observed when organic materials were incorporated into the soil (Akhtar, 1995b; Akhtar and Mahmood, 1993a). Predatory nematodes are comparatively large and therefore favour coarse soil with high organic matter.

Invertebrates

A number of soil invertebrates are known as biological control agents of nematodes. Some of the better known predators belong to the phylum Tardigrada and micro-arthropods (collembolans and mites) (Brown, 1954; Doncaster and Hooper, 1961; Hutchinson and Streu, 1960; Imabriani and Mankau, 1983). Frequent interactions of some invertebrates with nematodes probably occur in sandy soils, in well structured soils with a high organic matter content, and in the rhizosphere (Curl and Truelove, 1986). A turbellarian predator (*Adenoplea* sp.) isolated from greenhouse soil was able to consume several nematode species (Sayre and Powers, 1966). Tardigrades are not common in mineral soils and reports of their predation on nematodes are rare. Enchytraeids and earthworms usually feed on decaying organic matter in soil, but they may also ingest nematodes during feeding (Yeates, 1981). Plant-parasitic nematodes are likely to be consumed when oligochaetes feed on decomposing roots infected with endoparasitic nematodes. It may also be expected that a sizeable number of ectoparasitic plant nematodes are consumed during feeding and lead to nematode control, although studies on this aspect are too few to draw any sound conclusion. Some arthropods in soil, particularly mites, are known to consume considerable numbers of nematodes and could be considered as possible agents for nematode control (Muraoka and Ishibashi, 1976). Many nematophagous mites develop more rapidly, with higher reproductive rates when nematodes are included in their diet (Walter et al., 1987). At present it appears to be rather difficult to artificially produce or introduce invertebrate nematode into the soil.

FACTORS AFFECTING NATURAL PRODUCTS AND SOIL ORGANISMS

Many factors influence development of nematode populations including temperature, soil aeration, moisture, pH, exudates from host plants and other organic compounds, and inorganic ions (Clarke and Perry, 1977). Several environmental factors affect biological control agents and natural products in nematode control. The release of nutrients such as nitrogen, phosphorus and potassium depends on the decomposition and the demands made by heterogeneous populations of soil organisms. The addition of organic matter to soil stimulate the activity of bacteria, fungi, algae, and other soil-organisms. Increased microbial activity in amended soil causes enhanced enzymatic activities (Rodriguez-Kabana et al., 1983) and accumulation of decomposition end products and microbial metabolites, which may be detrimental to plant-parasitic nematodes (Johnson, 1959; Mankau and Minteer, 1962; Rodriguez-Kabana et al., 1987). The use of green manure crops, widely practised in some countries, may provide a simple and convenient method of enhancing the activities of nematophagous fungi in soil (Schlang et al., 1988).

Soil

Soil is a dynamic and complicated physical, chemical, and biological medium and organic compounds released into it may be altered and transported by biological, chemical and physical means (Halbrendt, 1996). Net mineralization of soil organic matter was found to be more rapid in sandy soils than in clay soils (Catroux et al., 1987; Hassink et al., 1991). The lower net mineralization rate in clay soils is assumed to be partly by a higher physical protection of soil organic matter, i.e., the organic matter is located in places relatively inaccessible to microbes (Vanveen and Kuikeman, 1990). Nematodes are restricted to pores of 30 mm diameter (Jones, 1982). Thus, tropic structure and available pore space have been found to influence decomposition and mineralization rate of organic matter (Elliot et al., 1980; Vanveen and Kuikeman, 1990). Christensen et al. (1989) demonstrated that the distribution of decomposable organic matter in the soil is strongly related to the particle size of the soil fraction, greater decomposition rate being associated with the larger sand-sized particles. Moreover the addition of organic materials usually improves soil structure and consequently the capacity of the soil to hold water and exchange ions so that, together with the nutrients released by the organic matter, the improved soil structure promotes root growth of plants. Predatory nematodes such as *Mononchus* spp. are comparatively large and are therefore favored by coarse soils of high organic matter content.

C:N Ratios

Microbial decomposition of organic matter and subsequent use of nitrogen by higher plants is dependent on the ratio of C:N in the organic matter.

With organic matter of a C:N ratio greater than 20:1, N will temporarily be immobilized in microbial tissue and this will create N-deficient soil for any plants grown following the addition of such organic wastes. For wastes or residues with a C:N ratio less than 20:1, N will be mineralized in the form of NH_4^+ or NO_3^- for absorption and uptake by plant roots (Jones, 1982). Miller et al. (1968, 1973) considered the C:N ratios of organic amendments and concluded that the availability of more nitrogen enhances the ability of amendment to control nematodes. Mian and Rodriguez-Kabana (1982) reported that the nematode control potential of on organic materials is directly related to N content or inversely related to the C:N ratio. The application of organic matter to soil sometimes creates problems because different types of material have different and often unique C:N ratios.

RELEASE OF NUTRIENTS FROM NATURAL PRODUCTS

Soil fauna can enhance soil organic matter decomposition and nutrient mineralization. Free-living (microbivorous) nematodes contribute to both decomposition processes of organic soil amendments and increase the mineralization of nitrogen and phosphorus (nutrient cycling) and may have indirect beneficial effect on plant growth (Yeates and Coleman, 1982). Abrams and Mitchell (1980) and Griffiths (1986) suggested that free-living nematodes accelerate the decomposition of organic soil amendments and increase the mineralization of nitrogen and phosphorus. Moreover, high levels of inorganic nitrogen fertilizer increased the number of bacterium *P. penetrans* endospores poroduced per nematode female and decreased the density of root-knot nematode *M. incognita* (Chen et al., 1994).

Nitrogen is the single most important fertilizer input and is required in the largest quantities for crop production. Nitrogen may also be a factor involved in controlling plant-parasitic nematodes as a result of the application of natural bio-resources. Rodriguez-Kabana (1986) reported that nitrogenous amendment to soil released ammonical nitrogen and this form of nitrogen can suppress plant-parasitic nematode population. In aerobic soil, ammonia produced by ammonifying bacteria during the natural decomposition of nitrogenous organic material has been implicated in plant-parasitic nematode suppression (Rodriguez-Kabana, 1986). This nematode suppression effect has been confirmed by many investigators (Eno et al., 1955; Vassalo, 1967; Rodriguez-Kabana et al., 1981, 1982). The nematicidal efficacy of additions of urea to soil, which is readily converted to ammonia by urease present in soil, was dosage dependent (Rodriguez-Kabana and King, 1980; Huebner et al., 1983; Akhtar and Mahmood, 1994d, 1996b). Release of nitrogen resulted in microbial activity in amended soil is known to bring about increased conversion of N to nitrate (nitrification). High rates of nitrogen are required for satisfactory control, but accumulation of nitrate and ammonical nitrogen in the soil can be phytotoxic (Huebner et al., 1983). In some experiments, equal amounts were added in all treatments, but the availability of nitrogen was different, because different substrate have different mineralization rates, so that equal amounts of N in organic form will not lead to equal amounts of inorganic N. Other nutrients were present in different quantities in the different substrates, which may have affected the plant growth responses. In this way the nematicidal effect of an amendment was, therefore, not only dependent on nitrogen, but also carbon content. Amendments with narrow C:N ratios, such as animal manure, oilcake and green manure, result in better nematode control than those with wide ratios (grassy hay, stubbles and cellulosic materials such as paper and sawdust). Chen et al. (1994) reported that some nitrogen levels may increase the density of *Pasteuria penetrans* on *Meloidogyne* spp. infecting a resistant tobacco cultivars.

RELEASE OF COMPOUNDS TOXIC TO NEMATODES

Organic materials release some chemicals into the soil that are directly responsible for nematode control. Recent work has shown that ricin, a protein derived from castor bean, has nematotoxic potential (Rich et al., 1989). The neem tree (*Azadirachta indica*) contains a group of chemicals known as limonoids, and these compounds have proven to be highly effective chemicals in nematode control. Phenols and tannins (the best known constituents of neem oilcake) are nematicidal at certain concentrations and, since some amendments contain high levels of these compounds, they may have a direct effect on nematode mortality (Badra and Elgindi, 1979).

Metabolic by-products, enzymes and toxin produced by microbes during decomposition of organic matter, can also be detrimental to plant-parasitic nematodes. Ammonia, nitrites, hydrogen sulphide, organic acids, and other chemicals that are produced from organic matter may be directly nematicidal or affect egg-hatch, or the motility of juveniles (Sayre et al., 1964, 1965; Hollis and Rodriguez-Kabana, 1966; Elmiligy and Norton, 1973; Badra and Elgindi, 1979). Tannic acid and its analogous are abundant in many plants products, and tannins have been reported to have antimicrobial properties (Scalbert, 1991). This compound reduced infection of tomato by the root-knot nematode in greenhouse and microplots (Hewlett et al., 1995). Avermectins are highly toxic to nematode parasites of plants. Several researchers have shown that root-dip, bulb-dip, and soil applications of avermectins were effective at controlling plant-parasitic nematodes on certain crops (Roberts and Matthews, 1995). Rodriguez-Kabana et al. (1987) suggested that the plant-parasitic nematode control effects of organic additives depend on their chemical compositions and the composition and activity of decomposer community that develop during degradation. For example, it has been reported that plants growing in amended soil contained greater concentrations of phenols than those growing in unamended soil and that this may induce disease resistance in roots (Badra et al., 1979).

Plant products may be released through volatilization, exudation from roots, leaching from plants or residues, and decomposition of residues (Pawlowski and Bachthaler, 1939; Lewis and Papavizas, 1971; Halbrendt, 1996). The prospect of exploiting naturally occurring plant constituents for nematode control has advantage over the current use of toxic chemicals and there have been many attempts to utilize this approach by rotation, intercropping, or green manure treatments (Halbrendt, 1995), although, crop rotation and green manure treatments do not have an immediate effect on population dynamics and their effects must be evaluated over time (Halbrendt, 1995). Naturally occurring biochemicals and plant allelochemicals can achieve effective reductions in population of target phytopathogens while minimizing environmental risk but non-target organisms are often negatively affected too.

In fugnal antagonists, nematode suppression was due to enzymes produced by the fungus, namely chitinase, collagenase, kerastase, and elastase which disintegrate nematode cuticle. Some bacteria and fungi produce metabolic by-products which interfere with nematode behavior and many soil organisms parasitize or prey on nematodes.

PLANT GROWTH AND YIELD

Organic materials represent a very important manurial resource for the improvement of soil fertility because decomposed materials ultimately serve as nutrients for plants and thereby improve crop production. In many reports, yields were increased following the application of organic matter to the soil, and the beneficial effects of organic materials are generally assumed to be due to the provision of extra nutrients of some sort to the crops. Page (1966) reported that plants grown in plots receiving organic manures were always larger than those receiving inorganic fertilizers. Very often, when there is a decrease in the population of soil pathogen(s), there is a consequent increase in crop yield. However, nematode control is by no means always followed by increased yields (Muller and Gooch, 1982).

The improvement in plant growth in the fungal treatment could not be attributed to control of nematodes because the total nematode densities were low and there were no differences in growth between plants with or without nematodes. The greater plant growth in fungus-treated soil may be attributed to either decomposition of the carrier, which may have provided nutrients, or fungi, such as *P. lilacinus* that produce substances that stimulated plant growth (Dickson et al., 1994).

CONCLUSIONS

Various non-chemical alternatives are available for the control of plant-parasitic nematodes. However, some non-chemical methods may not be particularly effective when used alone, making integration of methods necessary to achieve optimal nematode control, particularly sustainable agriculture. Application of natural plant-products have been evaluated for use in nematode control. Crop rotation, mixed cropping and green manuring have been a successful methods for suppressing plant-parasitic nematodes.

The plant-parasitic nematode population levels under natural products treatments may change for many reasons, including changes in soil properties, nutrients released to plants, increase in predators or parasitic microrganisms, toxic metabolites released from organic amendment breakdown or health of the host crop. However, changes in nematode populations are still dependent on C/N ratio of organic amendments, and the level of amendment needed for nematode control. However, there was insufficient evidence in these studies to conclude nematode suppression was due to these described reasons.

Soil fertilization is commonly employed by farmers to increase crop yield, particularly as land usage for agriculture intensifies. Proper re-cycloing of organic materials in agricultural endeavours can greatly improve soil tilth, fertility, water holding capacity and unfavourable habitat for nematodes. Most organic matters contain several nutrients essential for plants, so that levels of nitrogen, phosphorus, potassium and other essential elements are usually increased when organic materials are added to soil. Triterpenes compounds in neem products inhibit the nitrification process and provide more nitrogen in the form of ammonium to the plants for the same amount of nitrogen applied by the amendments.

Organic materials decompose slowly, thereby protecting the crop from pests for quite a long duration. Such a control tactic not only increases yield but also deters the spread of new infestations. As a result, the investment cost of treatment is reduced to a profitable level, and the next season's crop in the same plots without receiving any additional fertilizer may obtain benefits from the previous treatments. Further research on the quality and quantity of products applied still is needed to stimulate their widespread use.

Biological control offers an alternative or supplemental control tactic to chemical and cultural control of nematode pests. It is promising because bacterial parasites and numerous fungal parasites have been observed infecting nematodes in agricultural crops. *Pasteuria penetrans* has great potential in the near future for biological control of nematode pests, because it is widely distributed in agricultural soil throughout the world and contributes to natural and induced nematode control.

There is increasing interest in expanding information on the biological control agents such that future use of these organisms may provide an additional tool for the management of plant-parasitic nematodes. Currently, nematologists involved in the biological control of nematodes have adopted two main strategies: the manipulation of natural enemies in soil and the introduction of selected agents as inundative treatments. Organic matter can be modified by addition of specific compounds or by inoculation with particular microbial species to produce an amendment that will induce suppressiveness. However, despite considerable research effort, no organism is yet routinely used for the biological control of a nematode pest of any crop.

It is now possible to consider the role of natural enemies in control systems because most parasites and predators of plant-parasitic nematodes with biological control potential have a restricted host range and should not pose much of a threat to non-target organisms. Since biological control involves the use of organisms that are a natural part of the environment, it is largely non-polluting.

REFERENCES

Abrams, B.I. and M.J., Mitchell (1980). Role of nematode-bacterial interactions in hetrotrophic systems with emphasis on sewage-sludge decomposition. *Oikos* 34: 404-410.

Akhtar, M. (1993). Utilisation of plant-origin waste materials for the control of parasitic nematodes. *Bioresource Technology* 46: 255-257.

Akhtar, M. (1995a). Economic aspects of nematode disease-crop loss management. *Everyman's Science* 30: 71-75.

Akhtar, M. (1995b). Biological control of the root-knot nematode *Meloidogyne incognita* in tomato by the predatory nematode *Mononchus aquatic. International Pest Control*: 18-19.

Akhtar, M. (1997). Current options in integrated management of plant-parasitic nematodes. *International Pest Management reviews* 2:187-197.

Al:htar, M. (1998a). Biological control of plant-parasitic nematodes by neem products in agricultural soil. *Applied Soil Ecology* 7: 219-223.

Akhtar, M. (1998b). Effect of two compositae plant species and two types of fertilizer on nematodes in an alluvial soil, India. *Applied Soil Ecology* 10: 21-25.

Akhtar, M. (1999a). Biological control of plant-parasitic nematodes in pigeonpea field crops using neem based products and manurial treatments. *Applied Soil Ecology* 12: 191-195.

Akhtar, M. (1999b). Plant growth and nematode dynamics in response to soil amendments with neem-products, urea and compost. *Bioresource Technology* 69: 181-183.

Akhtar, M. (1999c). Effect of organic and urea amendments in soil on nematode communities and plant growth. *Soil Biology & Biochemistry* (in press).

Akhtar, M. and M.M. Alam (1990a). Control of plant-parasitic nematodes with agrowastes soil amendments. *Pakistan Journal of Nematology* 8: 25-28.

Akhtar, M. and M.M. Alam (1991). Integrated control of plant-parasitic nematodes on potato with organic amendments, nematicide and mixed cropping with mustard. *Nematologia Mediterranea* 19: 169-171.

Akhtar, M. and M.M. Alam (1992). Effect of crop residues amendments to soil for the control of plant-parasitic nematodes. *Bioresource Technology* 41: 81-83.

Akhtar, M. and M.M. Alam (1993a). Control of plant-parasitic nematodes by 'Nimin'–an urea-coating agent and some plant oils. *Zetschrifi fur Pflanzenkrankhei ten und Pflanzenschutz* 100: 337-342.

Akhtar, M. and M.M. Alam (1993b). Utilimtion of waste materials in nematode control: A Review. *Bioresource Technology* 45: 1-7.

Akhtar, M. and I. Mahmood (1993a). Effect of *Mononchus aguaticus* and organic amendments on *Meloidogyne incognita* development on chilli. *Nematologia Mediterranea* 21: 251-252.

Akhtar, M. and I Mahmood (1993b). Control of plant-parasitic nematodes with 'Nimin' and some plant oils by bare root-dip treatment. *Nematologia Mediterranea* 21: 89-92.

Akhtar, M. and I. Mahmood (1993c). Utilimtion of fallen leaves as soil amendments for the control of plant-parasitic nematodes. *Pakistan Journal of Nematology* 11: 131-138.

Akhtar, M. and I. Mahmood (1994a). Potentiality of phytochemicals in nematode control–A Review. *Bioresource Technology* 48: 189-201.

Akhtar, M. and I. Mahmood (1994b). Control of root-knot nematode by bare-root dip in decomposed and decomposed extracts of neem cake and leaf. *Nematologia Mediteranea* 22: 55-57.

Akhtar, M. and I. Mahmood (1994c). Prophylactic and therapeutic use of oilcakes and leaves of neem and castor extracts for the control of root-knot nematode in chilli. *Nematologia Mediterranea* 22: 107-129.

Akhtar, M. and I. Mahmood (1994d). Nematode populations and short-term tomato growth in response to neem-based products and other soil amendments. *Nematropica* 24: 169-173.

Akhtar, M. and I. Mahmood (1995a). Control of root-knot nematode *Meloidogyne incognita* in tomato plants by seed coating with Achook and neem oil. *International Pest Control* 37: 86-87.

Akhtar, M. and I. Mahmood (1 995b). Evaluation of a neem based product against rootknot nematode *Meloidogyne incognita*. *Tests of Agrochemicals and Cultivars (Supplement to Annals of Applied Biology)* 16: 6-7.

Akhtar, M. and I. Mahmood (1996a). Control of plant-parasitic nematodes with organic and inorganic amendments in agricultural soil. *Applied Soil Ecology* 4: 243-247.

Akhtar, M. and I. Mahmood (1996b). Integrated nematode control in potato *Solanum tubrosum*. *International Pest Control* 38: 62-64.

Aklitar, M. and I. Mahmood (1996c). Effect of a plant-based product–Nimin and some plant oils on nematodes. *Nematologia Mediterranea* 24: 3-5.

Akhtar, M. and I. Mahhmood (1996d). Evaluation of nematicidal properties of a biopesticide against root-knot nematode, *Meloidogyne incognita. Tests of Agrochemicals and Cultivar, (Supplement to Annals of Applied Biology)* 127: 104-105.

Akhtar, M. and I. Mahmood (1997a). Impact of organic and inorganic management and plant based products on plant-parasitic and microbivorous nematode communities. *Nematologia Mediterranea* 25: 21-23.

Akhtar, M. and I. Mahmood (1997b). Control of root-knot nematode, *Meloidogyne incognita* in tomato plants by seed coating with suneem and neem oil. *Journal of Pesticide Science* 22: 37-38.

Akhtar, M., S. Anver, and A. Yadav (1990). Effects of organic amendments to soil as nematode suppressants. *International Nematology Network Newsletter* 7: 21-22.

Badra, T. and D.M. Eligindi (1979). The relationship between phenolic content and *Tylenchulus semipenetrans* populations in nitrogen amended citrus plants. *Revue de Nematologie* 2: 161-164.

Badra, T., M.A. Saleh, and B.A. Oteifa (1979). Nematicidal activity and composition of some organic fertilizers and amendments. *Revue de Nematologie* 2: 29-36.

Bird, A.F. and P.G. Brisbane (1988). The influence of *Pasteuria penetrans* in field soils on the reproduction of root-knot nematodes. *Revue de Nematologie* 11: 75-81.

Boag, B., and L.V. Lopez-Llorea (1989). Nematodes and nematophagous fungi associated with cereal fields and permanent pasture in eastern Scotland. *Crop Research* 29: 1-10.

Bora, B.C. and P.N. Phukan (1983). Organic amendments for the control of root-knot nematodes on jute. *Journal of Assam Agriculture Unit* 4: 50-54.

Bridge, J. (1996). Nematode management in sustainable and subsistence agriculture. *Annual Review of Phtytopathology* 34: 201-225.

Brown, S.M., J.L. Kepner, and G.C. Smart Jr. (1985). Increased crop yields following application of *Bacillus penetrans* to field plots infested with *Meloidogyne incognita. Soil Biology & Biochemistry* 17: 483-486.

Brown, W.L. (1954). Collembola feeding on nematodes. *Ecology* 35: 421.

Burg, R.W., B. M. Miller, E.E. Baker, J. Birbaum, S.A. Currie, R. Hartma, Y. Kong, R.L. Monghan, G. Olson, I. Putter, J.B. Tunac, K. Wallicl, E.O. Stanley, R. Oiwa, and S. Omura (1979). Avermectins, a new family of potent anthelmintic agents: Producing organism and fermentation. *Antimicrobial Agents and Chemotherapy* 15: 361-367.

Carris, L.M., D.A. Glawe, C.A. Smyth, and D.I. Edwards (1989). Fungi associated with populations of *Heterodera glycines*, in two Illinois Soybean fields. *Mycologia* 81: 66-75.

Catroux, G., R. Chaussod, and B. Nicolardot (1987). Assessment of nitrogen supply from the soil. *Comptes Rendus Academic Agriculture Francais* 3: 71-79.

Cayrol, J.C. (1983). Lutte biologique contre les *Meladogyne* en moyend' *Arthrobotrys irregularis. Revue de nematologie* 6: 265-273.

Chen, S.Y., D.W. Dickson, J.W. Kimprough, R. McSorley, and D.J. Mitchell (1994).

Fungi associated with females and cysts of *Heterodera glycines* in a Florida soybean field. *Journal of Nematology* 26: 296-303.
Chen, S.Y., D.W. Dickson and D.J. Mitchell (1996a). Pathogenicity of fungi to eggs of *Heterodera glycines*. *Journal of Nematology* 28: 148-158.
Chen, S.Y., D.W. Dickson, and D.J. Mitchell (1996b). Population development of *Heterodera glycines* in response to mycoflora in soil from Florida. *Biological Control* 6: 226-231.
Chen, S.Y., D.W. Dickson, R. McSorley, D.J. Mitchell, and T.E. Hemlett (1996c). Suppression of *Meloidogyne areanaria* race 1 by soil application of endospores of Pasteuriapenetrans. *Journal of Nematology* 28: 159-168.
Christensen, H., D. Funck-Jensen, and A. Kjoller (1989). Growth rate of rhizosphere bacterid measured directly by the tritiated thymidene incorporation technique. *Soil Biology & Biochemistry* 21: 113-117.
Ciancio, A., A. Logrieco, and F. Lamberti (1986). Parasitism of *Xiphinema diversicaudatum* by the fungus *Hirsutella rhossiliensis*. *Nematologia Mediterranea* 14: 187-192.
Clarke, A.J. and R.N. Perry (1977). Hatching of cyst nematodes. *Nematologica* 23:350-368.
Cobb, H.A. (1917). The mononchus (*Mononchus* Bastian) a genus of free-living nematodes. *Soil Science* 3:431-486.
Culbreath, A.K., R. Rodriguez-Kabaa, and G. Morgan-Jones (1985). The use of hemicellulosic waste matter for reduction of phytotoxic effects of chitin and control of root-knot nematodes. *Nematropica* 15: 49-75.
Culbreath, A.K., R. Rodriguez-Kabana, and G. Morgan Jones (1986). Chitin and *Paecilomyces lilacinus* for control of *Meloidogyne arenaria*. *Nematropica* 16: 1253-166.
Curl, E.A. and B. Truelove (1986). *The Rhizosphere*. Springer Verlay, Berlin and New York.
Davide, R.G. and R.A. Zorilla (1983). Evaluation of a fungus, *Paecilomyces lilacinus* (Thom) Samson, for the biological control of potato cyst nematode, *Globodera rostochiensis* Woll, as compared with some nematicide. *Phillippine Agriculture* 66: 13-77.
Derrico, F.P. and F.D. Maio (1980). Effect of some organic materials on root-knot nematodes on tomato in field preliminary experiments. *Nematologia Mediterranea* 8: 107-111.
Dickson, D.W., D.J. Mitchell, T.E. Hewlett, M. Oostendrop, and M.E. Kanvischer-Mitchell (1991). Nematode suppressive soil from a peanut field. *Journal of Nematology* 23: 526. (Abst.)
Dickson, D.W., M. Oostendorp, R.M. Giblin-Davis, and D.J. Mitchell (1994). Control of plant parastic nematodes by biological antagonists. In: *Pest Management in the Sub-tropics: Biological Control, a Florida Perspective*. Eds., D. Rosen, P.D. Bennett, J.L. Jr. Capinera, Intercept Ltd. U.K.
Doncaster, C.C. and D.J. Hopper (1961). Nematodes attacked by protozoa and tardigrades. *Nematologica* 6: 333-335.
Dube, B. and G.C. Smart (1987). Biological control of *Meloidogyne incognita* by

Paecilomyces lilacinus and *Pasteuria penetrans*. *Journal of Nematology* 19: 222-227.

Duddington, C.L. and C.M.G. Duthoit (1960). Green manuring and cereal root eelworm. *Plant Pathology* 9: 7-9.

Duddington, C.L., C.O.R. Everard, and C.M.G. Duttoit (1961). Effect of green manuring and a predacious fungus on cereal root eelworm in oats. *Plant Pathology* 10: 108-109.

Duncan, L.W. (1991). Current optios for nematode management. *Annual Review of Phytopathology* 29: 467-490.

Dunn, M.T., R.M. Sayre, A. Carell, and W.P. Wergin (1982). Colonization of nematode eggs by *Paecilomyces lilacinus* (Thom) Samson as observed with the scaning electron microscope. *Scanning Electron Microscopy* 198: 1351-1357.

Egunjobi, O.A. and S.O. Afolami (1976). Effect of neem (*Azadirachta indica*) leaf extracts on populations of *Prarylenchus brachyurus* and on the growth and yield of maize. *Nematologica* 22, 125-132.

Egunjobi, O.A., P.T. Akonde, and F.E. Caveness (1986). Interaction between *Prarylenchus sefaensis*, *Meloidogyne javanica* and *Rorylenchulus reniformis* in sole and mixed crops of maize and cowpea. *Revue de Nematologie* 9: 61-70.

Ellenby, C. (1945). The influence of crucifers and mustard oil on the emergence of larvae of the potato root eelworm, *Heterodera rostochiensis* wollenweber. *Annals of Biology* 32: 67-70.

Elliot, E.T., R.V. Anderson, D.C. Coleman, and C.V. Cole (1980). Habitable pore space and microbial trophic interactions. *Oikos* 35: 327-335.

Elmiligy, I.A. and D.C. Norton (1973). Survival and reproduction of some nematodes as affected by muck and organic acids. *Journal of Nematology* 5: 50-54.

Eno, C.F., W.G. Blue, and J.M. Good (1955). The effect of ahydrous ammonia on nematodes, fungi, bacteria and nitrification in some Florida soils. *Proceedings of Soil Science Sociery of America* 19: 55-58.

Esser, R.P. (1987). Biological control of nematodes by nematodes I. Dorylaims (Nematoda: Dorylaimida). *Nematology Circular No.* 144: pp. 1-4.

Galper, S., E. Cohn, Y., Spiegel, and I. Chet (1990). Nematicidal effects of collagenamended soil and the influence of protease and colleagenase. *Revue de Nematologie* 13: 67-71.

Galper, S., E. Cohn, Y. Spiegel, and I. Chet (1991). A Collagenolytic fungus, *Cunninghamella elegans*, for biological control of plant-parasitic nematodes. *Journal of Nematology* 23: 269-274.

Gintis, B.O., G. Morgan-Jones, and R. Rodriguez-Kabana (1982). Mycoflora of young cysts of *Heterodera glycines* in North Carolina soils. *Nematropica* 12: 295-303.

Godoy, G., R. Rodriguez-Kabana, and G. Morgan-Jones (1983). Fungal parasties of *Meloidogyne arenaria* in Alabama soil. A mycological survey and green house studies. *Nematropica* 13: 201-213.

Gonzalez, A. and M. Canto-Saenz (1993). Comparison of five organic amendments for the control of *Globadera pallida* in microplots in Peru. *Nematropica* 23: 133-139.

Griffiths, B.S. (1986). Mineralization of nitrogen and phosphorus by mixed culture of the ciliate protozoan *Colpoda steinii*, the nematode *Rhabditis* sp. and the bacterium *Pseudomonas fluorescens*. *Soil Biology & Biochemistry* 18: 637-642.

Hackney, R.W. and O.J. Diskerson (1975). Marigold, castorbean and chrysanthemum as controls of *Meloidogyne incognita* and *Prarylenchus alleni*. *Journal of Nematology* 7: 84-90.

Halbrendt, J.M. (1995). Progress report : efficacy of crop rotahon green manure, and oxamyl as preplant nematode treatments. *Pennsylvania Fruit News* 75: 44-45.

Halbrendt, J.M. (1996). Allelopathy in the management of plant parasitic nematodes. *Journal of Nematology* 28: 8-14.

Hassink, J., G. Lebbink and J.A. Van Veen (1991). Microbial biomass and activity of a reclaimed-polder soil under a conventional or reduced-input farming system. *Soil Biology and Biochemistry* 23: 505-517.

Hewlett, T.E., R. Cox, D.W. Dickson, and R.A. Dunn (1994). Occurrence of *Pasteuria* spp. in Florida. *Journal of Nematology* 26: 616-619.

Hewlett, T.E., E.M. Hewlett, and D.W. Dickson (1995). Response of *Meloidogyne* spp. and *Heterodera glycines* to tannic acid. *Journal of Nematology* 27: 502.

Hollis, J.P. and R. Rodriguez-Kabana (1966). Rapid kill of nematodes in flooded soil. *Phytopathology* 56: 1015-1019.

Hollis J.P. and R. Rodriguez-Kabana (1967). Fatty acids on Louisiana rice fields. *Phytopathology* 57:841-847.

Huebner, R.A., R. Rodriguez-Kabana, and R.M. Pattemson (1983). Hemicellulosic waste and urea for control of plant parasitic nematodes: Effect on soil enzyme activities. *Nematropica* 13: 37-54.

Hutchinson, M.T. and H.T. Streu (1960). Tardigrades attaching nematodes. *Nematologica* 5: 149.

Imbriani, I.L. and R. Makau (1983). Studies on *Lasioseius sculpatus* a neostiamatid mite predacious on nematodes. *Journal of Nematolagy* 15: 523-528.

Jaffee, B.A. and A.E. Muldoon (1989). Suppression of cyst nematodes by natural infestation of a nematophagous fungus. *Journal of Nematology* 21: 505-510.

Jaffee, B.A. and E.I. Zehr (1982). Parasitism of the nematode *Criconemella xenoplax* by the fungus *Hirsutella rhossiliensis*. *Phytopathology* 72: 1378-1381.

Jaffee, B.A., H. Ferris, J.J. Stapleton, V.K. Norton, and A.E. Muldoon (1994). Parasitism of nematodes by the fungus *Hirsutella rhossiliensis* as affected by certain organic amendments. *Journal of Nematology* 26: 152-161.

Jansson, H.B. and B. Nordbring-Hertz (1988). Infection events in the fungus nematode system. In: *Diseases of Nematodes* (Eds), G.O. Poinar, H.B. Jansson, H.B., Boca Raton, FL: CRC Press. 59 p.

Jatala, P. (1985). Biological control of nematodes. In: *An Advanced Treatise on Meloidagyne Biology and Control*, Vol. I, (Eds) J.N. Sasser, C.C. Carter, N.C. State University, Department of Plant Pathology and USAID, Raleigh, NC. pp. 303-308.

Jatala, P. (1986). Biological control of plant-parasitic nematodes. *Annual Review of Phytopathology* 24:453-489.

Jatala, P., R. Kaltenbach, M. Bocangel, A.J. Devaux, and R. Campos (1980). Field application of *Pasteuria penetrans* for controlling *Meloidogyne incognita* on potatoes. *Journal of Nematology* 12: 226-227.

Jing, G.N. (1994). Evaluation of the nematicidal properties of Cruciferous plants for nematode management in replant orchards. Ph.D. Disseration. The Pennsylvania State University Park.

Johnson, T.M. (1959). Effect of fatty acid mixtures on the rice stylet nematode (*Tylenchorhynchus martini*) Fielding, 1959, Nature 183, 1392.

Jones, F.G.W. (1982). The soil plant environment. In: *Plant Nematology*. J.F. Southey (Ed.), Her Majesty's Stationery Office, London.

Kaplan, M. and J.P. Moe (1993). Effects of chicken excrement amendments on *Meloidogyne arenaria*. *Journal of Nematolagy* 27: 71-77.

Kerry, B.R. (1975). Fungi and the decrease of cereal cyst-nematode population in cereal monoculture. *Bulletin of European Mediterranean Plant Protection Organization* 5: 353-361.

Kerry, B.R. (1990). An assessment of progress towards microbial control of plant-parasitic nematodes. *Journal of Nematology* 22: 621-631.

Kerry, B.R., D.H. Crump, and L.A. Mullen (1982). Natural control of the cereal cyst nematode, *Heterodera avenue* Woll. by soil fungi at three sites. *Crop Protection* 1: 99-109.

Kerry, B.R. (1989). Fungi as biological control agents for plant parasitic nematodes. In: *Biotechnology of Fungi for Improving Plant Growth* Ed. J.M. Whipps and R.D. Lumsden, Cambridge, Cambridge University Press.

Lal, A., K.C. Sanwal, and V.K. Mathur (1983). Changes in the nematode population in undisturbed land with the introduction of land development practices and cropping sequences. *Indian Journal of Nematology* 13: 133-140.

Lear, B. (1959). Application of castor pomace and cropping of castor beans to soil to reduce nematode populations. *Plant Disease Reporter* 43: 459-460.

Lewis, J.A. and G.C. Papavizas (1971). Effect of sulphur-contarning volatile compounds and vapors from cabbage decomposition on *Apnanomyces euteiches*. *Phytopathology* 61: 208-214.

Linford, M.B. (1937). Stimulated activity of natural enemies of nematodes. *Science* 85: 123-124.

Linford, M.B. and F. Yap (1939). Root-knot nematode injury restricted bya fungus. *Phytopathology* 29:596-608.

Linford, M.B., F. Yap, and J.M. Oliviera (1938). Reduction of soil populations of rootknot nematode during decomposition of organic matter. *Soil Science* 45: 127-141.

Mankau, R. (1980). Biological control of nematode pests by natural enemies. *Annual Review of Phytopathology* 18: 415-440.

Mankau, R. and S. Das (1969). The influence of chitin amendments on *Melaidogyne incognita*. *Journal of Nematology* 1: 15-16.

Mankau, R. and R.J. Minteer (1962). Reduction of soil populations of the citrus nematode by the addition of organic materials. *Plant Disease Reporter* 46, 375-378.

McSorley, R. and L.W. Duncan (1995). Economic thresholds and nematode management. In: *Advances in Plant Pathology*, Vol. 11, Academic Press Ltd. pp. 147-171.

McSorley, R. and J.J. Frederick (1995). Responses of some common Cruciferae to root-knot nematodes. *Journal of Nematology* 27: 550-554.

McSorley, R. and R.N. Gallaher (1995). Effect of yard waste compost on plant-parasitic nematode densities in vegetable crops. *Journal of Nematology* 27: 545-549.

Mian, I.H. and R. Rodriguez-Kabana (1982). Orgaic amendments with high tannin and phenolic contents for control of *Meloidogyne arenaria* in infested soil. *Nematropica* 12: 221-234.

Mian, I.H., G. Godoy, R.A. Shelby, R. Rodriguez-Kabana, and G. Morgan-Jones (1982). Chitin arnenoments for control of *Meloioogyne arenaria* in infested soil. *Nematropica* 12: 71-84.

Miller, P.M. (1976). Effects of some nitrogenous materials and wetting agents on survival in soil of lesion. stylet and lance nematodes. *Phytopathology* 66: 798-800.

Miller, P.M., D.C. Sands, and S. Rich (1973). Effects of industrial residues, wood fibre wastes, and chitin on plant parasitic nematodes and some soil borne disease. *Plant Disease Reporter* 57: 438-442.

Miller, P.M., G.S. Taylor, and S.E. Wihrheim (1968). Effect of cellulosic amendments and fertilizers on *Heterodera tabacum. Plant Disease Reporter* 52: 441-445.

Minogue, M.J., I.C. Frances, P. Quatermass, M.B. Kappagoda, R. Bradbury R.S. Walls, and I. Motum (1984). Successful treatment of fungal keratitis caused by *Paecilomyces lilaciniis. Amertfa Journal of Ophthalmology* 98: 625-626.

Minton, H.A. and R.M. Sayre (1989). Suppressive influence of *Pasteuria penetrans* in Georgia soils on reproduction of *Meloidogyne arenaria. Journal of Nematology* 21: 574-575.

Mojtahedi, H., G.S. Santo, A.N. Hang, and J.H. Wilson (1991). Suppression of root-knot nematode populations with selected rapeseed cultivars in green manure. *Journal of Nematology* 23: 170-174.

Mojtahedi, H., G.S. Santo, J.H. Wilson, and A.N. Hang (1993). Managing *Meloidogyne chitwoodi* on potato with rapeseed as green manure. *Plant Disease* 77: 42-46.

Morgan-Jones, G., G.F. White, and R. Rodriguez-Kabana (1984). Fungal parasites of *Meloidogyne incognita* in an Alabama soybean field soil. *Nematropica* 14: 93-96.

Muller, R. and P.S. Gooch (1982). Organic amendments in nematode control. An examination of the literature. *Nematropica* 12: 319-326.

Muraoka, M. and M. Ishibashi (1976). Nematode feeding mites and their feeding behaviour. *Applied Entomology and Zoology* 11: 1-7.

Nusbaum C.J. and H. Ferris (1973). The role of cropping systems in nematode population management. *Annual Review of Phytopathology* 11: 424-440.

Okopnyi, N. S., Y.E. Gutsu, and N.A. Barba (1980). Nematicidal activity of some isothiocynates and their derivatives containting thiomide groups, *Izvestiya Akadenoii Nauk* 3: 84-85.

Oostenbrink, M., K. Kuiper, and J.J. Jacob (1957). *Tagetes* als feindpflanzen von *Prarylenchus*-Arten. *Nematologica* 2: 423-424.

Oosterdorp, M. and R.A. Sikora (1989). Seed treatment with antagonistic rhizobacteria for the suppression of *Heterodera schachtii* early root infection of sugar beet. *Review de Nematologie* 12: 77-83.

Oosterdorp, M., D.W. Dickson, and D.J. Mitchell (1990). Host range and ecology of isolates of *Pasteuria* spp. from the southeastern United States. *Journal of Nematology* 22: 525-531.

Osman, A.A. and D.R. Vigiierchio (1988). Efficacy of biologically active agents as nontraditional nematicides for M*eloidogyne Javanica*. *Revue de Nematologie* 11: 93-98.

Page, E.R. (1966). The micronutrient content of young vegetable plants as affected by FYM. *Journal of Ho rticultural Science* 41: 257-261.

Pawloski, L. and G. Bachthaler (1989). Allelopathic interrlations for weed control on arable land. Components Technical Notes. U.S. Sustainable Agricultural Research and Education Program. California.

Poinar, G.O. and E.L. Hansen (1986). Association between nematodes and bacteria. *Helminthological Abstract* 55: 61-81.

Poinar, G.O. Jr., and K.S. Jansson (1988). *Diseases of Nematodes*, Vol. 1, Vol. II, 150 p., CRC Press Boca Raton, FL, USA.

Prot, J.C. and J.M. Kornprobst (1983). Effects of *Azadirachta indica*, *Hannoa undulata* and *Hannoa klaineana* seed extracts on the ability of *Meloidogyne javanica* juveniles to penetrate tomato roots. *Revue de Nematologie* 6: 330-332.

Reddy, K.C., A.R. Stoffes, G.M. Prine, and R.A. Dunn (1986). Tropical legurnes for green manure Il. Nematode population and their effects on succeeding crop yields. *Agronomy Journal* 78: 5-10.

Rich, J.R., G.S. Rahi, C.H. Opperinan, and E.L. Davis (1989). Influence of the castor bean (*Ricinus communis*) lectin (Ricin) on motiliry of *Meloidogyne incognita*. *Nematropica* 19: 99-101.

Roberts, P.A. and W.C. Matthews (1995). Disinfection alternatives for control of *Dirylenchus dipsaci* in garlic seed cloves. *Journal of Nematology* 27:448-456.

Roccuzzo, G., A. Ciancio, and R. Bonsignore (1993). Population density and soil antagonists of *Meloidogyne hapla* infecting kiwi in southern Italy. *Fundamental and Applied Nematology* 16: 151-154.

Rodriguez-Kabana, R. (1986). Organic and inorganic amendments to soil as nematode suppressants. *Journal of Nematology* 18: 129-135.

Rodriguez-Kabana, R., J.W. Jordan, and J.P. Hollis (1965). Nematodes: Biological control in rice fields: Role of hydrogen sulflde. *Science* 148: 524-526.

Rodriguez-Kabana, R. and P.S. King (1980). Use of mixtures of urea and blackstrap molasses for control of root-knot nematodes in soil. *Nematropica* 10: 38-44.

Rodriguez-Kabana, R. and G. Morgan-Jones (1988). Potential for nematode control by mycofloras endmic in the tropics. *Journal of Nematology* 20: 191-203.

Rodriguez-Kabana, R., P.S. King, and M.H. Pope (1981). Combinations of anhydrous ammonia and ethylene dibromide for control of nematodes parasitic on soybeans. *Nematropica* 11: 27-41.

Rodriguez-Kabana, R., G. Godoy, G. Morgan-Jones, and R.A. Shelby (1983). The determination of soil chitinase activity: conditions for assay and econological studies. *Plant and Soil* 75: 95-106.

Rodriguez-Kabana, R., G. Moigan-Jones and B. Ownley-Gintis (1984). Effects of chitin amendments to soil on *Heterodera glycines*, microbial populations, and colonization of cysts by fungi. *Nematropica* 14: 10-25.

Rodrigues-Kabana, R., G. Morgan-Jones, and I. Chet (1987). Biological control of nematodes: soil amendments and microbial antagonists. *Plant and Soil* 10: 237-247.

Rodriguez-Kabana, R., R.A. Shelby, P.S. King, and M.H. Pope (1982). Combinations of anhydrous ammonia and 1,3-dichloropropenes for control of root-knot nematodes in soybean. *Nematropica* 12: 61-69.

Rodrigues-Kabana, R., D.G. Robertson, P.T. Backman, and H. Ivey (1988a). Soybean-peanut rotations for the management of *Meloidogyne arenaria*. *Journal of Nematology* 20: 81-85.

Rodriguez-Kabana, R., D.G. Robertson, L. Wells, and R.W. Young (1988b). Hairy indigo for the management of *Meloidogyne arenaria* in peanut. *Nematropica* 18: 137-142.

Sayre, R.M. (1988). Bacterial diseases of nematodes and their role in controlling nematode populations. *Agriculture, Ecosystems and Environment* 24: 263-279.

Sayre, R.M., Z.A. Patrick, and H.J. Thorpe (1964). Substances toxic to plant-parasitic nematodes in decomposing plant residue. *Phytopathology* 54; 205.

Sayre, R.M., Z.A. Patrick, and H.I. Thorpe (1965). Identification of a selective nematicidal component in extracts of plant residues decomposing in soil. *Nematologica* 12: 263-268.

Sayre, R.M. and E.M. Powers (1966). A predacious soil turbellarian that feeds on free-living and plant-parasitic nematodes. *Nematologica* 12: 619-629.

Sayre, R.M. and M.P. Start (1988). Bacterial diseases as antagonists of nematodes. In: *Diseases of Nematodes*. Vol. I, (Eds.) G.O. Poinar, H.B. Jansson. CRCP, Boca Raton, FL pp. 70-101.

Sayre, R.M. and D.E. Walter (1991). Factors affecting the efficacy of natural enemies of nematodes. *Annual Review of Phytopathology* 29: 149-166.

Scalbert, A. (1991). Antimicrobial properties of tannin. *Phytochemistry* 30: 3875-3883.

Schlang, J., W. Steudel, and J. Muller (1988). Influence of resistant green manure crops on the population dynamics of *Heterodera schachtii* and its fungal egg parasites. *Proceedings of European Societv of Nematology, 19th International Nematology Symposium*, Uppsala, Sweden, pp. 69 (Abstr).

Sharma, R.D. (1992). Biocontrol efficiency of *Pasteuria penetrans* against *Meloidogyne javanica*. *Ciencia Biological Ecology Systematics*. 12: 43-47.

Siddiqui, M.A. and M.M. Alam (1987a). Control of plant-parasitic nematodes by intercropping with *Tagetes minuta*. *Nematologia Mediterranea* 15: 205-211.

Siddiqui, M.A. and M.M. Alam (1987b). Utilization of marigold plant wastes for the control of plant-parasitic nematodes. *Biological Wastes* 21: 221-229.

Siddiqui, M.A. and M.M. Alam (1988). Control of plant-parasitic nematodes with *Tagetes terifolia*. *Revue de Nematologie*, 11: 369-370.

Singh, R.S. and K. Sitaramaiah (1970). Control of plant-parasitic nematodes with organic soil amendments. *PANS* 16: 287-297.

Small, R.W. (1987). A review of the prey of predatory soil nematodes. *Pedobiologia* 30: 179-206.

Spiegel, Y., E. Cohn and I. Chet (1986). Use of chitin for controlling plant-parasitic nematodes. I. Direct effects on nematode reproduction and plant performance. *Plant and Soil* 95: 87-95.

Spiegel, Y., I. Chet and E. Cohn (1987). Use of chitin for controlling plant-parasitic nematodes. II. Mode of action. *Plant and Soil* 98: 337-345.

Spiegel, Y., I. Chet, E. Cohn, S. Galper, and E. Sharon (1988). Use of chitin for controlling plant-parasitic nematodes III. Influence of temperature on nematicidal effect, mineralization and microbial population build-up. *Plant and Soil* 109: 251-256.

Spiegel, Y., E. Cohn, and I. Chet (1989). Use of chitin for controlling *Heterodera avenae* and *Tylenchulus semipenetrans*. *Journal of Nematology* 21: 419-422.

Steiner, G. (1941). Mematodes parasitic on and associated with roots of marigold (*Tagetes hybrida*). *Proceeding of the Biological Sociery of Washington.* 54: 31-34.

Stirling, G.R. (1991). Conservation and enhancement of naturally occurring antagonists and the role of organic matter. In: *Biological Control of Plant Parasitic Mematodes. Progress, Problems and Prospects*. CAB International, Wallingford, U.K.

Stirling, G.R. and B.R. Kerry (1983). Antagonists of the cereal cyst nematode *Heterodera avenae* Woll. In Australian Soils, Australian Journal Experimental. *Agriculture and Animal Husbandry* 23: 378-324.

Stirling, G.R. and R. Mankau (1978). Parasitism of *Meloidogyne* eggs by a new fungal parasite. *Journal of Nematology* 10: 236-240.

Stirling, G.R. and M.G. Wachtel (1980). Mass production of *Bacillus penetrans* for the biological control of root-knot nematodes. Nematologica 26: 308-312.

Stirling, G.R. and A.M. White (1982). Distribution of parasites of root-knot nematodes in a south Australian Vineyard. *Plant Disease* 66: 52-53.

Sukul, N.C. (1992). Plant antagonistic to plant-parasitic nematodes. *Indian Review of Life Science* 12: 23-52.

Thakur, R.S., B.S. Singh, and A. Goswami (1981). Review Article. E. *Azadirachta indica* A. Juss. *CROMAP*, 3.

Thorne, G. (1940). *Duboscquia penetrans* n.sp. (Sporozoa, Microsporidia Nosematidae), a parasite of the nematode *Prarylenchulus pratensis* (de Man) Filipjev, *Proceeding of Helminthological Society of Washington* 7: 51-53.

Triffitt, M.I. (1929). Preliminary researches on mustard as a factor inhibiting cyst formation in *Heterodera schachtii*. *Journal of Helminthology* 7: 81-82.

Tyler, I. (1938). Proceedings of the root-knot nematode conference held at Atlanta, Georgia, February 4, 1938. *Plant Disease Reporter* 109: 133-151.

Tzortzakakis, E.A. and S.R. Gowen (1994). Evaluation of *Pasteuria penetrans* alone and in combination with oxamyl, plant resistance and solarization for control of *Meloidogyne* spp. on vegetables grown in greenhouse in Crete. *Crop Protection* 13: 455-461.

Uhlenbroek, I.H. and I.D. Bijloo (1958). Investigations on nematicides. I. Isolation and structure of nematicidal principle occurring in *Tagetes* roots. *Recueil des Travaux Chimiques des Pays-Bas et de la Belgique* 77: 1004-1009.

Van der Laan, P.A. (1956). The influence of organic manuring on the development of the potato eelworm, *Heterodera rostochiensis*. *Nematologica* 1: 112-115.

Van Veen, I.A. and P.I. Kuikeman (1990). Soil structural aspects of decomposition of organic matter by micro-organisms. *Biogeochemistry* 11: 213-233.

Vassalo, M.A. (1967). The nematicidal power of ammonia. *Nematologica* 13: 155 (Abst.).

Villanueva, L.M. and R.G. Davide (1984). Evaluation of several isolates of soil fungi for biological control of root-knot nematodes. *The Phillipine Agriculture* 67: 361-371.

Walter, D.E., H.W. Hunt, and F.T. Elliott (1987). The influence of prey type on the development and reproduction of some predatory soil mites. *Pedobiologia* 30: 419-424.

Webster, J.M. (1972). Nematode and biological control. In: *Economic Nematology*, (Ed.), J.M. Webster, Academic Press, New York, 563 p.

Weibelzabl-Fulton, E., D.W. Dickson and E.B. Wbitty (1996). Suppression of *Meloidogyne incognita* and *M javanica* by *Pasteuria penetrans* in field soil. *Journal of Nematology* 28: 43-49.

Yeates, G.W. (1969). Predation by *Mononchoides potohikus* (Nematoda: Diplogasteridae) in laboratory culture. *Nematologica* 15: 1-9.

Yeates, G.W. (1981). Soil nematode populations depressed in the presence of earthworm. *Pedobiologia* 22: 191-201.

Yeates, G.W. and D.C. Coleman. (1982). Nematodes and decomposition. In: *Nematodes in Soil Ecosystems* Ed. D.W. Freckman. Austin. University of Iexas Press, pp. 55-82.

Yuhara, I. (1971). Effect of soil application of chopped plant material of Crotalaria or marigold on *Meloidogyne hapla* population. *Annual Report of Society of Plant Protection* 22: 62.

Zechineister, L. and J.W. Sease (1941). A blue-florescing compound terthienyl, isolated from marigold. *Journal of American Chemical Society* 69: 273-275.

Index

